Apollo 12 astronaut drives a core tube into lunar surface for scientific exploration of the Moon.

NASA SP-235

APOLLO 12

Preliminary Science Report

PREPARED BY
NASA MANNED SPACECRAFT CENTER

Scientific and Technical Information Division
OFFICE OF TECHNOLOGY UTILIZATION 1970
NATIONAL AERONAUTICS AND SPACE ADMINISTRATION
Washington, D.C.

ACKNOWLEDGMENT

The material submitted for the "Apollo 12 Preliminary Science Report" was reviewed by a NASA Manned Spacecraft Center Editorial Review Board consisting of the following members: Anthony J. Calio (Chairman), John W. Harris, John H. Langford, R. Mercer, Jamie L. Moon, Scott H. Simpkinson, William K. Stephenson, Jeffrey L. Warner, and Julian M. West.

Foreword

OUR SECOND JOURNEY to the Moon opened the new age of extraterrestrial scientific exploration by man. Going beyond Apollo 11, which demonstrated to an eager world that astronauts can set foot on a celestial body and return safely to Earth, Apollo 12 concentrated on a systematic scientific sampling designed to help unlock some of the secrets of the solar system's origin and early history.

At Apollo 12's destination we had in the spring of 1967 landed Surveyor 3, a spiderlike mechanical creature with three legs, a clawlike arm, and a roving TV eye. Less than 3 years later, Pete Conrad and Alan Bean landed their lunar module *Intrepid* with precision a few thousand feet from Surveyor 3, enabling them to disassemble parts exposed to the lunar environment for a known time for later analysis back on Earth. This was but a single task in a long series of tasks the astronauts had to perform, but to me this precise landing so close to a preselected site a quarter of a million miles from Earth points up the marvelous inseparability of mechanical and scientific capabilities in space exploration.

This document, like the initial report of the first manned landing, relates the preliminary scientific observations resulting from the mission of Apollo 12. Further study and more detailed analysis will undoubtedly produce additional significant results, just as additional manned landings will undoubtedly reveal new mechanical and scientific marvels.

THOMAS O. PAINE
Administrator
National Aeronautics and Space Administration

JUNE 1, 1970

Contents

		PAGE
	INTRODUCTION *A. J. Calio*	ix
	MISSION DESCRIPTION *W. K. Stephenson*	xi
	SUMMARY OF SCIENTIFIC RESULTS *Gene Simmons and A. J. Calio*	1
1	PHOTOGRAPHIC SUMMARY OF THE APOLLO 12 MISSION *L. C. Wade*	7
2	CREW OBSERVATIONS *Alan L. Bean, Charles Conrad, Jr., and Richard F. Gordon*	29
3	PASSIVE SEISMIC EXPERIMENT *Gary V. Latham, Maurice Ewing, Frank Press, George Sutton, James Dorman, Yosio Nakamura, Nafi Toksoz, Ralph Wiggins, and Robert Kovach*	39
4	LUNAR SURFACE MAGNETOMETER EXPERIMENT *P. Dyal, C. W. Parkin, and C. P. Sonett*	55
5	THE SOLAR-WIND SPECTROMETER EXPERIMENT *Conway W. Snyder, Douglas R. Clay, and Marcia Neugebauer*	75
6	SUPRATHERMAL ION DETECTOR EXPERIMENT (LUNAR IONOSPHERE DETECTOR) *J. W. Freeman, Jr., H. Balsiger, and H. K. Hills*	83
7	COLD CATHODE GAGE (LUNAR ATMOSPHERE DETECTOR) *F. S. Johnson, D. E. Evans, and J. M. Carroll*	93
8	THE SOLAR-WIND COMPOSITION EXPERIMENT *J. Geiss, P. Eberhardt, P. Signer, F. Buehler, and J. Meister*	99
9	APOLLO 12 MULTISPECTRAL PHOTOGRAPHY EXPERIMENT *A. F. H. Goetz, F. C. Billingsley, E. Yost, and T. B. McCord*	103
10	PRELIMINARY GEOLOGIC INVESTIGATION OF THE APOLLO 12 LANDING SITE	113
	PART A. GEOLOGY OF THE APOLLO 12 LANDING SITE *E. M. Shoemaker, R. M. Batson, A. L. Bean, C. Conrad, Jr., D. H. Dahlem, E. N. Goddard, M. H. Hait, K. B. Larson, G. G. Schaber, D. L. Schleicher, R. L. Sutton, G. A. Swann, and A. C. Waters*	113
	PART B. PHOTOMETRIC AND POLARIMETRIC PROPERTIES OF THE LUNAR REGOLITH *H. E. Holt and J. J. Rennilson*	157

	PAGE
PART C. MECHANICAL PROPERTIES OF THE LUNAR REGOLITH R. F. Scott, W. D. Carrier, N. C. Costes, and J. K. Mitchell	161
11 LUNAR SURFACE CLOSEUP STEREOSCOPIC PHOTOGRAPHY T. Gold, F. Pearce, and R. Jones	183
12 PRELIMINARY EXAMINATION OF LUNAR SAMPLES The Lunar Sample Preliminary Examination Team	189
13 PRELIMINARY RESULTS FROM SURVEYOR 3 ANALYSIS R. E. Benson, B. G. Cour-Palais, L. E. Giddings, Jr., Stephen Jacobs, P. H. Johnson, J. R. Martin, F. J. Mitchell, and K. A. Richardson	217
APPENDIX A – GLOSSARY OF TERMS	225
APPENDIX B – ACRONYMS	227

Introduction

THE APOLLO 11 MISSION, primarily designed to land men on the Moon and return them safely to Earth, signaled a new phase of the manned space program. Based on the success of Apollo 11, the first of a series of missions designed for the systematic exploration of the Moon was successfully accomplished on Apollo 12. The fact that the Apollo 12 astronauts were able to achieve a pinpoint landing at a preselected site, and then spend an extended time on the lunar surface, graphically illustrates the rapid progress of the Apollo program.

The Apollo 12 mission added significantly to man's knowledge of the Moon. The precise landing capability allowed the crew to accomplish a wide variety of preplanned tasks and paved the way for planning future missions to smaller, more selected landing areas with the possibility of significant scientific returns.

The Apollo 12 mission also benefited lunar orbital science. By changing the orbital plane of the command and service module (CSM) twice, once for rendezvous and once to accomplish photographic tasks, the crew demonstrated the capability to explore new areas of the lunar surface during orbital operations. Future flights will take advantage of this capability to photograph additional potential landing sites and to make scientific observations of the surface, both visually and photographically. The success of the Apollo 12 crew in lunar orbit allowed an increase in the planned orbital activities for the Apollo 13 mission.

The large quantity of lunar soil and rocks brought back by the Apollo 12 astronauts will add to the detailed scientific information already obtained from the Apollo 11 samples. The emplaced scientific experiments have yielded considerable geophysical data that were unavailable prior to the Apollo 12 mission. With the landing of future missions, a network of scientific instruments will be created that will greatly enhance the gathering of data. This report is preliminary and covers only the initial scientific results of the Apollo 12 mission. A great deal of work remains for the scientists involved to interpret and understand the returned lunar material and the data being constantly transmitted from the Apollo 12 scientific experiments on the Moon.

A. J. CALIO
NASA Manned Spacecraft Center

Mission Description

W. K. Stephenson[a]

The Apollo 12 mission was the second manned lunar landing mission. Its objective was to perform detailed scientific lunar exploration.

The space vehicle with a crew of Charles Conrad, Jr., the commander; Richard F. Gordon, the command module pilot; and Alan L. Bean, the lunar module pilot, was launched from Kennedy Space Center, Fla., at 11:22:00 a.m. e.s.t. (16:22:00 G.m.t.) on November 14, 1969. The activities during Earth-orbital checkout, translunar injection, and translunar coast were similar to those of Apollo 11, except for the special attention given to verifying all spacecraft systems as a result of the electrostatic discharges shortly after launch at 36.5 and 52 sec ground elapsed time (g.e.t.). All equipment checked perfectly, so permission to proceed was given. Only one midcourse correction, applied at 31 hr g.e.t. to place the spacecraft on the planned non-free-return trajectory, was required prior to lunar orbit insertion.

The spacecraft was inserted into an elliptical lunar orbit at 83.5 hr g.e.t., and the orbit was circularized two revolutions later. Following undocking of the command and service module and the lunar module, the descent orbit began at 109.5 hr g.e.t. One hour later, a precision landing was made using automatic guidance, with only small manual corrections required in the final phases of descent. The touchdown occurred at 110.5 hr g.e.t. at a point only 600 ft from the target point, the Surveyor 3 spacecraft. The landing coordinates were 3.2° south latitude and 23.4° west longitude in the Ocean of Storms. This precision landing is of great significance to the future lunar exploration program, because landing points in rough terrain of great scientific interest may now be targeted.

[a] NASA Manned Spacecraft Center.

The first of the two planned extravehicular activity periods began at 115 hr g.e.t. A color television camera mounted on the descent stage provided live television coverage of the descent of both astronauts to the lunar surface. A contingency sample of lunar soil was collected by the commander and placed aboard the lunar module prior to the descent of the lunar module pilot. Live television coverage was subsequently lost because of the inadvertent pointing of the camera at the Sun. The crew emplaced the U.S. flag and the solar-wind composition experiment. Then, the Apollo lunar surface experiments were deployed at a safe distance away from the lunar module. Additional lunar surface samples, including core-tube specimens, were collected. The first extravehicular activity period, lasting approximately 4 hr, was recorded with the color still cameras and the sequence camera and by stereoscopic and panoramic views taken with the still cameras. Upon return to the lunar module, the crew replenished the portable life-support systems and, after a planned 7-hr rest period, prepared for the second extravehicular activity period.

The second extravehicular activity period began at 131.5 hr g.e.t. with the descent of the commander to the lunar surface. The lunar module pilot followed shortly thereafter, and the two astronauts started the geology traverse. The traverse covered a distance of approximately 4300 ft and lasted 3 hr 50 min. During the traverse, documented samples, core-tube samples, trench site samples, and gas analysis samples were collected. The crew photographed Surveyor 3, which landed on the lunar surface in April 1967. Also, for scientific and engineering evaluation of the long-term effects of the lunar environment on Earth-made materials, the crew retrieved a painted tube, an unpainted tube, the television camera, and the Surveyor 3 scoop. Stereoscopic and black-and-white still cameras were used to document sample collection and the traverse. As on Apollo 11, crew mobility was excellent throughout the total 7-hr-and-46-min extravehicular activity period.

Another rest period and a final checkout preceded the liftoff of the lunar module ascent stage from the lunar surface at 142 hr g.e.t. After a nominal rendezvous sequence, the two spacecraft were docked at 145.5 hr g.e.t. Following crew transfer, the ascent stage was remotely guided to impact on the lunar surface to provide an active seismic source for the passive seismic experiment. Impact occurred at 150 hr g.e.t., approximately 40 miles from the seismic equipment that had been emplaced during the first extravehicular activity period.

Extensive landmark tracking and photography were performed in lunar orbit. The lunar orbit photography was conducted using a 500-mm long-range lens to obtain mapping and training data for future missions.

Transearth injection was accomplished with the service propulsion engine at 172.5 hr g.e.t. During transearth coast, two small midcourse corrections were executed, and the entry sequence was normal. The command module landed in the Pacific Ocean at 244.5 hr g.e.t. The landing coordinates, as determined from the onboard computer, were 15°47' south latitude and 165°11' west longitude.

Summary of Scientific Results

Gene Simmons[a] and A. J. Calio[b]

The Apollo 12 mission provided the first opportunity in the scientific exploration of the Moon to sample extensively the rocks within a radius of ½ km of the landing site, to obtain geologic data from firsthand observations made on the Moon, to measure on the surface of the Moon the vector components of the lunar magnetic field, to measure the pressure of the lunar atmosphere, and to collect seismic data on the interior of the Moon from the impact of the lunar module (LM) ascent stage. During the two extravehicular activity (EVA) periods, a total duration of 7.5 hr, the astronauts collected three core tubes of lunar soil and additional surface samples along a geologic traverse. They obtained material from the bottom of a shallow trench and brought back several items from the Surveyor 3 spacecraft. The astronauts caught some of the solar wind in an aluminum foil, and they obtained extensive photographs of the lunar surface and of crew activities by using 70-mm Hasselblad cameras and a closeup stereoscopic camera.

The Apollo 12 LM landed on the northwest rim of the 200-m-diameter Surveyor Crater in the Ocean of Storms. The landing site was at 23.4° west longitude and 3.2° south latitude, approximately 120 km southeast of the crater Lansberg and due north of the center of Mare Cognitum. The landing site is near a ray associated with the crater Copernicus, which is situated approximately 370 km to the north. The landing site is characterized by a distinctive cluster of craters that range in diameter from 50 to 400 m. The traverses during the two EVA periods were generally made on or near the rims of these craters and on deposits of ejecta from them. Therefore, the samples returned to Earth contain a variety of material ejected from local craters. Some of the fine-grained material was derived locally, and some probably from distant sources.

The Apollo 12 results obtained to date are summarized in this section. The reader should understand that these results are preliminary and that the interpretation, especially, is likely to change in the future.

Surface Experiments

Geology

Igneous rocks, breccias, and soils were collected from a variety of local geologic features that included a mound and several craters. Along several parts of the traverse made during the second EVA period, the astronauts found fine-grained material of relatively high albedo that, at some places, was buried in the shallow subsurface and, at other places, was situated on the surface. This light-gray material was specifically reported to be at the surface near Sharp Crater and a few centimeters below the surface near Head, Bench, and Block Craters. It is possible that some of this light-gray material may constitute a discontinuous deposit that is observed telescopically as a Copernican ray.

Small linear patterns similar to those at the Apollo 11 site were noted in the surface. These patterns are probably caused by drainage of fine-grained material into fractures in the underlying bedrock. This interpretation implies northeast- and northwest-trending joint sets in the bedrock of the Apollo 11 site and north- and east-trending joint sets in bedrock of the Apollo 12 site. The lineated strips of ground reported by the crew probably reflect joint sets within larger fracture zones in the bedrock.

[a] NASA Manned Spacecraft Center and Massachusetts Institute of Technology.
[b] NASA Manned Spacecraft Center.

Darker regolith material that generally overlies the light-gray material is only a few centimeters thick at some places, but probably thickens greatly on the rims of some craters. The regolith varies locally in the size, shape, and abundance of constituent particles and in the presence or absence of patterned ground. Most of the local differences are probably the result of local cratering events.

The Apollo 12 site is younger than the Apollo 11 landing site as suggested by the fewer number of kilometer-size craters. The Apollo 11 site has about 2.37 times as many kilometer-size craters as does the Apollo 12 site.

One of the notable differences between the set of rocks collected at the Apollo 12 site and that collected at the Apollo 11 site (Tranquility Base) is the ratio of crystalline rocks to breccias. At the Apollo 12 site the rocks collected are predominantly crystalline, whereas at Tranquility Base, approximately half the rocks collected are crystalline and half are microbreccia. This difference exists probably because the rocks from the Apollo 12 landing site were collected primarily on or near crater rims. The regolith is thin on the crater rims, and many of the rocks are probably derived from craters that have been excavated in the bedrock that lies well below the regolith. Tranquility Base was on a thick, mature regolith, where many of the observed rock fragments were produced by shock lithification of regolith material and were ejected from craters too shallow to excavate bedrock.

Two mounds were situated in the area north of Head Crater. Both mounds are visible on the high-resolution Lunar Orbiter photographs. These mounds are probably clumps of regolith material that were slightly indurated by impact and ejected from one of the nearby craters—possibly from Head Crater. Bombardment by meteoritic material and by secondary impacts and, possibly, the effects of diurnal temperature changes have probably caused sloughing of the sides of the mounds, which has resulted in their present rather smooth form.

The lunar surface materials near the Surveyor 3 spacecraft were examined for measurable changes that might have occurred in photometric properties during the 30 months since the Surveyor landing. None occurred, within the limits of the measurements. The optical properties indicate that the lunar surface in the area of the Surveyor spacecraft has not received a new covering of dust nor been mechanically altered by the lunar environment during the 30 months. A significant change occurred in the reflectance of the Surveyor footpad imprint over the 30-month span; the change may have been caused by microscopic mechanical alteration of the compressed surface.

In spite of local variations in soil texture, color, grain size, compactness, and consistency, the soil at the Apollo 12 site is similar in appearance and behavior to the soils encountered at the Apollo 11 and Surveyor 3 equatorial landing sites. Although the deformation behavior of the surface material involves both compression and shear effects, the conclusion drawn earlier from the Surveyor 3 mission results—that the soil at the Surveyor 3 landing site is essentially incompressible—is consistent with the consistency, compactness, and average grain size of the soil at the Surveyor 3 site, as assessed during the Apollo 12 EVA. There appears to be no direct correlation between crater slope angle and consistency of soil cover. The latter depends mainly on the geologic history of the terrain feature and on the local environmental conditions.

Seismology

No seismic signals with characteristics similar to terrestrial signals have been observed for the Moon. This fact is a major scientific result. The high sensitivity at which the lunar instruments were operated would have resulted in the detection of many such signals if the Moon were as seismically active as the Earth and had the same transmission characteristics as the Earth. Thus, the data obtained to date indicate that either seismic energy release is far less for the Moon than for the Earth or the interior of the Moon is highly attenuating for seismic waves.

The LM ascent stage was impacted approximately 80 km E 24° S from the seismometer. The event was recorded on all three long-period seismometers. The signal amplitude increased slowly and then decreased slowly. The signal continued for approximately an hour. The coherence between various signal components was quite low. None of the signal features was typical of analogous terrestrial events. The velocity of the first arrival was 3.1 to 3.5 km/sec.

The passive seismic experiment has recorded 30 prolonged signals with a gradual buildup and then a slow decrease in signal amplitude. Signals with these characteristics may imply transmissions with very low attenuation and intense wave scattering—conditions that are mutually exclusive on Earth. Because of the similarity with the signal from the impact of the LM ascent stage, the 30 recorded signals are thought to be produced by meteoroid impacts of shallow moonquakes. Most of the events that produced these signals appear to have originated within 100 km of the Apollo lunar surface experiments package (ALSEP). The occurrence of signals with similar characteristics during both the Apollo 11 and 12 missions greatly strengthens the present belief that at least some signals observed, including the 30 signals recorded by the seismometer, are likely to be of natural origin.

Magnetometer

The Apollo 12 magnetometer is a very sophisticated three-component fluxgate instrument. It is the first magnetometer to be operated on the lunar surface. A permanent magnetic field of 36 gammas and a gradient of 4×10^{-3} gammas/cm were measured at the Apollo 12 site. The magnitude of the gradient is interpreted to mean that the local magnetic body must be at least 0.2 km in size. The largest transient magnetic field measured in space at distances greater than a few Earth radii, approximately 96 gammas, was recorded on November 26, 1969, when the Moon was in the vicinity of the Earth magnetohydrodynamic bow shock.

Solar-Wind Spectrometer

Examination of the data obtained from the solar-wind spectrometer during the first 35 days of operation indicates that the solar plasma at the lunar surface is superficially indistinguishable from the solar plasma at some distance from the Moon, both when the Moon is ahead of and when the Moon is behind the plasma bow shock of the Earth. No detectable plasma appears to exist in the magnetospheric tail of the Earth or in the shadow of the Moon. Times of passage through the bow shock or through the magnetospheric-tail boundary, as indicated by the solar-wind spectrometer and by the lunar surface magnetometer, are in agreement when comparison of data has been possible. Highly variable spectra that may involve unexpected phenomena were observed on November 27, 1969, and at lunar sunrise; otherwise, observations have been as expected.

Suprathermal Ion Detector Experiment

Preliminary analysis of data from the suprathermal ion detector reveals the following features: a concentration of ions in the 18- to 50-amu/q (mass-per-unit-charge) range, the frequent appearance of ions in the 10- to several-hundred eV range, the sporadic appearance of 1- to 3-keV ions early in the lunar night, and the presence of solar-wind ions on the nightside of the Moon approximately 4 days before lunar sunrise. Energetic ion fluxes correlate well with the impact of the LM ascent stage into the lunar surface. There is a strong suggestion that the impact-released gases have been ionized and accelerated by the solar wind. High background count rates observed during the second lunar day may be indicative of large quantities of gas escaping impulsively from the LM descent-stage tanks.

Cold Cathode Gage

The ambient lunar atmospheric pressure is less than 8×10^{-9} torr. The gas cloud around an astronaut on the lunar surface exceeds the upper range of the gage (approximately 10^{-6} torr) when the gage is a distance of several meters from the astronaut; however, no perceptible residual contamination at the 10^{-8} torr level remains around the gage for longer than a few minutes after the departure of the astronaut.

Multispectral Photography Experiment

The astronauts obtained a total of 142 black-and-white photographs, taken with blue-, green-, red-, and infrared-filtered cameras, that are suitable for color-difference analysis. Two existing image data-reduction methods are being expanded to produce images that display greatly enhanced three-color contrast. Two-color difference pictures have been produced, and the method is effective. The color enhancement of the Apollo 13 landing-site frame shows a lack of color variation. The frame containing Lalande η exhibits color differences, the first such differences to be detected in high-resolution photography of the lunar surface, that can probably be attributed to compositional variations.

Lunar Surface Closeup Stereoscopic Photography

The almost complete absence of dust on the surfaces of rocks, clearly evident in several of the Apollo lunar surface stereoscopic camera photographs taken during the Apollo 12 mission and in several similar photographs taken during the Apollo 11 mission, is most remarkable. The absence of dust cannot be attributed to any cleansing effect of the exhaust gases from the descent engine because shadowing would have to be evident, and shadowing is not evident in any of the photographs. During the time required to form the many impact holes on the surface of the rock, a similar number of impacts on the neighboring powdery ground would have scattered much powder; and the average condition, if impacts were the only process, would have to be a substantial blanket of dust so that the loss from the dust blanket by impacts equaled, in the long run, the gain from material scattered from nearby impacts. The almost complete absence of dust on the rocks requires an explanation other than such an equilibrium. It must be assumed that there is either a general removal of dust from the lunar surface that dominates all other processes that distribute dust or that there is a dust-transportation process over the lunar surface that has a strong tendency for downhill flow and in which the particles are generally not lifted as high (i.e., more than 5 or 10 cm) as the surfaces of the rocks that exhibit the clean areas. The latter possibility is more in accord with other observations, such as the scarcity of trenches adjoining rocks whose distribution clearly indicates that they fell to their present positions. The trench and pileup that must have been common in the soft soil surrounding a fallen rock must thus be eradicated, yet at the same time, no significant amount of material must be deposited on the tops of the rocks. This is a strong indication for a process of surface creep that may be a major process in the long-term evolution of the lunar surface.

Experiments on Returned Materials

Returned Lunar Samples

Three categories of samples were collected on the lunar surface. The contingency sample was collected early in the first EVA period in the vicinity of the LM. The selected sample was collected, after deployment of the ALSEP, in the vicinity of the mounds and near Middle Crescent Crater; and a core tube was driven into the surface near the LM late in the first EVA period. The documented sample, collected along the geologic traverse during the second EVA period, included a variety of rock and soil samples, one single- and one double-core tube, the special environment and gas analysis samples, and several totebag samples that were brought back in a totebag. One sample was grapefruit size.

Igneous rocks and breccias were collected on this mission. The igneous rocks are basaltic and vary widely in both texture and modal composition. Most of the igneous rocks fit a fractional crystallization sequence that indicates either that they represent parts of a single intrusive sequence or that they are samples of a number of similar sequences.

The breccias and fines have a higher carbon content than the crystalline rocks, which is presumably largely due to contributions of meteoritic material and the solar wind. The level of indigenous organic material capable of volatilization or pyrolysis, or both, appears to be extremely low ($\leqslant 10$ to 200 ppb).

The content of noble gas of solar-wind origin is less in the fines and breccias of the Apollo 12 rocks than in similar material from Tranquility Base. The breccias contain less solar-wind contribution than the fines, which indicates either that the breccias were formed from fines that were lower in solar-wind noble-gas content than the fines presently at the surface or that the gases escaped during the process of formation.

The presence of nuclides produced by cosmic rays shows that the rocks have been within 1 m of the lunar surface for 1 to 200 million years. The preliminary ^{40}K-^{40}Ar measurements on igneous rocks show that they crystallized 1.7 to 2.7 billion years ago.

The Apollo 12 breccias and fines are chemically similar and contain only half the titanium content of the Apollo 11 fines. The composition of the crystalline rocks is distinct from that of the fines material in containing less nickel, potassium, rubidium, zirconium, uranium, and thorium. The Apollo 12 rocks contain less titanium, zirconium, potassium, and rubidium and contain more iron, magnesium, and nickel than the Apollo 11 samples. Systematic variations among the magnesium, nickel, and chromium contents occur in the crystalline rocks, but there are only small differences in the potassium and rubidium contents.

Comparison of the Apollo 12 samples from Oceanus Procellarum with the Apollo 11 samples from Mare Tranquillitatis shows that the chemistry at the two mare sites is clearly related. Both sites show the distinctive features of high concentrations of refractory elements and low contents of volatile elements; these two features most clearly distinguish lunar material from other material. This overall similarity indicates that the Apollo 11 sample composition is not unique. Taken in conjunction with the Surveyor 5 and 6 chemical data, this similarity is suggestive of a similar chemistry for the maria basin fill. Unlike the Tranquility Base samples, the element abundances in the fines of the Apollo 12 samples display a generally more fractionated character than the rocks. The fine material and the breccias are generally quite similar in composition and could not have formed directly from the large crystalline rock samples. The chemistry of the fine material is not uniform in the different maria.

The overall geochemical behavior of the lunar rocks is consistent with the patterns observed during fractional crystallization in terrestrial igneous rocks—a process that involves separation of olivine and pyroxene; depletion in the silicate melt of the elements such as nickel and chromium, which preferentially enter these mineral phases; and enrichment of the residual melt in such elements as barium and potassium, which are excluded from the early crystal fractions. The slight degree of enrichment of barium and potassium indicates an early stage of the fractional crystallization process. Whether these rocks form a related sequence or are a heterogeneous collection of similar origins cannot be answered from the chemical evidence.

The chemistry of the Apollo 12 samples does not resemble that of any known meteorite because nickel, in particular, is strikingly depleted. The Apollo 12 sample chemistry, however, has interesting similarities with the eucrites; and there now seems to be a fairly good possibility that rocks of chemistry similar to the eucritic meteorites are present on the Moon.

The Apollo 12 material is enriched in many elements by 1 to 2 orders of magnitude in comparison with estimates of cosmic abundances. The mare material is clearly strongly fractionated relative to most models of the composition of the primitive solar nebula.

The K-Ar age of the Apollo 12 rocks is most interesting scientifically. The ^{40}K-^{40}Ar age of these rocks reinforces the possibility recognized from the data obtained on the Apollo 11 rocks, that is, that the lunar maria are geologically very old. If the minimum ages established by the K-Ar method are indicative of the true age of the Apollo 12 rocks, then the mare material in Oceanus Procellarum at the Apollo 12 site is approximately 1 billion years younger than that at the Apollo 11 site. Although this K-Ar age is subject to various uncertainties, the younger age for the Apollo 12 material is consistent with geological observations. This large age difference indicates a prolonged period of mare filling.

Solar-Wind Composition Experiment

The Apollo 12 foil had the same dimensions, general makeup, and trapping properties as the Apollo 11 foil. Three small foil pieces were decontaminated by ultrasonic means. It was found that a piece of the foil that had been shielded from the solar wind had a ^4He concentration per unit area that was less than 1 percent of the concentrations found in the foil pieces exposed to the solar wind. There is good agreement between the concentrations and the $^4He/^3He$ ratios measured in two exposed foil pieces.

From the first two Apollo 12 foil pieces analyzed, the $^4HE/^3HE$ ratio is 2600 ± 200 for the Apollo 12 exposure period. This value is higher than the $^4He/^3He$ ratio obtained thus far from analyses of pieces of the Apollo 11 foil. Comparative analyses of pieces of foils from the two flights are being continued to confirm this difference. Actually, time variations in isotopic ratios in the solar wind can be expected, and the $^4He/^3He$ ratio has to be determined repeatedly to assess the range of occurring variations before an average for the present-day solar wind can be established. This average is of high astrophysical significance, because it can be compared with ancient $^4He/^3He$ ratios derived from solar-wind gases trapped in the lunar surface or in meteorites.

Preliminary results from the investigation of the returned Surveyor 3 components indicate a surprisingly low number (0 to 3) of high-velocity-impact pits of meteoritic origin. The majority of low-velocity-impact pits have been identified as resulting from the LM descent. The Surveyor 3 spacecraft was found to be covered with a light-brown coating that has been identified as lunar dust. Other investigations in progress are engineering (cold welding) investigations, radioactivity analysis of the Surveyor 3 TV camera, dust analysis, and alpha-particle measurements.

1. Photographic Summary of the Apollo 12 Mission

L. C. Wade[a]

The scientific, geologic, and photographic objectives of the Apollo 12 mission were designed to achieve the maximum return of lunar data. The staytime on the lunar surface was increased to accommodate two extravehicular activity (EVA) periods. The photography accomplished during the Apollo 12 mission was designed to document and augment the experimental, observational, and geologic data obtained from the extended lunar surface activities and from the scientific instruments placed on the lunar surface. A further photographic objective was to obtain, from lunar orbit, photographs of future landing sites. Orbital photographs will aid in the planning of pinpoint landings in lunar terrain more rugged and of even greater scientific interest than the mare-type terrain at the Apollo 11 and 12 landing sites.

This chapter is a brief description of the Apollo 12 mission and is illustrated with a small sample of the hundreds of photographs taken by the astronauts during the mission. These photographs will require years to analyze completely and, thus, will be a continuing contribution to the improvement of man's knowledge of the Moon. The lunar multispectral photography experiment may aid in the determination of subtle color or tone changes on the lunar surface. The photographs taken during the EVA periods will aid in identifying the original locations and positions of many of the returned lunar samples. Photographs of Surveyor 3, when examined in conjunction with the parts of the spacecraft that were returned to Earth, will provide engineering data on the effects of Earth-made materials of long-duration exposure to the lunar environment.

During the early part of the lunar orbital phase of the mission, the crew took many photographs of the lunar surface. A series of photographs of particular interest was that series of the Fra Mauro area taken during revolution 10. These photographs of the Apollo 13 landing site showed the site at a 7° Sun elevation angle and will be used to train the Apollo 13 crew for their pinpoint landing, which will occur at approximately that Sun angle.

The lunar multispectral photography experiment was performed by Astronaut Gordon, the command module (CM) pilot, while the lunar module (LM) was on the lunar surface. Astronaut Gordon took a series of stereoscopic strip photographs with the four-camera array during two orbital revolutions and photographed selected target areas on a third revolution.

On revolution 39, the command and service module (CSM) accomplished an orbital plane change to prepare for extensive photography of future landing sites. The plane change opened up a considerable amount of new area on the lunar surface to both photographic and visual observation and placed the orbital trace of the CSM over the three potential landing sites to be photographed—the crater Lalande, an area north of the crater Descartes, and the Apollo 13 landing site in the Fra Mauro region.

On revolution 40, a terminator-to-terminator strip of photographs was taken with the bracket-mounted Hasselblad camera using the 80-mm lens. An intervalometer was used to trigger one frame every 20 sec with sufficient overlap to provide stereoscopic photography. By using this strip of stereoscopic photographs, the approach terrain into the landing sites can be better defined, which will be important in designing the profiles for the three sites.

On revolution 41, high-resolution photographs of two potential landing site areas were taken using the 500-mm lens on the Hasselblad camera.

[a] NASA Manned Spacecraft Center.

Unfortunately, a film magazine malfunctioned, and these photographs were partially lost. The photographs of the two areas were successfully retaken on revolution 43, with another film magazine; however, this resulted in the inability to complete a second terminator-to-terminator stereoscopic strip of photographs that had been planned for revolution 44. The crater Lalande was successfully photographed on revolution 45. In addition to these photographs, the crew took several photographs of various areas using the 250-mm lens on the Hasselblad camera. The Davy Rille was one such area photographed. The Apollo 12 photographs, combined with earlier Lunar Orbiter photographs, have kindled scientific interest in the Davy Rille area, and it is planned to rephotograph this area on the Apollo 13 mission to determine if it may be selected as a future landing site.

After separation, the LM began a series of maneuvers that resulted in the successful lunar landing. Unlike the Apollo 11 LM, the Apollo 12 LM flew the entire descent phase of the mission in the heads-up position; therefore, photographic documentation of the Apollo 12 LM descent phase started with the pitchover maneuver. At this point, the 16-mm data acquisition camera vividly recorded the scene as described by Astronaut Conrad during the actual landing. The "Snowman" formation, for which the crew targeted, is clearly visible during most of the descent film. Astronaut Conrad's landing maneuvers are also evident, as is the actual moment of touchdown, which occurs within an extensive dust cloud. This film has been correlated with the telemetered data to document the final landing maneuvers performed by the LM crew.

Following touchdown, photographs of the immediate vicinity of the LM were taken through the LM windows. Panoramic mosaics of these photographs are presented in chapter 10 of this document and show the lunar surface before and after both EVA periods. Initial activities on the first EVA period were recorded by the Hasselblad cameras and by a color television camera. However, shortly after both astronauts had egressed the LM, the television camera was inadvertently pointed directly at the Sun, and television coverage of the mission was lost. Thus, the photographs taken by the crew are the only visual record of their lunar surface activities.

The 16-mm data acquisition camera was used by Astronaut Bean to record Astronaut Conrad's descent down the ladder and his early activities on the lunar surface. Photographs of Astronaut Conrad picking up the contingency sample provided a record of the sample location and illustrated Astronaut Conrad's movements in the 1/6g environment on the lunar surface.

Both crewmen took hundreds of photographs while on the surface, including many panoramas. They photographed the unloading of the Apollo lunar surface experiments package (ALSEP) and its deployment on the lunar surface. Some unusual features, which the crew called mounds, were seen and photographed on the surface. These features can be located on the Lunar Orbiter photography of the areas and, thus, can be used as a key to interpreting the Lunar Orbiter photography. After completion of ALSEP deployment, and at the request of geologists on Earth, the crew made a traverse to a large subdued crater approximately 300 m in diameter.

During the second EVA period, the first photographic activity after departure from the LM was to take black-and-white photographs, with polarizing filters on the Hasselblad cameras. After taking these photographs, the filters were discarded, and the astronauts began the documented geological traverse. During the course of the traverse, the crew photographed the rock samples collected, the core samples taken, and the craters visited.

The final leg of the traverse on the second EVA period took the crew to the crater in which Surveyor 3 landed in April 1967. The crew recorded photographically their activities in the vicinity of Surveyor 3; photographed the Surveyor 3 spacecraft; and with their cameras, duplicated many of the scenes first recorded by the Surveyor television camera.

Near the completion of the second EVA period, Astronaut Bean took several stereoscopic pairs of photographs with the Apollo lunar surface closeup camera. These photographs show, in exceptional detail, the fine grain structure of the lunar surface material.

On their return from the Moon, the crew followed the practice of previous Apollo crews and took a series of photographs of the Moon after transearth injection. Several of these photographs were planned to provide slightly different lunar aerial coverage as an aid in solving selenodetic problems of the Moon.

As the spacecraft neared the Earth, the crew reported and photographed the impressive sight of the Sun being eclipsed by the Earth. This unusual sighting was reported to be one of the most spectacular views observed during the entire mission. The eclipse was photographed using color film in the 16-mm data acquisition camera and black-and-white film in the Hasselblad camera.

Reentry of the Apollo 12 spacecraft over the Pacific Ocean occurred later in the day than had previous Apollo flights, and the crew recorded an excellent sequence of reentry data on 16-mm film. The film data cover the period from the first evidence of heat-shield burning to the deployment of the three main parachutes.

Overall, the crew of Apollo 12 took many excellent photographs of scientific interest, provided extremely valuable photographic information concerning possible future landing sites, and photographically recorded many items of engineering and scientific interest. Cameras, film types, and usage are listed in tables 1-I and 1-II.

TABLE 1-I. *Apollo 12 Photographic Equipment Used in CM*

Camera	Features	Film	Remarks
Hasselblad	Electric; with 80-mm lens, 250-mm lens, and 500-mm lens	70-mm, type SO-368 Ektachrome MS color-reversal film, with a normal ASA of 64; and 70-mm, type SO-3400 Panatomic X black-and-white film, with a normal ASA of 80	Used in lunar orbit to make stereoscopic strip photographs of potential landing sites
Data acquisition camera	With 5-mm lens, 18-mm lens, and 75-mm lens	16-mm, type SO-368 film; and 16-mm, type SO-164 Panatomic X black-and-white film, with a normal ASA of 80. Also, 16-mm, type SO-168 Ektachrome EF color-reversal film, with a normal ASA of 160	Type SO-168 film exposed and developed with an ASA of 1000
Multispectral photography experiment array	See section 9	See section 9	

* For complete information, see section 10 of this document.

TABLE 1-II. *Apollo 12 Photographic Equipment Used in LM and During EVA Periods*

Camera	Features	Film	Remarks
Hasselblad (2)	Electric with 60-mm lens and reseau plate	70-mm, type SO-168 film (first EVA period) and 70-mm, type SO-267 plus XX black-and-white film, with a normal ASA of 278	
Data acquisition camera	With 10-mm lens	16-mm, type SO-368 film	Mounted behind right window LM; recorded LM descent (from approximately 6 km) and lunar surface activities
Apollo lunar surface closeup camera	Stereoscopic; has 46-mm M-39 lens with aperture of f/22.6 and built-in light source	35-mm, type SO-368 film	Stereoscopic convergent angle of 9°

FIGURE 1-1. — On November 14, 1969, at 11:22 a.m. e.s.t., the second U.S. lunar landing mission, Apollo 12, lifted off from Cape Kennedy, Fla. On board were Astronauts Charles Conrad, Jr., Richard F. Gordon, and Alan L. Bean. (NASA photograph S-69-58883)

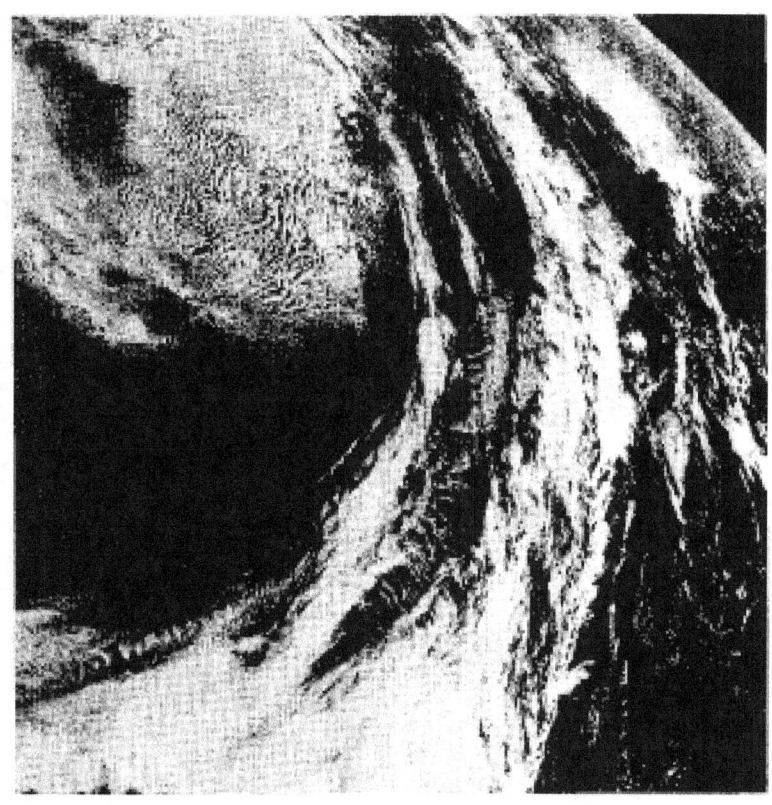

FIGURE 1-2. — Shortly after translunar injection, the central portion of the United States was photographed. Lake Michigan can be seen in the upper left corner of the photograph. The river at the center of the photograph is the Red River, and the point of land visible in the lower right corner is the Yucatan Peninsula. The cloud formation suggests a frontal passage that is in the Gulf of Mexico. (NASA photograph AS12-50-7325)

FIGURE 1-3. — Made during the translunar coast phase of the Apollo 12 mission, this view overlooks a great expanse of the Pacific Ocean. Baja California and Mexico can be seen in the lower right corner of the photograph, and the Yucatan Peninsula and Central America are visible in the lower left center. The object near the center of the photograph and above the clouds is one of the four panels that protect the LM during launch and that are jettisoned at the time of CSM separation from the SIVB stage of the launch vehicle. (NASA photograph AS12-50-7326)

FIGURE 1-4.—The LM is shown in place on the SIVB stage shortly after separation of the CSM and the SIVB stage. (NASA photograph AS12-50-7329)

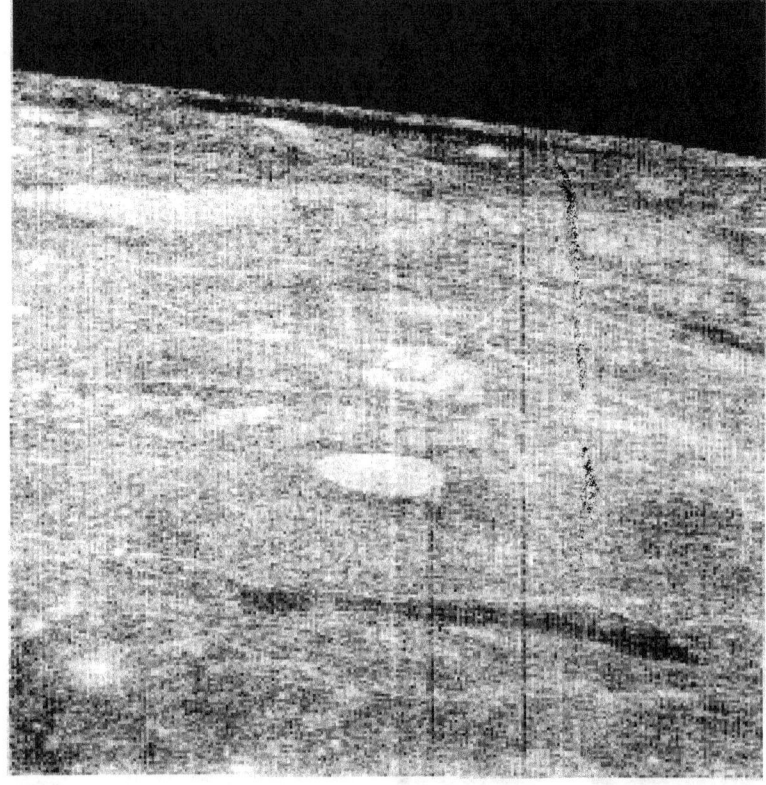

FIGURE 1-5.—Looking back to the southeast on the lunar surface, the crew took this dramatic view of the crater Humboldt. The central peaks appear "snow" white because of the high-Sun elevation angle. The arcuate fractures within Humboldt are evidence of the forces working on the surface of the Moon to change the lunar topography. (NASA photograph AS12-50-7416)

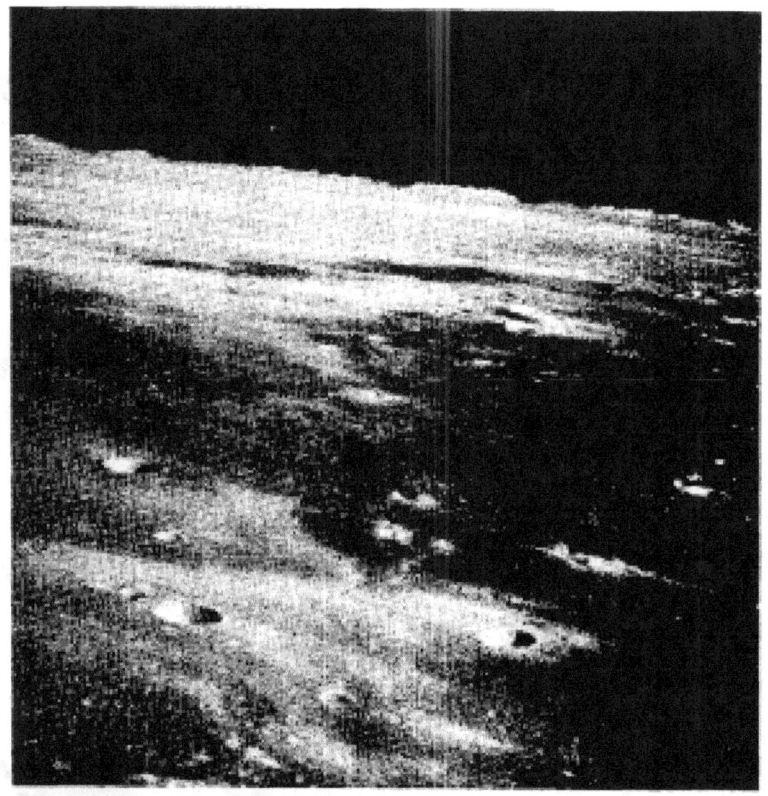

FIGURE 1-6. — Looking into Sinus Aestuum. A striking example of the differences in albedo of the lunar surface. (NASA photograph AS12-52-7733)

FIGURE 1-7. — Alphonsus, the target point of Ranger 9. The dark "haloed" areas on the floor of Alphonsus are distinctly evident. These areas are of high geologic interest. (NASA photograph AS12-51-7580)

FIGURE 1-8.—Looking into the terminator. The crater Kepler can be seen (at the center of the photograph) just to the east of the terminator. On the dark side of the terminator, the crew could distinguish surface detail in earthshine. (NASA photograph AS12-51-7547)

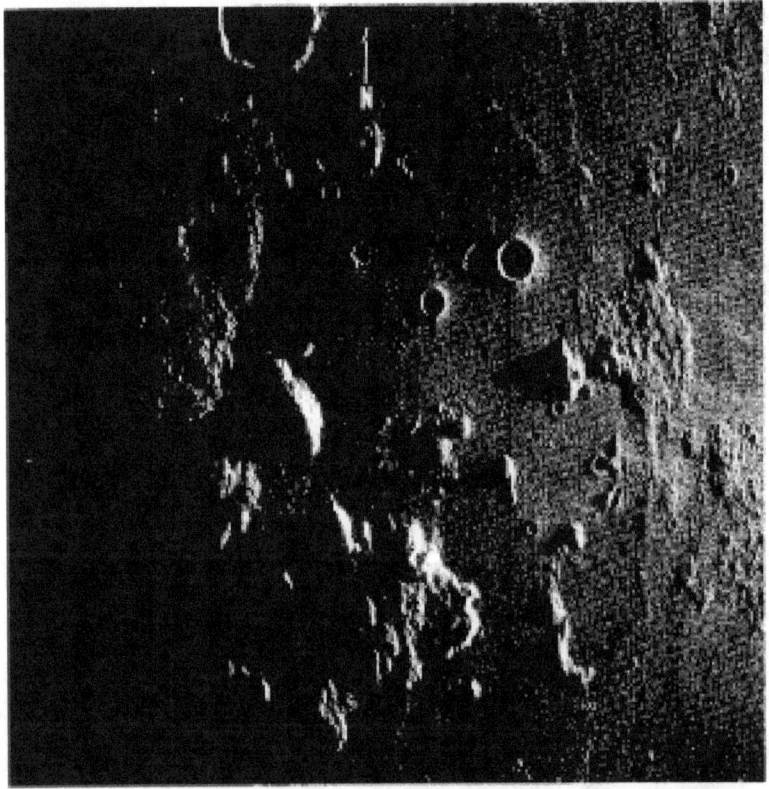

FIGURE 1-9.—The difference between day and night on the Moon. The terminator is just to the west of the crater Gambart, which is at the extreme north in the photograph. This view is representative of the lunar surface as seen by the crew as they crossed the terminator twice every orbit. The Apollo 13 landing site is located in the highland region in the shadow area of the photograph. (NASA photograph AS12-50-7438)

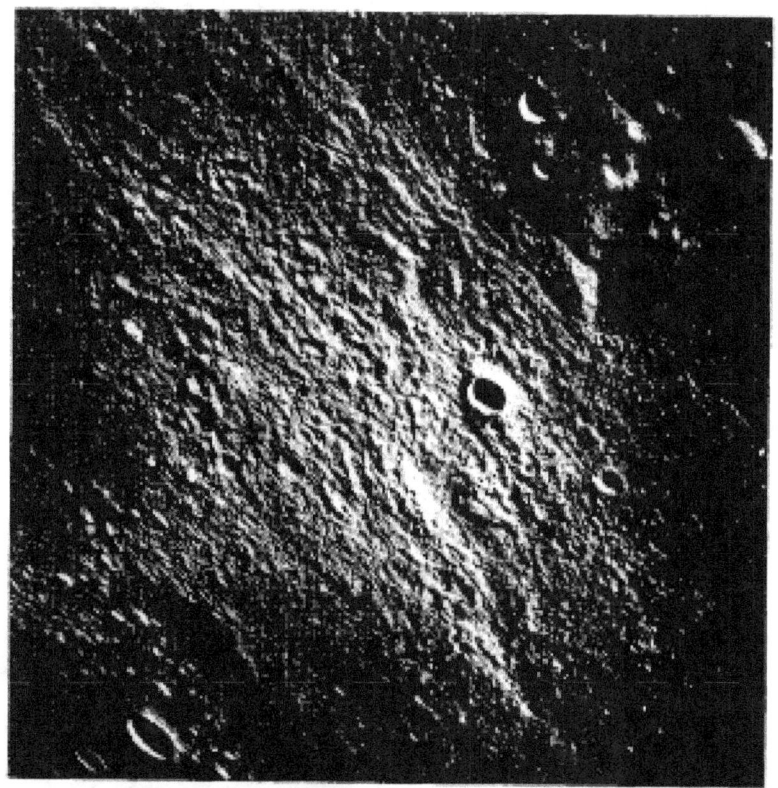

FIGURE 1-10.— An oblique view looking northwest at the highland area north of the crater Fra Mauro. This photograph was taken from the CM at an orbital altitude of 60 n. mi. The Sun elevation angle was 7°. The Sun will be only slightly higher when Apollo 13 lands in this area. (Arrow indicates landing site.) Although taken early in the mission, this photograph supplements the site photography taken later. (NASA photograph AS12-52-7597)

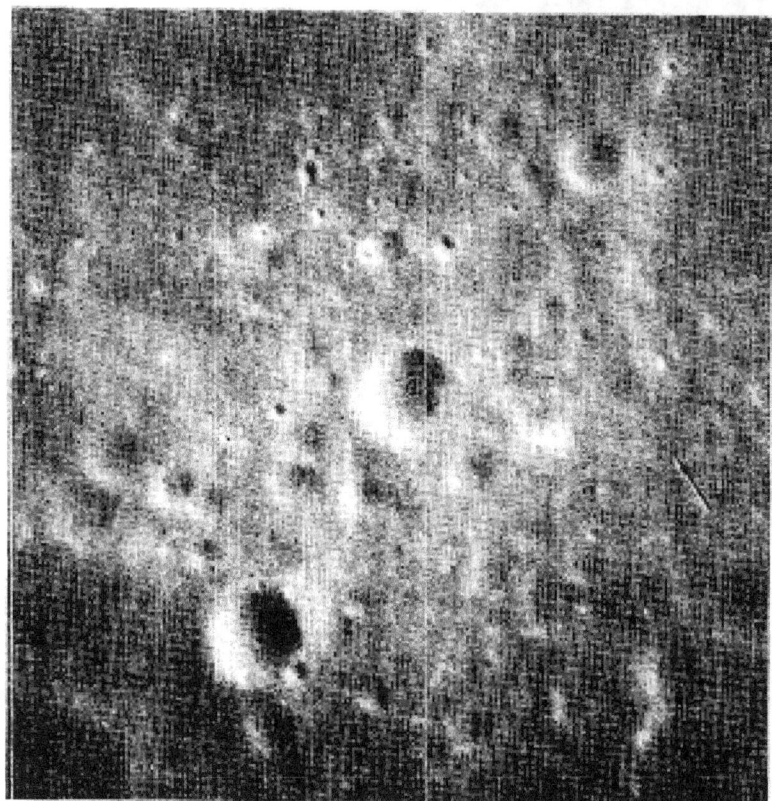

FIGURE 1-11.— The Fra Mauro area. This photograph was taken with a 500-mm lens from the CM at an approximate orbital altitude of 60 n. mi. This photograph is representative of the photography taken during the final orbital phase of the mission. The arrow indicates a hilly area slightly north of Fra Mauro Crater that has been selected as the landing site for the Apollo 13 lunar mission. This photograph was taken with a 40° Sun elevation angle. (Compare with fig. 1-10.) To take this and other orbital photographs, Astronaut Gordon constantly had to adjust the pitch of the spacecraft to accomplish image motion compensation. (NASA photograph AS12-53-7833)

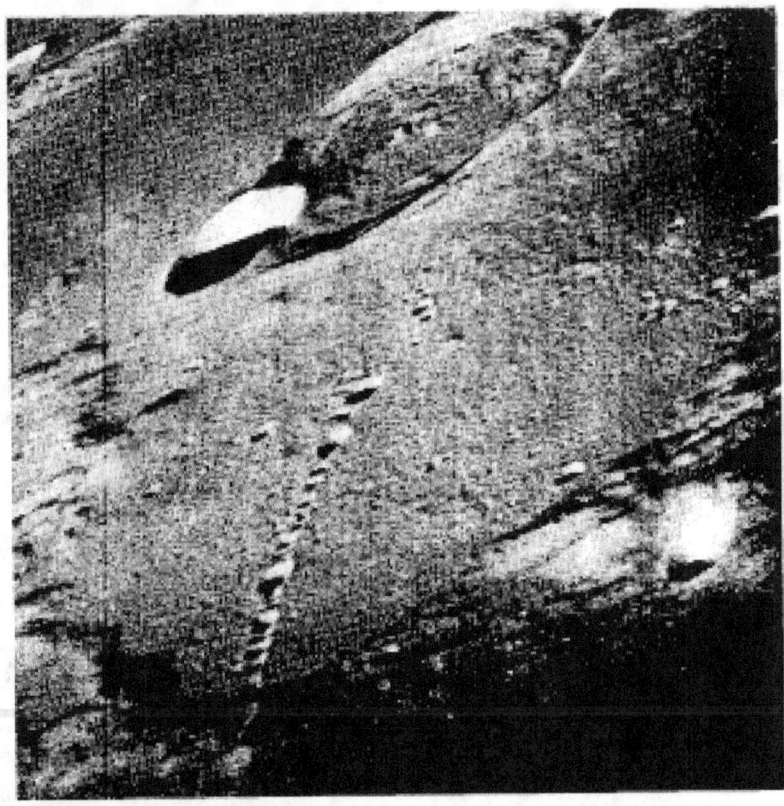

FIGURE 1-12.— Davy Rille, potential landing site for a future lunar mission. The proposed landing site is at the point where the rille touches the highlands in the east. North is at the right of the photograph. (NASA photograph AS12-51-7485)

FIGURE 1-13.— The LM is above the floor of the giant crater Ptolemaeus in this westward-looking oblique. The LM appears to be close to the surface, even though it is some 60 n. mi. high. This was one of Astronaut Gordon's last views of the LM as he began his 1½ days of solo flight around the Moon. (NASA photograph AS12-51-7507)

Figure 1-14.— A high-oblique view looking northeast. This photograph was taken from the LM. The large crater Copernicus is in full view, and the Carpathian Mountain Range is visible on the horizon. The stark lunar relief is accented by the low-Sun elevation angle. (NASA photograph AS12-47-6875)

Figure 1-15.— Earthrise as seen from the LM. This photograph was taken before the LM started its final descent to the lunar surface. (NASA photograph AS12-47-6880)

FIGURE I-16.—After collecting the contingency sample, Astronaut Conrad took several panoramas in the vicinity of the LM. In this panorama, the LM, Surveyor Crater, the television camera, and the S-band antenna are visible. The Surveyor 3 spacecraft can barely be distinguished in the shadow of the crater. (NASA photograph S-70-22360)

FIGURE I-17.—The crew erected the American flag after landing and collecting the contingency sample. The long shadow of the LM and the bleak lunar surface serve as a fitting background. (NASA photograph AS12-47-6897)

Figure 1-18. — Astronaut Bean unloading equipment from Intrepid. Astronaut Conrad was standing to the north of Intrepid when he took this photograph. Both the edge and the interior of Surveyor Crater are just to the east. (NASA photograph AS12-46-6749)

Figure 1-19. — After unloading the ALSEP, Astronaut Bean used the "barbell" carry to take the ALSEP to its deployment site. (NASA photograph AS12-46-6807)

FIGURE 1-20.— The deployed ALSEP is visible in this northwest-looking view. The lunar surface magnetometer experiment is in the foreground, and the other objects on the lunar surface (from left to right) are: a discarded subpallet and cover; the solar-wind spectrometer experiment; the suprathermal ion detector experiment, with the darker radioisotope thermoelectric generator behind it; the ALSEP central station, with an astronaut adjusting the antenna; the passive seismic experiment (PSE); and the discarded PSE girdle. The deployment of the ALSEP was one of the major tasks of the first EVA period. (NASA photograph AS12-47-6921)

FIGURE 1-21.— Looking back at the LM from the ALSEP deployment site. The ALSEP central station is approximately 690 ft from the LM. The magnetometer and the passive seismometer are clearly visible. Only a small part of the rim of Surveyor Crater, which is located behind the LM, can be seen at this distance. (NASA photograph AS12-47-6928)

FIGURE 1-22.—After deploying the S-band antenna and the solar-wind experiment and after erecting the flag, the crew moved around the LM and photographed their spacecraft on the edge of the Surveyor Crater. Surveyor 3 is below the rim of the crater and cannot be seen in this view. (NASA photograph AS12-47-6899)

FIGURE 1-23.—This mound on the lunar surface was photographed looking towards the southwest. This type of feature and its formation have generated considerable interest among lunar geologists. (NASA photograph AS12-46-6795)

FIGURE 1-24.— Two panoramas of the large subdued crater visited during the first EVA period. The crater is more than 300 m across and was visited near the end of the EVA period. (NASA photograph S-70-22361)

FIGURE 1-25.— The tools of the lunar geologist. The gnomon, the core tube, and the tool carrier were carried on the documented geology traverse during the second EVA period. (NASA photograph AS12-49-7320)

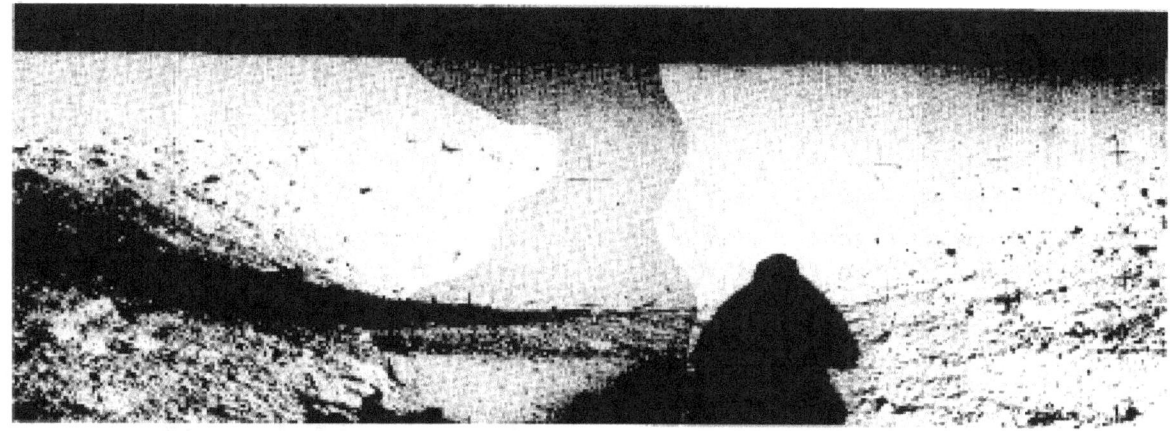

FIGURE 1-26.—A panorama looking southwest across Head Crater. Astronaut Conrad rolled a rock down the side of this crater. Head Crater was visited during the early part of the second EVA period. (NASA photograph S-70-24309)

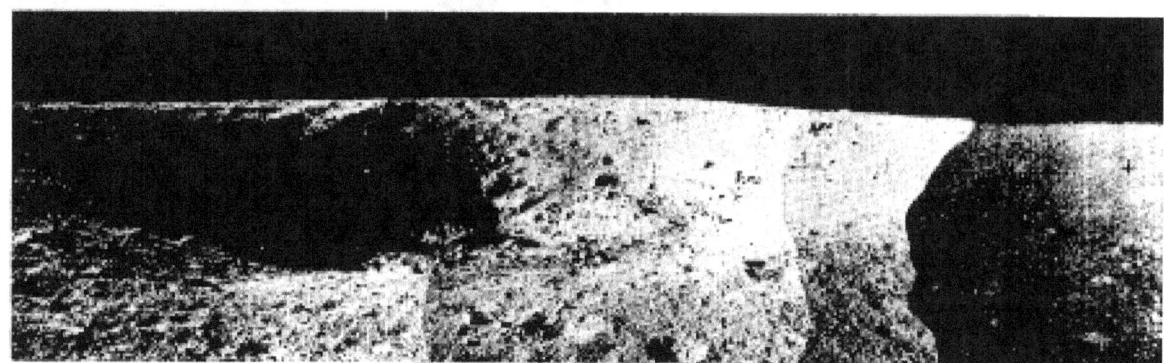

FIGURE 1-27.—A panorama of Bench Crater. This crater was selected by the geologists as a key crater to visit during the second EVA period. (NASA photograph S-70-24311)

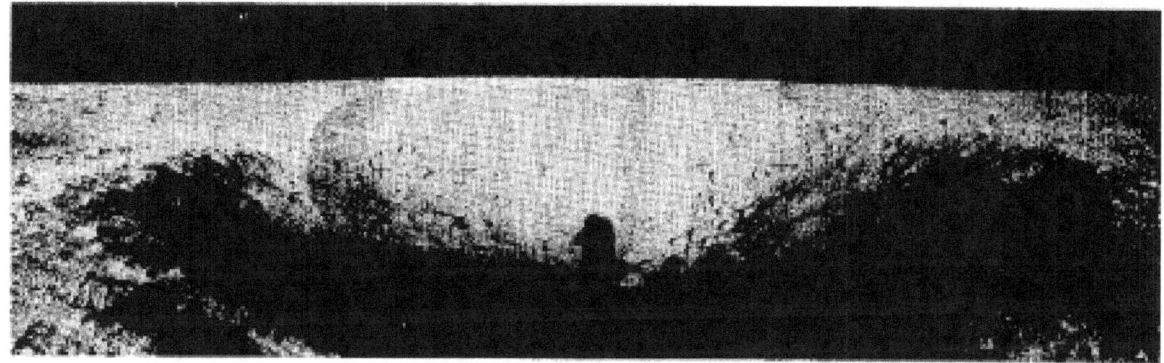

FIGURE 1-28.—A panorama of Sharp Crater. This crater was at the outermost point of the crew's traverse during the second EVA period. (NASA photograph S-70-22363)

FIGURE 1-29.—The LM is visible at the extreme right center of the photograph, with Head Crater visible in the foreground. The view, taken during the second EVA period, is toward the east-northeast. The astronaut is moving away from Head Crater. (NASA photograph AS12-49-7213)

FIGURE 1-30.—The core-sample tool embedded in the lunar surface near Halo Crater. In this southeast-looking view, the loosely compacted lunar soil is clearly visible, and Halo Crater can be seen in the background. (NASA photograph AS12-49-7288)

FIGURE 1-31.—The LM on the lunar surface near the rim of Surveyor Crater. A portion of Surveyor Crater is visible in the foreground. The flag is visible just to the left of the LM, and the ALSEP can be seen at left center of the photograph. (NASA photograph AS12-49-7317)

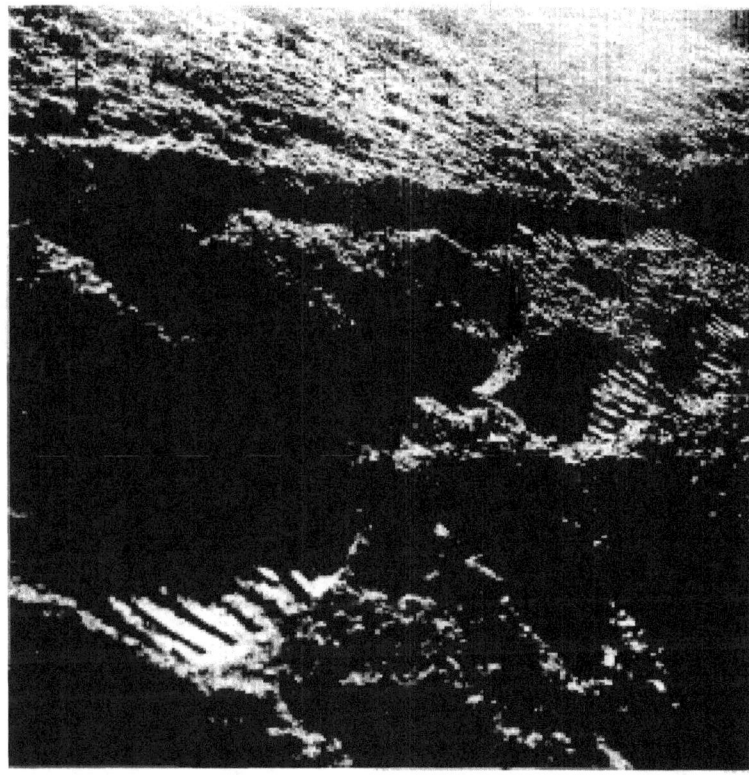

FIGURE 1-32.—Footprints left by the astronauts walking on the interior slope of Surveyor Crater. The footprints are sharp, and a slight break in the lunar soil can be seen. (NASA photograph AS12-48-7098)

FIGURE 1-33. — A closeup of the lower part of Surveyor 3. Those parts that were not returned to Earth were photographed extensively by the Apollo 12 crew. (NASA photograph AS12-48-7138)

FIGURE 1-34. — Astronaut Bean and two U.S. spacecraft on the surface of the Moon. This photograph and figure 1-29 clearly show how close Astronaut Conrad landed the LM to its predesignated landing point. After completing their work around Surveyor 3, the crew moved back toward the LM. (NASA photograph AS12-48-7133)

FIGURE 1-35.—The Moon, photographed after transearth injection. The geographic areas visible include Smyth's Sea, the Sea of Crises, the Sea of Fertility, and the Sea of Tranquility. The areas of coverage on this photograph are from approximately 110° east longitude to 25° east longitude. (NASA photograph AS12-55-8221)

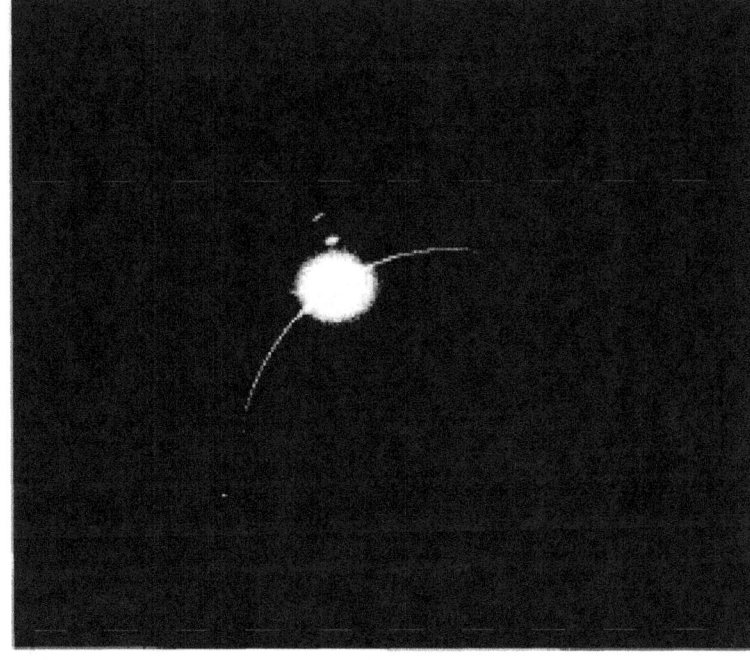

FIGURE 1-36.—A photograph of the solar eclipse, taken from the CM during transearth coast shortly before reentry. (NASA photograph AS12-53-7917)

2. Crew Observations

Alan L. Bean,[a] Charles Conrad, Jr.,[a] and Richard F. Gordon[a]

Apollo 12 was the first in a series of missions designed to take advantage of the unique capabilities of the trained observer in combination with a carefully designed instrument package to insure the maximum possible return of scientific data about the Moon. Apollo 11 proved the capabilities for a manned lunar landing, for working in the lunar environment on the lunar surface, for successfully deploying an instrument package for long-term data retrieval, and for a safe return to Earth. The operational use of these capabilities on Apollo 12 has been an important first step in the scientific exploration of the Moon.

This section is a summary of the scientific observations of the Apollo 12 crew from lunar orbit and from the lunar surface. Included in the section are comments on the appearance of the Earth and Moon from orbit, observations from the lunar surface on the appearance of lunar terrain and features, descriptions of crew adaptability to the lunar environment, and geologic and soil mechanics observations made during the two extravehicular activity periods. The section is concluded with a short discussion of the solar eclipse witnessed by the crew during transearth coast.

Observations of Earth From Space During Translunar Flight

After about 8 hr into the mission, the Earth appeared to be about volleyball size, and nothing was readily discernible but clouds and water. There appeared to be a glare point on Earth similar to the zero-phase point. On the dark side of the Earth, it was difficult to tell where the Earth stopped and space began.

There was no noticeable rotation or movement of the Earth after about 11 hr into the mission.

[a] NASA Manned Spacecraft Center.

It was a strange feeling; the Earth just seemed to hang in black space. The subsolar point was over the water and caused the surface to glint, somewhat like a light on a billiard ball. One spot in particular showed a specular reflection with a bright halo. The red Earth colors were easy to distinguish, but the greens and grays were difficult to distinguish from the blues.

After about 32 hr into the mission, the Earth appeared to be about the size of a golfball held at arm's length. Even though landmasses seemed to be brown, it was difficult to distinguish landmasses from water. Later in the mission, it was noted that the vegetation on Earth made a kind of gray-green color area that tended to blend into the ocean.

As we got farther away from Earth, only areas that contrasted strongly with the blue-gray and blue-green areas of Earth were noticeable. Overall, the Earth seemed to be a pretty blue and white, and our planet looked like an oasis.

Lunar Surface Features as Seen From Lunar Orbit

We began to get our first good look at the lunar surface when the spacecraft went into lunar orbit. The mare areas were generally smooth, but there were several long ridges running through these areas, and several isolated hills were scattered throughout the smooth maria. At first glance, these hills resembled cumulus clouds because the low sun angle caused them to be extremely bright on top. By using the monocular, though, it was possible to see that the hills were high and isolated. While using the monocular, we also noticed three or four bright craters that had very dark-gray material streaking down the crater sides, like landslides. The gray streaks had the appearance of something like a liquid spill; the streaks were completely different from anything else we saw. The Ocean of Storms looked much like a desert area, but we could not think of a place on Earth that is similar to it.

In contrast to the smooth maria, the surface in the terra was covered with rounded knolls and seemed to be quite a bit rougher. We also saw several areas that seemed to indicate that the lunar surface has been involved in some volcanic action.

Most of the lunar craters that we observed from lunar orbit appeared to be caused by impacts on the lunar surface. One fresh-looking crater had a collapsed side; another crater had a fantastic ejecta blanket that must have gone out to a distance of 50 to 60 times the crater diameter. Two young craters in the mare area were particularly startling because of their perfectly circular ray patterns.

In the crater Humboldt, we saw great fracture marks that looked like crisscrossed roads in the desert. Big, black boulders appeared to dot the surface of the central peak of the crater Langrenus. Boulders were also observed on the central peak of Theophilus, and on this central peak there were some rather well-defined ridge lines. Upon looking closely, we saw what appeared to be a mild terracing effect that ran parallel to the ridge lines and also some rilles that ran perpendicular to the ridge lines.

We also saw rilles as we passed over the Sea of Serenity. In this sea, there were two parallel rilles (long narrow valleys) and two parallel grabens (depressed segments of the lunar crust bounded on two sides by faults) with associated crater chains.

As for the color of the Moon, we noticed very little change as we orbited. At first, the surface had a very light gray-white concrete appearance. On later orbits, we began to notice a touch of brown in the surface features. One interesting contrast was that of the Moon to the black sky. In comparison to the coal blackness of the sky, the Moon appeared to be a very light concrete color. The terra was a gray color, and as we passed over the Sea of Fertility, the sea appeared to be a slightly darker gray than the terra. We commented at the time that the Sea of Fertility resembled wet beach sand.

We also got a look at the back side of the Moon during lunar orbit. The command module pilot commented that he was more impressed by the lunar back side.

One especially impressive sight was the Moon in earthshine. The Moon is fairly easy to see in earthshine; it is quite beautiful and very soft looking, and the earthshine gives the Moon a greenish tinge, making it look gray-green.

Landing and Observations From the Lunar Module

Landing

As soon as the pitchover started, the computer was used to enable the landing point designator (LPD), and we immediately looked out the window. For the first couple of seconds none of the lunar surface features were recognizable, although the visibility was excellent. The field of view and the lunar surface detail were also excellent, but photographs do not provide adequate preparation for the first look out the window. However, when we looked down the 40° line on the LPD, our five-crater chain and the "Snowman" became very obvious. No early landing site redesignations were required. We were targeted in the middle of Surveyor Crater.

We made a steeper-than-planned descent. At approximately 700 ft, we went to manual control and reduced our descent rate to enable us to get a better look around. As we passed over Surveyor Crater, at approximately 500 ft, a suitable landing area was apparent between Surveyor and Head Craters. It was necessary to fly around Surveyor Crater to get to this area, which appeared to be perfectly smooth. The vertical descent was begun at approximately 300 ft, and the descent rate was relatively low. At 175 ft, we began to pick up dust. The dust went out as far as we could see in any direction and continued to build up until the lunar surface was completely obscured during the last 40 ft of descent. The reduced visibility forced us to get our attitude references from the instruments. Then, the lunar contact light came on, and the descent engine was shut down. We dropped approximately 2 or 3 ft to a gentle touchdown on a slight slope.

Color and Contact

At first glance out of the spacecraft window, there was absolutely no distinguishable color difference. About the only difference we noticed was in looking cross-Sun versus looking down-Sun. Looking down-Sun, the surface material looks the same, but it appears to be of different colors. Some of the rocks must definitely have had different colors and different textures, but viewed from the spacecraft, they did not appear to be different. All the materials on the horizon and the blocks on the horizon appeared to be of the same material, and when using the monocular with the low Sun angle, they all appeared to be white.

From the lunar module (LM), there did not seem to be any possibility of seeing a contact between different colored surfaces. We hoped that we could discern the contacts or different materials by looking at the textures during the extravehicular activity (EVA) periods. For example, the area that was described as being directly in front of the LM had north-south lines running through it. Other than those lines, the area looked like one uniform surface with many craters in it. There were no immediately apparent white rim craters near us. Most of the craters we saw from the LM window did not have a raised rim, and neither did they have any particular elongation. The craters seemed to be the same texture as the areas surrounding them. This lack of contact was verified during the EVA. We found that all the material looked the same until we were very close to the individual rocks.

First EVA Period

Lunar Surface Visibility

We were positive of where we were, but it was difficult to pinpoint our exact location because of the limited field of view out the LM windows, because of the general tendency to underestimate distances (sometimes by as much as 100 percent), and because of the difficulty of seeing even large craters beyond a distance of about 100 ft. None of the shadows that were visible in flight, the ones in the bottoms of craters, were visible after touchdown. The bright part of the landscape could always be seen, but it was difficult to find the craters. When we were on the ground, things that were far away looked closer than they really were. After egress to the lunar surface, however, it was possible to get an accurate position for the LM because it was apparent that we were close to the edge of Surveyor Crater, approximately 600 ft from the Surveyor spacecraft.

Lunar surface visibility was not too unlike Earth visibility, with the exception that the Sun was extremely bright. Cross-Sun and down-Sun viewing were not hindered to any great degree, except that shadows were visible only when viewing cross-Sun. It was difficult to view down-Sun exactly along the zero-phase direction, but this deficiency did not hinder normal lunar surface operations because the eyes could be scanned back and forth across these bright zones for visual assimilation. Objects in shadows could be seen with only a slight amount of dark adaptation. The only difficulty in seeing in a shadow occurred when some object was reflecting sunlight into the helmet visor while we were trying to see into the shadow. Once we entered the shadow, it was possible to see well.

Mobility

Our mobility and stability were generally the same as reported by the Apollo 11 crew. Running on the lunar surface, moving from side to side, hopping, and so forth, felt almost exactly as it had during simulations on Earth. There was no noticeable tendency for our boots to slip on the lunar surface.

Moving around on the lunar surface proved to be no special problem, although, at first, we moved slowly because of the required adjustment to a different center of gravity. It was difficult for us to walk "heel-toe, heel-toe" as we would have in a normal walk on Earth. There is really no such thing as walking on the lunar surface; it takes more energy to move slowly and take a normal Earth step than it does to move at a lope. Because of the reduced gravity, there is a brief period when both feet are off the ground at the same time. This condition gave us the impression that we were moving rapidly, although in reality, we were moving at about a normal Earth walking pace. Loping actually seemed to be the most natural way to move around.

The fall experienced by the commander proved to be no problem. When a fall begins, you first lose your balance rather quickly, particularly if you try to back up, because the ground is uneven and there is a possibility of stepping in holes or on rocks. The fall progresses so slowly, though, that there is plenty of time to almost turn around or to catch your footing before you actually fall to the surface. Because a fall begins so slowly on the Moon, it is usually possible to spin around, bend the knees, and recover. Several times, in trying to bend over to get something, we would start to fall over, but the fall progressed so slowly that we could start moving our feet and keep moving until they came back under us again.

Working on the lunar surface was not excessively strenuous. The metabolic rate was 900 Btu/hr for the commander and 1000 Btu/hr for the lunar module pilot. We did not feel too tired, even wearing the pressurized suit that tends to tire the wearer even during simulations. On the Moon, in the same light gravity as experienced in the simulations, with the same suit, and with the same weight, the wearer's legs never seem to get tired. The problem with the suit is that it does not always bend as the wearer wants to bend. For example, the suit bends fairly well in the knees and in the ankles, but it does not want to bend near the top of the thigh. This one area of inflexibility results in loping in a stiff-legged fashion — running with straight legs, landing flatfooted, and then pushing off with the toes. We expected to become tired loping in this manner, but apparently it takes much less force to walk this way on the Moon than on Earth, and as a result, the legs do not seem to tire.

Soil Mechanics

Our first impression of the lunar soil was that it was soft and queasy; however, we did not sink in too far. Our boot prints left a very shiny, compacted surface. We guessed that the particles were small and cohesive, because the prints were so extremely well defined and because no grains could be detected in the boot prints. Also, when walking in the bottoms of small craters, we seemed to sink in deeper. We believe that the bottoms of those small craters contain a softer dust than is found on the crater rim. The difference in softness is not great, but it is noticeable. The sides of the craters did not appear to be particularly slippery. When deploying the experiments, we noticed that tamping of the lunar surface material seemed to do little good in emplacing an experiment. The core tube was a little difficult to drive in, but if it were augered first, it could be pounded in to full length.

Upon inspection of the LM, we found that the Intrepid footpads sank in a little farther than did the Apollo 11 LM footpads. Our estimated depth of penetration was 1½ to 2 in. As observed on the Apollo 11 mission, the LM descent propulsion system (DPS) engine did not dig a crater, but the surface under the engine was clean. There were no loose dust particles, but a number of small, dirtlike clods seemed to be strewn out radially from the skirt of the engine. We did notice one unusual thing, though: there was an angular rock that was sitting approximately 6 in. from the engine exhaust skirt. It was approximately 3½ by 3½ in., and it was not stuck in the ground — it was just sitting there loosely. The ground around it was glassy clean, yet the DPS engine exhaust did not blow the rock away.

While sampling some of the lunar rocks and soil, we frequently noticed linear patterns that extended from north to south. These patterns seemed to indicate that an impact had occurred not long ago at the particular location.

Experiment Deployment

Several minor problems arose during deployment of the experiments. The Apollo lunar surface experiments package (ALSEP) fuel element would not come out of its cask easily, and several minutes were spent working with the delicate element before it was removed satisfactorily. We also noticed that while carrying the ALSEP in the ⅙g environment, the whole pallet tended to rotate, especially the pallet containing the radioisotope thermoelectric generator.

During deployment of the passive seismic experiment, the Mylar skirt on the experiment would not lie flat. We believe that the skirt had been folded for so long that it had an "elastic memory," and it wanted to go back to its original folded shape because of this memory. To make the skirt lie flat, we had to put lunar soil and bolts along the skirt edges.

There was also a problem with the deployment of the cold cathode ion gage (CCIG). The gage tended to undeploy itself, but we finally got it to lie down while pointing upward at about a 60° angle. The problem with the CCIG was caused by the cable, which kept pulling back on the instrument and causing the gage to tip over.

Generally, deployment of the experiments was accomplished with no major problems. However, there seems to be no way to avoid getting dust on the experiments during unloading, transport, and deployment. The experiments got dusty from dust splattered off the lunar surface by our boots, and they got dusty when we put them down on the lunar surface. Because there seems to be no operationally simple means of alleviating this dust condition, the presence of dust should be considered during the design of future lunar surface experiments. Of all the experiments deployed on this mission, the solar-wind composition experiment was the cleanest, and the passive seismic experiment was the next cleanest.

After lunar liftoff, when we were again in a 0g environment, a great quantity of dust floated free within the cabin. This dust made breathing without the helmet difficult, and enough particles were present in the cabin atmosphere to affect our vision. The use of a whisk broom prior to ingress would probably not be satisfactory in solving the dust problem, because the dust tends to rub deeper into the garment rather than to brush off.

During the transearth coast phase, we noticed that much of the dust that had adhered to the camera magazines and other equipment on the lunar surface had floated free in the 0g condition, leaving the equipment relatively clean. This fact was also true of the spacesuits, which were not as dusty after flight as they were on the surface after final ingress.

Second EVA Period

Color Observations

One of the more difficult aspects of the traverses was the fact that there did not appear to be any difference in color among either the rocks or the soils; they all looked about the same. The first day, everything appeared to be dull gray. If we looked very closely, of course, now and then it was possible to observe a white rock or, in an area where we had disturbed the soil, perhaps a slightly different shade of gray. Between the first and second days, definite color change accompanied the Sun-angle change. On the second day, everything that had appeared to be gray on the first day started looking either a dark or a tannish brown.

One of the most interesting aspects of the lunar surface operations was how much color change resulted from only about a 7° Sun-angle change. Because the Sun angle and the angle of viewing had such a pronounced effect on color, minerals in the rocks were very difficult to identify, even when we held the rocks in our hand to get the best lighting. Geological operations on the Moon are probably going to be more difficult than on Earth, because the color cues are not going to be there. The lunar geologist is going to have to look for texture, fracture, and luster, among other things, to aid in determining differences in rocks and minerals.

There might be a subtle enough distinction in colors from a distance, perhaps sufficient to distinguish the ray patterns around a crater; but up close, there was not sufficient apparent difference in color to distinguish ray material from the surrounding surface material. At one point on the lunar surface, we talked about seeing large white boulders in the distance and, on other occasions, about seeing other white-appearing objects. We talked about this again when we were back in orbit. At the high-Sun angles, the ground looked white; we concluded that because the Sun, almost directly behind the spacecraft, is so bright at high-Sun angles, everything viewed at a distance appears to be almost chalk white.

The one place where we found a fairly definite color difference in the soils was out on the northwest rim of Head Crater. About 50 ft inside the upper rim of the crater, our boots scraped away the surface layer (about an eighth of an inch thick) to reveal a much lighter cement-colored soil right underneath. We trenched down about 6 in., and the soil remained that light-gray color.

In moving away from the area, we turned over one rock that had a much whiter bottom than that of any other rocks on the traverse, and we think this lighter color was caused by contact of the rock with the lighter gray material beneath the surface. The soil in this area around Head Crater was different from the soil around the LM. We kicked up quite a bit of soil around the LM, and it was all the same dark-gray color. We are almost sure that the light-gray color observed in the area around Head Crater was true soil color and not just due to the Sun angle.

Observations on the Geological Traverse

We collected a sample of anything that was different in texture, differently weathered, interesting in its location, or in any way unusual. On the geological traverse, the first significant thing observed was at Head Crater, where there was a difference in soil color just under the lunar surface. The next thing of significance we noted was the difference in the textures of the rocks in the bottom of Bench Crater. We were fairly certain that there were big fragments of bedrock visible, but we also noted the fact that some of the rocks were rounded on top and not as jagged, almost as though they had been melted on top. The melted material had a lavalike appearance, but that is not to imply that we believe that it was volcanic in origin. The melted material looked more like the effects of some high-speed impact; it had a volcanic appearance only in that it resembled the melted knobby-looking mounds of basalt frequently associated with zones of weakness around volcanic-type craters on Earth.

Somewhere between Bench Crater and Sharp Crater, we obviously ran over what must have been a contact, in that the ground very definitely changed to a softer, finer dust. We sank in deeper not only at Sharp Crater but also in the surrounding area. It was not possible at the time to be certain that we had actually reached Halo Crater, which was the next stop on our traverse, but if not, we were close to it. As we came up to Halo Crater, it was apparent that we were on still another type of soil texture. The ground seemed to be very firm and was similar to the firmest ground we had traversed, which was around Surveyor Crater. We sank in less in Surveyor Crater than anywhere else on the traverse, both on the way to the Surveyor spacecraft and on the way toward the LM through Block Crater on the other side of Surveyor Crater.

This blocky crater inside Surveyor Crater was another interesting feature. Our impression was that Surveyor Crater was a very old crater that, in being formed, had exposed the bedrock. Then, after the crater had become rather smooth through weathering, another object impacted directly into the lightly covered bedrock in Surveyor Crater and really banged it out to form Block Crater. The blocks around this small blocky crater were the most angular and, we think, the most recently exposed blocks observed.

One thing that impressed us about the geological traverse was the fact that we went so far and found it as easy going as we did. The only thing that kept us from moving faster was that there was so much to see, and the only thing that kept us from studying more details at each site was the need to keep moving. We were not able to stay as long as we would have liked at any one site. The whole EVA period could have been spent in any one of the craters — trenching around, collecting different types and sizes of rocks, and examining the ejecta blanket to see if there were any apparent differences in texture, etc. — but there just was not enough time.

One contrast between what we were seeing as opposed to what the Apollo 11 crew had observed was that we never saw any vesicular material or any rocks with a vesicular type of structure. At one point, it was believed that we had found such a rock, but when we picked it up, we found that its pits were not vesicles. In moving across the lunar surface, we saw everything from fine- to coarse-grained basalts and even some reddish-gray colored rocks with a crystalline structure almost like granite.

The main objectives of the geological traverse were not just to grab samples and take photographs but to try to understand the morphology and stratigraphy of the area being investigated. However, except for deciding which craters were newer than others, we were not able to discern any of the little morphological clues that had been observed in training on Earth. The whole area on the Moon seemed to have been acted on by meteoroids or something, so that all the features that might have provided morphological clues were obscured. When we walked up to a crater, it was never possible to determine where the normal lunar surface ended and the ejecta blanket began. The only clues were a difference in slope or the fact that the ground became a little more powdery under foot, and these clues are a poor index.

Because we saw what was believed to be bedrock on the outside of some of the craters, we thought that when we got to them and looked in, a deep contact between regolith and the bedrock would be visible. However, the insides of the craters looked just about like the surface except for a few rocks resting on the walls and bottom and, occasionally, some exposed bedrock. Just as trees and grass and such things hide morphological clues on Earth, the morphology of the Moon is hidden from the lunar geologist. We did not see any places, regardless of the slope, that did not have a coating of the dust-type material on the lunar surface.

Two very interesting formations observed were in the form of mounds of material on the lunar surface. They were fairly small (4 ft high by about 5 ft across on about a 20-ft circular base as the contact with the surface), were oriented in an east-west direction, and were roughly triangular in shape. The mounds did not appear to be volcanic in origin; they appeared to be a big chunk of material that had been tossed into the area from one of the nearby craters or from something farther away. We looked around the mounds for vent holes or ejected pyroclastics that might have come from the mounds, but we did not find any of this kind of evidence.

Generally speaking, the glassy material seemed to be about evenly distributed over the entire surface and in the bottoms of even the smallest craters. Glass beads were all over the surface, and we found both beads and glass-covered rocks in the bottoms of craters 3 or 4 ft in diameter. The glass-covered rocks we saw and sampled were very similar to those returned by Apollo 11.

Soil Mechanics and Characteristics

Our observation that there were three distinct types of soil was more or less subjective in that there were no positive color differences to confirm the distinctions. It appeared from the way that we sank that some soil was firmer than other soil. The finer soil was generally softer. This type of soil was encountered near Sharp Crater, at the farthest point from the LM. The soil around the LM could be described to be of medium texture. The firmest texture we encountered was in Surveyor Crater, where the soil seemed to have the greatest bearing strength.

We traversed several areas of patterned ground that were covered with radial streaks. The streaks seemed to be about perpendicular to the direction of any such streaks that could have been caused by the LM exhaust plume. The streak pattern in Surveyor Crater seemed to run from the north or northeast to the southwest, and the streaks around the area of the LM appeared to run from the north or northwest toward the south. The patterns were made up of little hilly streaks that were about a sixteenth to an eighth of an inch wide; about a sixteenth of an inch, or less, high; and about three-eighths of an inch apart. One place in particular where we did not see any radial patterning was at Sharp Crater, where the ground was very soft. It may be a pertinent point that the firmer the ground, the more we saw the radial patterning. The firm ground was also the ground that had little blobs in it similar to a smooth, level dirt field after rain. The patterned ground seemed to extend in the direction of the grooves as far as the eye could see. We never saw any kind of contact along the transverse direction of the grooves in the vicinity of the LM, in Surveyor Crater, or in any of the other places where we found the streaks.

In the area of Halo Crater, we came upon a clear contact between two apparently different soils. The area was not so smooth and had a completely different texture. The surface material appeared to be finer in texture and more cohesive. At Sharp Crater, the soil around the sharp, white raised rim appeared to be very soft and had a color similar to the soil uncovered by our boots at Head Crater. There appeared to be blast-effect material in a radial spray pattern all around Sharp Crater. Although we looked for Copernican ray material around this area, we could discern only a very few distinct rays. For craters older than Sharp Crater, there did not seem to be any discernible differentiation or contact between surface material and ejected material or rays.

We made some interesting observations on the lunar surface by rolling rocks down the side of a crater. We had the idea that it would be quite easy to get a rock started, but it was not. Once we got one going, however, it went along in a sort of animated slow motion and kept going for a very long time. The rock bounced and slid as it would have done on Earth, but the whole process was very much prolonged. Without walking down into the crater, which we could not do, there was no way to determine what kind of track the rock made rolling down. Also, we did not notice any tracks in the craters that might have been made by rocks other than the ones we rolled down. If there were a lot of rock rolling that might have been causing some of the signals received by the passive seismic experiment, it was not obvious to us on the lunar surface. The majority of the rocks on the sides and in the bottoms of the craters were partially buried or had dust fillets around the bottom. None of these looked as if they had moved for a very long time or were likely to move in the near future. There were other rocks in the craters, however, that were positioned such that some thermal or seismic mechanism could possibly cause them to roll. A few of these, if they had been dislodged, would have had a drop of 2 or 3 ft to the surface of the crater slope and might have been given a sufficient impetus to have rolled. It is not possible to completely discount rock rolling as a possible seismic source.

We had no trouble driving the single-core tubes, and we did not have to auger them as was required for the double-core tubes. Friction steadily built up as the tube went into the surface, but there was never a significant increase in friction with depth. Driving the double-core tube required that we use harder hammer blows but not more blows. The holes made by the core tubes stayed pretty much uncollapsed except for the top inch or so. When the tubes were withdrawn from the holes, this top inch would fall in, but the sides were still relatively vertical. The same was true of the trenches. When we dug a trench, the sides would be almost 90°. As long as we did not touch the trench accidentally with the shovel or a boot, the angle of repose stayed about 85° to 90°. In fact, there seemed to be some angles greater than 90° that might imply layering or that the material had been built up over different time periods.

Surveyor Observations

During the first EVA period, the slope at the Surveyor location was in the shadow and appeared to have an inclination of approximately 35°. The next day, however, when the Sun had risen sufficiently to place the Surveyor slope in sunlight, the inclination appeared to be only 10° or 15°, as had been determined from the Surveyor data. When we walked around the crater and got to Surveyor, it was sitting on about a 12° slope and was situated firmly enough so that there was no possibility of it sliding down the slope on us. On the first day, while it was in the shadow, the Surveyor spacecraft looked white, but when we got close to it, the spacecraft was covered with a tenacious covering of tan-colored dust. The dust was not in a very thick layer, but it was very hard to rub off. The dust was on all sides of the Surveyor but was not uniform around any specific part. Generally, the dust was thickest on the areas that were most easily viewed when walking around the spacecraft. For example, the side of a tube or strut that faced the interior of the Surveyor was relatively clean as compared to a side facing out. The color of the surface around the Surveyor also had a brown appearance, but we are fairly sure that the dust on the Surveyor was not blown there by the LM exhaust. If the dust had been blown onto the Surveyor, we would have noticed a directional pattern to the dust, and there was no directional pattern. Except for the fact that it had changed color, the Surveyor spacecraft appeared to be in very good condition. The waffle imprint of the pads was still visible in the lunar surface.

Using the cutting tool to collect the television (TV) camera from the Surveyor spacecraft was not difficult. The aluminum tubing appeared to be more brittle and easier to cut than the tubes used in training. It was as though some process of crystallization had affected the tubes during their long exposure to the lunar environment. The insulation on the wire bundles had become very hard and dry and appeared to have the texture of old asbestos. The mirrors on the surfaces of the electronics packages were generally in good condition, with a few cracks but no serious debondings. The only mirrors that had become unbonded were those on the flight-control electronics package. The glass on the Surveyor thermal switchplate was bonded to metal, which we did not know, and the bonding was in such perfect condition that there was no way to get a sample of the glass. When we tried to get a sample, the glass shattered into tiny fragments, so we left it alone.

As a bonus, we retrieved the Surveyor scoop. Although the steel tape was too thick to cut, the end attached to the scoop became debonded when we twisted the tape with the cutting tool. We collected several rock samples in the field of view of the TV camera for comparison with original Surveyor photography, but the only soil sample from that area was a small amount that was in the bottom of the Surveyor scoop. The added weight of the Surveyor components and samples did not appear to affect either our stability or our mobility on the return traverse.

Apollo Lunar Surface Closeup Camera (ALSCC)

The ALSCC photographic activity suffered somewhat from the lack of time remaining to do it at the end of the EVA period. The areas photographed were mostly around the LM. We were able to get a few shots of some areas that we had not been in, but most of the photographs were of the bottoms of small craters, dust patterns on the lunar surface, rocks, and footprints in the lunar soil. Two photographs were taken in the vicinity near the LM descent engine. The entire sequence of ALSCC photographs was completed in about 5 min; this did not allow enough time to orient the camera properly and to document photographs by describing what was being taken.

Experiment Retrieval

During retrieval of the solar-wind composition experiment, the foil on the experiment rolled up approximately 1½ ft. After that point, the foil would crinkle rather than roll. Using great care, we tried to roll up the foil, but on about the fifth attempt to roll it up, a longitudinal crack appeared in the crinkle area. We finally had to use our hands to roll the foil up, and as a result, the foil was soiled by the lunar material adhering to our gloves. After the foil was rolled up, we discovered that the roll was too big to fit into the container that was to be used to return the foil to Earth. To make the foil fit into the container, we had to crush it with our hands. Perhaps one reason for these problems with the foil was the "setting" we had observed. On the second day, as we looked out the LM window, it appeared that the foil had wrapped itself around the pole on which it was mounted. Upon inspection during the second EVA period, we decided that the foil tended to set and that it did not want to roll up because the set was stronger than the spring tension of the roller.

Ascent From the Lunar Surface

Ascent from the lunar surface was smooth, and everything went as our training had conditioned us to expect. The ascent engine performed beautifully. The spacecraft pitched over, and we could look down directly at the landing site. The ALSEP did not appear to be affected by the liftoff, but a considerable amount of silver and gold Kapton insulation could be seen flying off the descent stage and traveling radially outward for considerable distance parallel to the lunar surface, as had been reported by the Apollo 11 crew.

Solar Eclipse

An event of significance that took place during transearth coast was the observation and photography of a solar eclipse that occurred when the Earth came between the spacecraft and the Sun. This was one of the most spectacular things we saw throughout the entire flight. The Earth was completely invisible except through the smoked glass, even when we used one hand to shade our eyes to block out the Sun. When the eclipse was almost total, the atmosphere of the Earth was completely illuminated, and the Earth was as black as space.

The coloring of the illuminated atmosphere was very interesting. The atmosphere was segmented in alternating color shades of blue and pink. Pinkish-gray clouds were visible over the darkened part of the Earth during part of the eclipse cycle, and from our vantage point of approximately 25 750 n. m., we could also see thunderstorms and lightning flashes over part of the Earth. We were not, however, able to distinguish landmasses.

3. Passive Seismic Experiment

Gary V. Latham,[a,f] *Maurice Ewing,*[a] *Frank Press,*[b] *George Sutton,*[c] *James Dorman,*[a] *Yosio Nakamura,*[d] *Nafi Toksoz,*[b] *Ralph Wiggins,*[b] *and Robert Kovach*[e]

The purpose of the passive seismic experiment (PSE) is to detect vibrations of the lunar surface and to use these data to determine the internal structure, physical state, and tectonic activity of the Moon. Sources of seismic energy may be internal (moonquakes) or external (meteoroid impacts and manmade impacts). A secondary objective of the experiment is the determination of the number and the mass of meteoroids that strike the lunar surface. The instrument is also capable of measuring tilts of the lunar surface and changes in gravity that occur at the instrument location. Detailed investigation of lunar structure must await the establishment of a network of seismic stations; however, a single, large, well-recorded seismic event can provide information of fundamental importance that could not be gained by any other method.

Since deployment and activation of the PSE on November 19, 1969, the instrument has operated as planned, except as noted in the paragraphs entitled "Instrument Description and Performance." The sensor was installed west-northwest from the lunar module (LM) at a distance of 130 m from the nearest LM footpad. With the successful installation and operation of the first Apollo lunar surface experiments package (ALSEP), the feasibility of using long-lived geophysical stations to study the Moon has been demonstrated.

Signals of 40 seismic events have been identified on the records for the 42-day period following LM ascent. Of these signals, 10 are thought to be produced by noise sources within the LM descent stage. The remaining 30 signals, classified as type L, are prolonged, with gradual buildup and decrease in signal amplitude. This signal character may imply transmission with very low attenuation and intense wave scattering — conditions that are mutually exclusive on Earth. Because of the similarity with the signal from the impact of the LM ascent stage, L-signals are thought to be produced by meteoroid impacts or shallow moonquakes. Most of the L-events appear to have originated within 100 km of the ALSEP. The occurrence of similar L-events during both the Apollo 11 and Apollo 12 missions greatly strengthens the present belief that, of the various types of signals observed, L-events are the most likely to be of natural origin.

The fact that no natural seismic signals with characteristics similar to those typically recorded on the Earth were observed during the combined recording period for Apollo 11 and 12 (63 days at this writing) is a major scientific result. The high sensitivity at which the lunar instruments were operated would have resulted in the detection of many such signals if the Moon were as seismically active as the Earth and had the same transmission characteristics as the Earth. Thus, the data obtained indicate that seismic energy release is either far less for the Moon than for the Earth or that the interior of the Moon is highly attenuating for seismic waves. Although the material of the outer region of the Moon (to depths of at least 20 km) appears to exhibit very low attenuation in the regions studied, the possibility of the existence of high attenuation at greater depths cannot presently be excluded. The

[a] Lamont-Doherty Geological Observatory.
[b] Massachusetts Institute of Technology.
[c] University of Hawaii.
[d] General Dynamics, Fort Worth, Tex.
[e] Stanford University.
[f] Principal investigator.

absence of significant seismic activity within the Moon, if verified by future data, would imply the absence of tectonic processes similar to those associated with major crustal movements on the Earth and would imply lower specific thermal energy in the lunar interior than is present in the interior of the Earth.

However, it is important to remember that all results obtained thus far pertain to mare regions. Quite different results may be obtained for nonmare regions in which the lunar structure may differ radically from that in the mare.

Instrument Description and Performance

A seismometer consists simply of a mass, free to move in one direction, that is suspended by means of a spring (or a combination of springs and hinges) from a framework. The suspended mass is provided with damping to suppress vibrations at the natural frequency of the system. The framework rests on the surface whose motions are to be studied and moves with the surface. The suspended mass tends to remain fixed in space because of its own inertia while the frame moves around the mass. The resulting relative motion between the mass and the frame can be recorded and used to calculate original ground motion if the instrumental constants are known. Conventional seismic instruments are prohibitively large and delicate for use on lunar missions. For example, a typical single-axis low-frequency seismometer designed for use on Earth weighs approximately 30 kg and occupies 1×10^6 cm^3.

The Apollo 12 PSE (ref. 3-1) consists of two main subsystems: the sensor unit and the electronics module. The sensor, shown schematically in figure 3-1, contains three matched long-period (LP) seismometers (with resonant periods of 15 sec) alined orthogonally to measure one vertical and two horizontal components of surface motion. The sensor also includes a single-axis short-period (SP) seismometer (with a resonant period of 1 sec) sensitive to vertical motion at higher frequencies.

The instrument is constructed principally of beryllium and weighs 11.5 kg, including the electronics module and thermal insulation. Without insulation, the sensor is 23 cm in diameter and 29 cm high. Total power drain varies between 4.3 and 7.4 W.

Instrument temperature control is provided by a 2.5-W heater, a proportional controller, and an insulating wrapping of aluminized Mylar. The insulating shroud is spread over the local surface to reduce temperature variations of the surface material. In this way, it is expected that thermally induced tilts of the local surface will be reduced to acceptable levels.

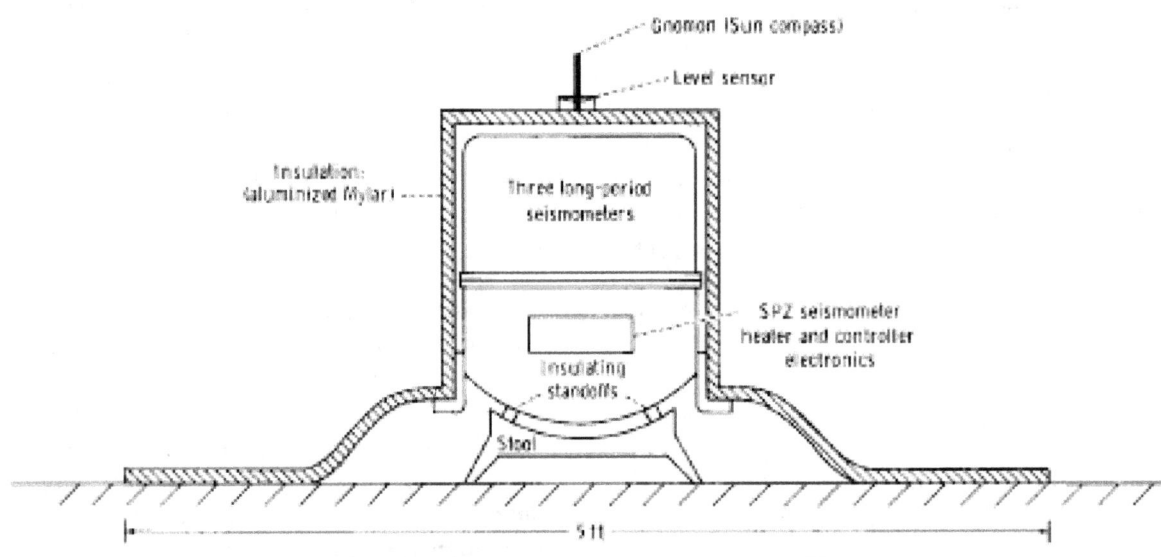

FIGURE 3-1. — Schematic diagram of the PSE sensor unit.

The LP seismometers will detect vibrations of the lunar surface in the frequency range from 0.004 to 2 Hz. The SP seismometer covers the band from 0.05 to 20 Hz. The LP seismometers can detect ground motions as small as 0.3 nm at maximum sensitivity; the SP seismometer can detect ground motions of 0.3 nm at 1 Hz.

The LP horizontal-component (LPX and LPY) seismometers are very sensitive to tilt and must be leveled to high accuracy. In the Apollo system, the seismometers are leveled by means of a two-axis motor-driven gimbal. A third motor adjusts the LP vertical-component (LPZ) seismometer in the vertical direction. Motor operation is controlled by command. These elements are shown schematically in figure 3-2. As shown in figure 3-2, the LP seismometers are mounted in crisscross fashion to reduce the required volume.

Calibration of the complete system is accomplished by applying an accurate increment or step of current to the coil of each of the four seismometers by transmission of a command from Earth. The current step is equivalent to a known step of ground acceleration.

A caging system is provided to secure all critical elements of the instrument against damage during the transport and deployment phases of the Apollo mission. In the present design, a pneumatic system is used in which pressurized bellows expand to clamp fragile parts in place. Uncaging is performed on command by piercing the connecting line by means of a small explosive device.

The seismometer system is controlled from Earth by a set of 15 commands that govern functions such as speed and direction of leveling motors, instrument gain, and calibration. The seismometer is shown fully deployed on the lunar surface in figure 3-3.

The PSE instrumentation has operated successfully throughout the first 42 days of the experiment, the time period discussed in this report. The instrument difficulties that have been observed are described in the following paragraphs.

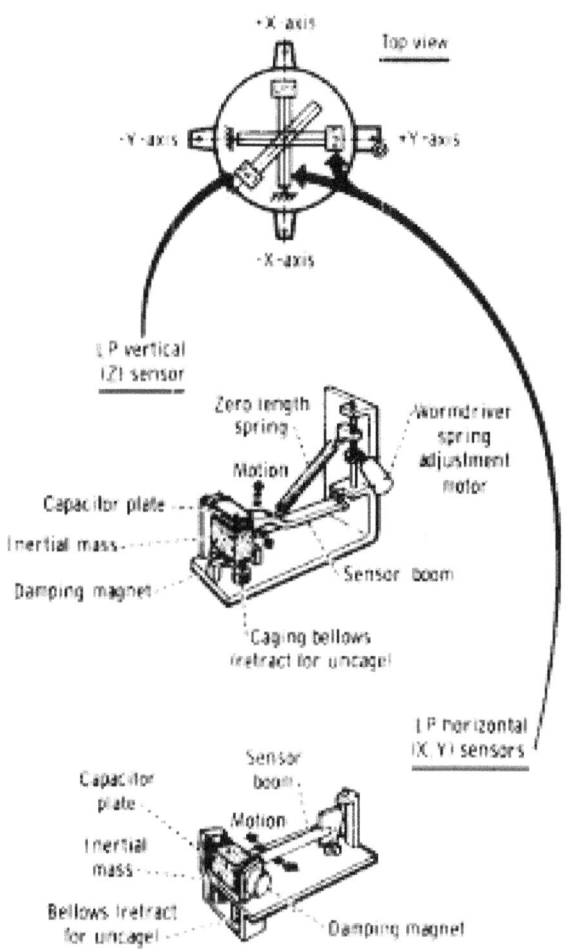

FIGURE 3-2. — Schematic diagram of the elements of LP seismometers.

FIGURE 3-3. — Photograph of the seismometer after deployment on the lunar surface.

Short-Period Seismometer

The SPZ seismometer appears to be operating at reduced gain. The first evidence of this problem appeared when the instrument failed to respond to calibration pulses. (No calibration pulses have been detected to date on the SPZ component.) Detailed comparison between signals observed on both LPZ and SPZ seismometers has led to the tentative conclusion that the inertial mass of the SP seismometer is rubbing slightly on the frame. Response, which is apparently normal, is observed for large signals, presumably because such signals produce forces large enough to exceed the static frictional restraining forces. Restraining forces introduced by sliding friction are apparently less important. The threshold ground-motion acceleration required to produce an observable signal cannot be determined accurately; however, the smallest signals observed correspond to a surface acceleration of 8×10^{-4} cm/sec^2 (peak-to-peak surface motion amplitude of 2 nm at a frequency of 10 Hz). Lunar surface accelerations less than this approximate threshold are apparently not detected by the SPZ seismometer.

A series of square-wave pulses observed on the SPZ seismometer trace began at 11:00 G.m.t. on December 2, 1969, and lasted approximately 13 hr. A similar "storm" commenced at approximately 08:00 G.m.t. on December 28, 1969, and ended some time between 01:00 and 23:00 G.m.t. on January 1, 1970. The pulse amplitude was constant and was approximately equal to a shift in the third least-significant bit of the 10-bit binary ALSEP data word. These pulses are also observable on the records from the LP seismometers, but the pulses have reduced amplitude. The source of these pulses has not yet been identified, but malfunction of the PSE analog-to-digital (A/D) converter or of the converter reference voltage is suspected. The fact that each of these "storms" occurred just subsequent to lunar noon, that the second storm was the stronger of the two, and that the seismometer temperature rose to 142° F at the second lunar noon as compared to 134° F at the first lunar noon suggests that the high temperatures in the seismometer may have caused the trouble.

Long-Period Seismometers

The response of the LPZ seismometer to a calibration pulse was observed to be oscillatory soon after activation. This effect gradually increased to the point of instability. In the presence of feedback, this tendency toward instability can be produced if the natural period of the seismometer is lengthened (or if the feedback-filter corner period is shortened) beyond the design value. It is considered most likely, at this point in the analysis, that vibration effects lengthened the natural period of the seismometer from 15 sec to approximately 60 sec. Acceptable seismometer operation has been achieved by removing the feedback filters from all three seismometers by command. In this configuration, the seismometers have responses equal to underdamped pendulums with natural periods of 2.2 sec. The instrument response curve corresponding to this mode of operation is shown in figure 3–4.

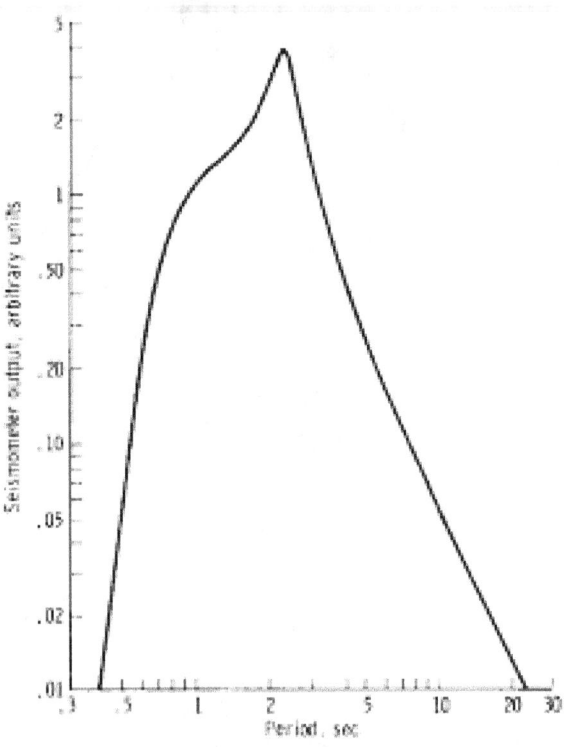

FIGURE 3-4. — Response of LP seismometers with feedback filters removed. Seismometer sensitivity for unity ordinate value is 5-mV output for 0.3-nm ground motion.

Thermal Control

The PSE active thermal control system was designed to maintain constant temperature to ±18° F about a set point of 125° F. The observed range is from +85° F (predicted by extrapolation) during the lunar night to +142° F during the lunar day. This temperature variation will not degrade the quality of seismic data but will greatly reduce the probability of obtaining useful tidal data. It appears that higher-than-anticipated heater output is required to maintain the temperature of the sensor unit within the ±18° F tolerance.

Deployment

Several points raised by the astronauts relative to the sensor emplacement are worthy of note. It appears that tamping the lunar surface material with the ribbed soles of the astronauts' boots is not an effective means of preparing the surface for experiment emplacement. The total compaction achieved by such tamping is reported to be small. Secondly, spreading the thermal shroud smoothly over the surface was difficult. The lightweight Mylar sheets of which the shroud is constructed appeared to resist a flat-lying configuration. It is not known whether this resistance is caused by electrical effects or by elastic memory within the Mylar.

Description of Recorded Seismic Signals

This experiment is a continuation of observations made during the Apollo 11 mission (refs. 3-2 and 3-3).

Preascent Period

Prior to LM ascent, many signals corresponding to various astronaut activities on the surface and within the LM were recorded, primarily on the LPZ component. The astronauts' footfalls were detectable at all points along their traverse (maximum range, approximately 360 m). Signals of particular interest were generated by test firings of the reaction control system (RCS) thrustors while the LM was on the lunar surface and by the LM ascent. Signals received from these sources are shown in figure 3-5. By measuring the elapsed time between engine ignition and signal arrival at the PSE for the RCS test firings and for the LM ascent, the compressional velocity of the lunar surface material has been determined to be approximately 108 m/sec. This value is consistent with estimates derived on the basis of mechanical properties measured by Surveyor, as presented in references 3-4 and 3-5 and by Sutton and Duennebier in "Seismic Characteristics for the Lunar Surface From Surveyor Spacecraft Data" (to be published).

Signals were also recorded from the impacts of the two portable life-support systems as they struck the lunar surface after ejection from the LM. Observed signal amplitudes from these sources are smaller by a factor of 80 than the signals observed from the same sources during Apollo 11. This reduced amplitude of the observed signals is due to the increased separation of the PSE and the LM during the Apollo 12 mission.

Postascent Period

Forty seismic signals of possible natural origin have been identified on the records for the 38-day period following LM ascent (the period for which data are available): 10 on the SPZ component and 30 on the LP components. All but one of the 10 high-frequency events detected by the SPZ component were recorded within 8 hr after LM ascent and may correspond to LM venting processes. This observation is in contrast to the thousands of signals assumed to be of LM origin that were recorded during the first 8 days of the Apollo 11 seismometer operation. This drastic reduction in the number of interfering noises from the LM is due, primarily, to the nearly eightfold increase in distance from the LM (130 m for Apollo 12 as compared to 16.8 m for Apollo 11). Part of the reduction may be attributed to the reduced sensitivity of the SPZ component to small signals.

Direct correlation has been made between signals recorded by the magnetometer (also on the lunar surface) and those recorded by the SPZ component. This correlation was particularly noticeable during passage of the ALSEP through the transition zone between the tail of the magnetic field of the Earth and interplanetary space, where rapid variations in the magnetic field were observed on the magnetometer record. It is assumed that detectable currents are induced in

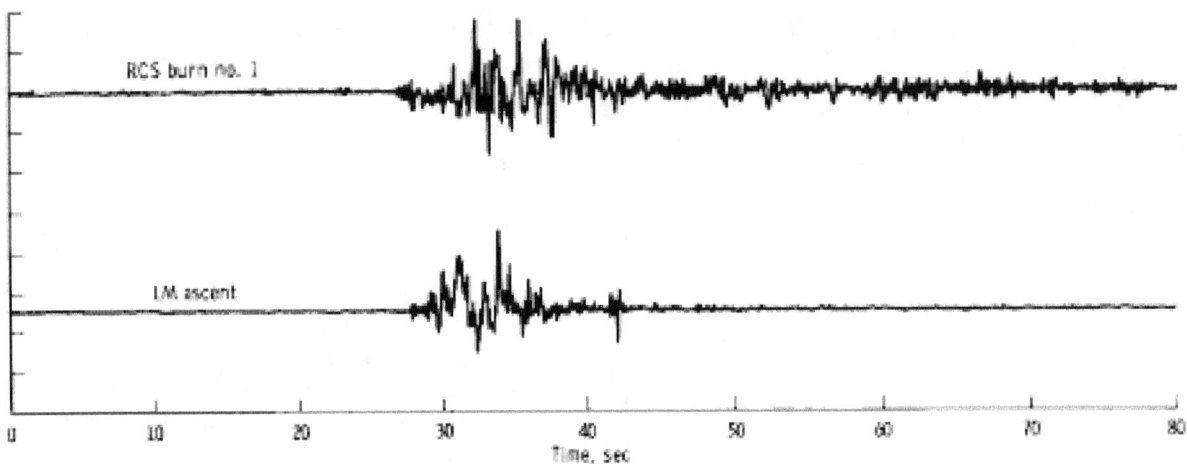

FIGURE 3-5.—Signals recorded during test fire of the RCS thrusters on the lunar surface and during LM ascension.

the main coil of the SPZ component by variations in magnetic flux. These signals will be studied with a view toward the possibility of extending the measurement of magnetic-field variations to higher frequencies than can be recorded with the magnetometer because of its lower data rate.

The two largest events recorded to date are shown on a compressed time scale in figure 3-6 along with the signal from the LM impact, which is to be discussed subsequently. The signals are prolonged (total durations are between 30 min and 1 hr), with the gradual increase and decrease in signal strength that is characteristic of all signals thus far recorded by the LP seismometers. These signals are classified as L-signals according to the nomenclature adopted in the "Apollo 11 Preliminary Science Report" (ref. 3-2). It should be noted that this designation does not refer to a particular type of seismic wave; the designation is simply a shorthand means of referring to the class of signals described. In general, L-signals are complex, with little correlation between the three LP seismometer channels. The familiar pattern of signals corresponding to the various body waves and surface waves typically observed from earthquakes is not observed in any of the recorded signals. However, there is some indication of the arrival of body waves (compressional (P) waves and shear (S) waves) in the early parts of the larger L-signals and in the LM impact signal.

It is expected that these phases, although indistinct, can be identified by use of more sophisticated analysis techniques and can be used to determine the velocity structure of the outer regions of the Moon.

Another interesting feature of L-signals is that the peak amplitudes recorded on all three LP seismometers are nearly the same in every case. The spectrum of the largest event (December 10, 1969) is shown in figure 3-7. The spectrum of the signal is broad, with maximum energy near 1.6 Hz. The very-low-frequency peak in the spectrum is produced by an oscillation present on the LPZ seismometer at the time of the event and is not related to the seismic signal.

A very significant event was recorded when the LM ascent stage impacted at a distance of 75.9 km from the ALSEP (azimuth from ALSEP, E 24° S). The angle between the LM trajectory and the mean lunar surface was 3.7° at the point of impact. The azimuth of the trajectory was 305.85°. Signal from the impact was recorded well on all three LP seismometers. The signal amplitude built up gradually to a maximum of 10 nm, peak to peak (all components), over a period of approximately 7 min and thereafter decreased gradually into the background. The total signal duration was approximately 55 min. The signal is shown with a compressed time scale in figures 3-6 and 3-8. Except for the very beginning of the wave train, distinct signals corresponding to various types of seismic waves

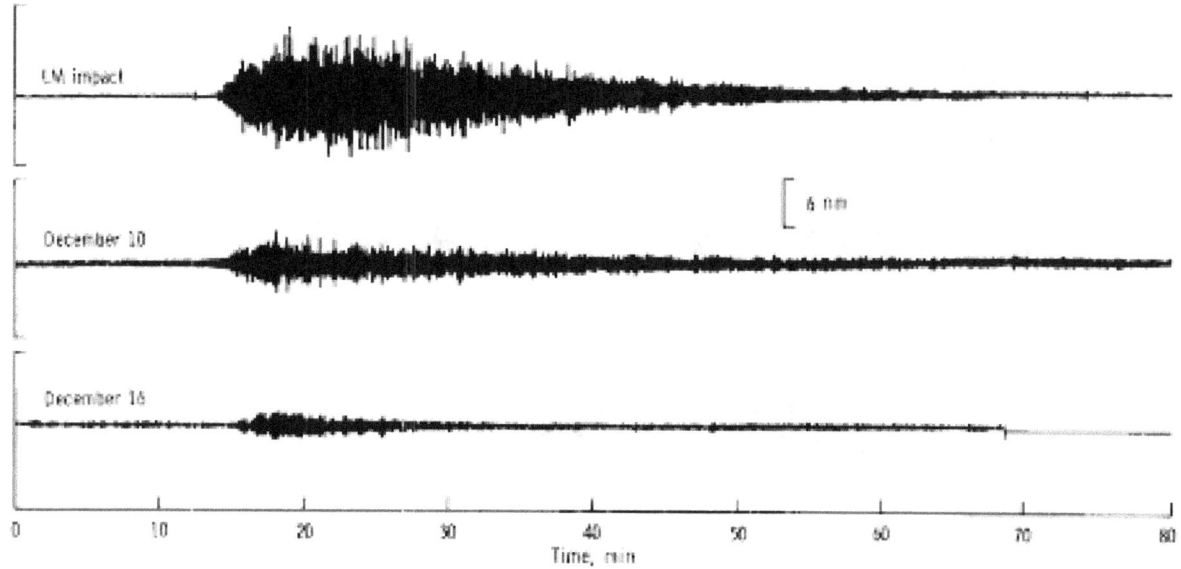

FIGURE 3-6. — The two largest seismic events recorded to date, in compressed time scale, compared with the signal received from the LM impact. Only LPZ components are shown.

(phases) are not apparent. The first few minutes of the filtered and rotated seismograms are shown with an expanded time scale in figure 3-9. The first signal, assumed to be the compressional wave P, is very small in amplitude, and it is difficult to specify the exact arrival time. The prominent signal that occurs near the beginning of the wave train was produced by a sudden tilt of the instrument. Such tilts occurred throughout the recording period, but they are especially prominent at times of terminator crossings. A possible second arrival, marked "S(?)" in figure 3-9, may be the arrival of the direct shear wave.

There is no visible correlation between components after the very first portion of the wave train. A test for coherence was carried out by generating the cross-spectral matrix for the X, Y, and Z components. Figure 3-10 shows power and cross spectra for the first 20 min of the LM impact signal. The data sample starts at 22 hr 17 min 36.43 sec (approximately the onset time of the emergent first arrival) and ends at 22 hr 38 min 12.9 sec. Spectra are normalized to the amplitude of the LPX spectrum, and the raw data were smoothed by the first finite difference (high-pass filtering) before the spectral estimates were computed.

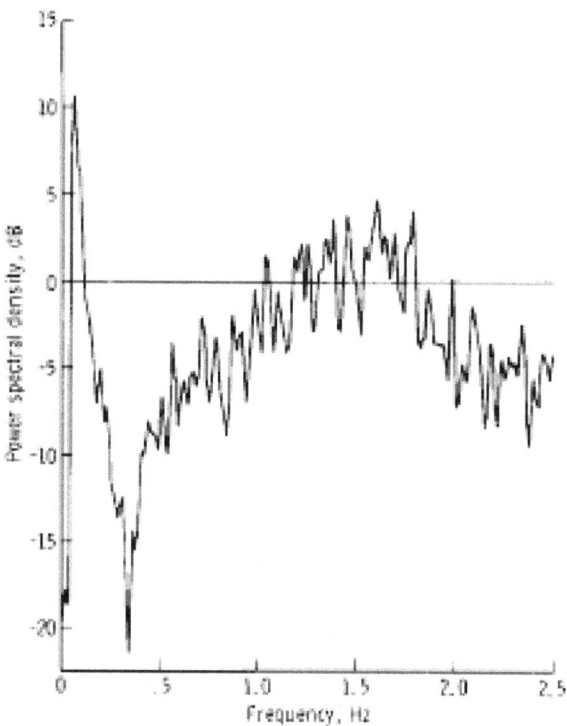

FIGURE 3-7. — Frequency spectrum of the event on December 10, 1969.

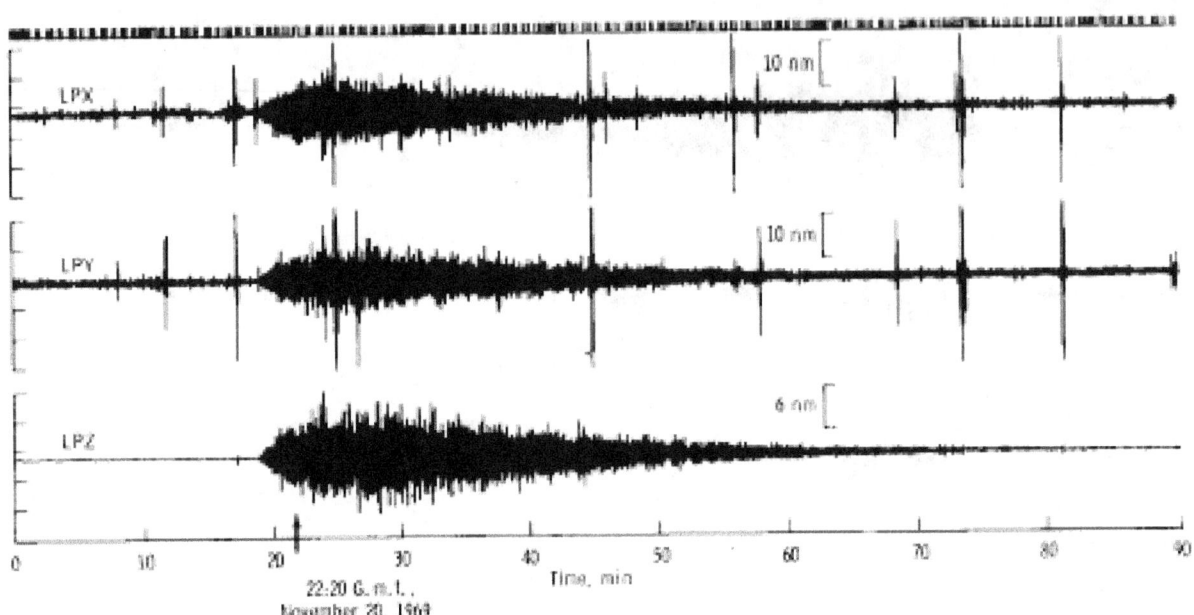

FIGURE 3-8. — Three LP components of the LM impact signal in compressed time scale.

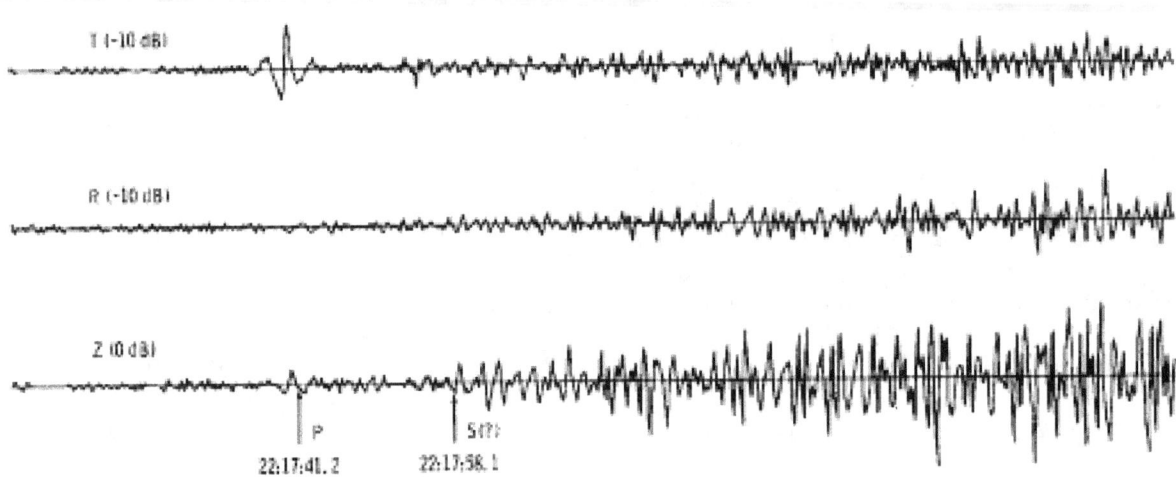

FIGURE 3-9. — Signals from the LM impact in expanded time scale. The data from the two LP horizontal-components seismometers are rotated to give transverse (T) and radial (R) components with respect to the impact point.

In general, the spectra for the LPX, LPY, and LPZ components are broadly centered at 1 Hz, with peaks at 0.95 and 1.2 Hz. The phases of the cospectra between the LPX-LPY and the LPX-LPZ components do not reveal consistent phase relationships between these components. Interestingly, there is some indication of a possible consistent phase relation between the LPY and LPZ components, although shorter time samples of data do not show this correlation. By using the real parts of power spectra and cospectra, the coherence between signal components was computed, using the definition

$$C_{XY} = \left(\frac{S_{XY} \cdot S_{YX}}{S_{XX} \cdot S_{YY}} \right)^{1/2}$$

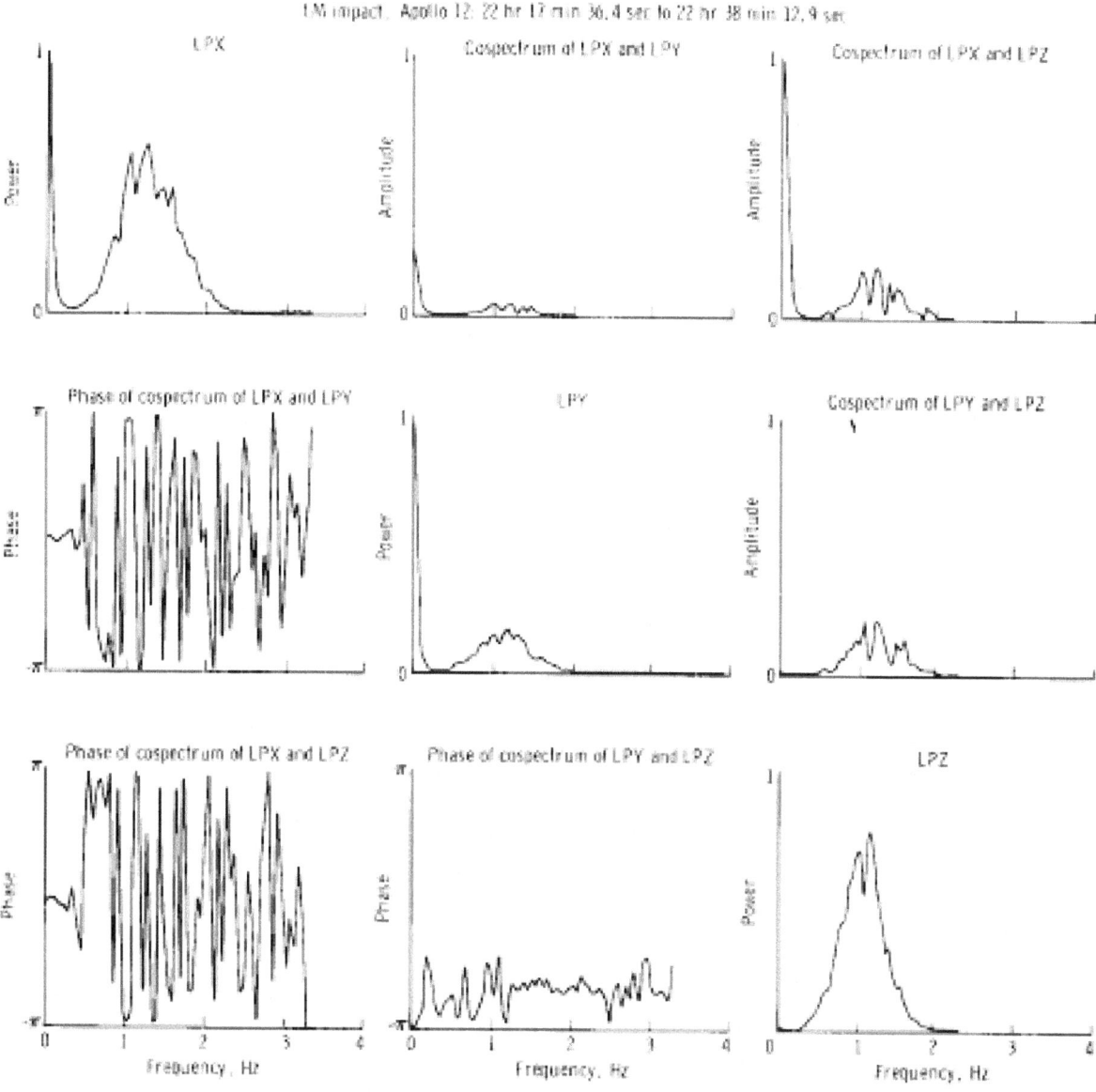

FIGURE 3-10. — Power spectra, cospectra, and phases of cospectra of the three LP components of the LM signal for the first 20 min. The spectral amplitudes are normalized to the LPX spectrum.

where C is coherence, S_{XY} and S_{YX} are cospectra, and S_{XX} and S_{YY} are spectra of X and Y components. In the frequency band of peak power ($f = 0.5$ to 1.5 Hz), the coherence between the LPX-LPY and LPX-LPZ seismometer pairs is quite low ($C < 0.5$). The coherence between LPY and LPZ components is somewhat higher ($C \approx 0.6$), indicating a better correlation. These results suggest that during the first 20 min of the impact wave train (1) the coherent motion was primarily P, SV, or Rayleigh type and (2) the primary direction of propagation was from the source (impact site) toward the instrument. The lack of strong coherence between any of the components indicates the presence of interference either between different wave types or between arrivals from different directions owing to reflections or to scattering. No signal was detectable on the SPZ component. This is attributed to the reduced sensitivity of the SPZ seismometer.

The seismic wave velocity corresponding to the first arrival ranges between 3.1 and 3.5 km/sec. The range in velocities is due to the uncertainty in the exact location of the emergent beginning of the wave.

Feedback (Tidal) Outputs and Instrument Temperature

The PSE LP seismometers are sensitive to tilt (horizontal components) and changes in gravity (vertical component). These data are transmitted through separate data channels referred to as "feedback" or "tidal" outputs. Large tilts (0.5 to 1 minute of arc) have been observed during passage of the terminator over the PSE site, when thermal changes are most rapid. When the change is from day to night, tilting begins several hours before local terminator crossing and lasts for several days. When the change is from night to day, tilting begins abruptly at the time of terminator passage and continues for several days. The close correlation between the time of terminator passage and the onset of tilting suggests that such tilts result from thermal effects on the instrument. Rapid heating and cooling of the Mylar thermal shroud is thought to be a major source of these disturbances.

Discussion

The background seismic noise at the Apollo 12 site is frequently not detectable by the PSE, that is, it is below 0.3 nm at 1 Hz. Thus, seismometers can be used on the Moon at much higher sensitivities than could be achieved on Earth. This result was also obtained for the Apollo 11 recording site (refs. 3–2 and 3–3).

The occurrence of similar L-events during both the Apollo 11 and 12 missions greatly strengthens the present belief that, of the various types of signals observed, L-events are most likely to be of natural origin. Equally important, the inclusion of the LM impact signal in this family shows that L-type signals can be generated by impulsive sources on the Moon. This observation suggests that all L-events were produced either by meteoroid impacts or by shallow moonquakes. By comparing the various L-phases with the LM impact signal, it may be concluded that few, if any, L-events originated significantly farther from the seismometer than the LM impact. The LM impact occurred at a distance of 75.9 km. Many, however, appear to have come from smaller distances.

To explain the unexpectedly long duration of the wave train, it must be assumed either that the effective source mechanism was prolonged in some manner or that the long duration of the wave is a propagation effect. An extended source from an impact might result from (1) triggering of rockslides within a crater located near the point of impact, (2) distribution of secondary impacts of ejecta that would presumably rain downrange (toward ALSEP) from the primary impact point, (3) disturbance by an expanding gas cloud consisting of residual LM fuel (180 kg) and volatilized ejecta, and (4) collapse of "fairy castle" or other fragile structures triggered by seismic waves. None of these mechanisms is considered likely, although the possible effects of secondary impacts deserve closer consideration. Since the signal maximum occurred approximately 7 min after impact, it is assumed that the main contribution from secondary impacts would correspond to a time of flight to 7 min for the ejecta. The range from the primary impact point to the secondary impact point can then be computed as a function of the velocity of the ejected particle. The results of this calculation are given in table 3–1. It can be seen from table 3–I that ejecta velocities would have to be less than 0.4 km/sec and the corresponding ejection angle greater than 54° to account for the arrival of

TABLE 3–I. *Distance From Primary Impact Point to Secondary Impact Point as a Function of Ejecta Velocity for a 7-Min Time of Flight*

Ejecta velocity, km/sec	Ejection angle (from horizontal), deg	Range, km
1.6	1.1	493
1.4	4.4	555
1.2	8.0	460
1.0	12.7	381
.8	19.4	299
.4	54.0	93

secondary impacts in the vicinity of the PSE (range=76 km) 7 min after the impact. These values are well outside the expected range for the shallow impact angle of the LM. (The LM struck the lunar surface traveling at 1.67 km/sec at an angle of 3.7° from the horizontal.) Since the signal produced by the LM impact began approximately 23.5 sec after impact, not all of the signal can be attributed to ejecta landing near the seismometers. The minimum time of flight at near-orbital velocity would be approximately 45 sec. The same remark applies to propellant gases released upon impact. Only seismic wave propagation permits signal velocities high enough to account for the beginning of the disturbance, regardless of the details of the mechanism postulated. In addition, the fact that the same signal character is observed for events of natural origin suggests that the signal character is produced primarily by propagation effects.

If the signal duration is a propagation effect, the attenuation of seismic waves in the lunar material through which these waves traveled must be extremely low. Attenuation of elastic energy in a vibrating system is frequently specified by the quantity Q (quality factor) for the system; or $1/Q$, the dissipation function, where $1/Q$ is the fractional loss of elastic energy per cycle of vibration of the system. Thus, a high Q implies low attenuation. The value of Q for the lunar material in the region of the Apollo 12 site ranges between 3000 and 5000. This range is in contrast with values of Q between 10 and 300 for most crustal materials on Earth.

One hypothesis that could explain the signal character of L-phases is that the Moon not only has a high Q but also is very heterogeneous, at least in its outer regions. Heterogeneity is also implied by surface evidence. The scattering of seismic waves that would occur through a highly heterogeneous material would tend to increase the duration of the observed seismic wave and to suppress the appearance of distinct phases within the wave train. A medium that shows high Q and high scattering efficiency is unlike anything observed within the Earth. Cold blocks of different composition in welded contact might show these properties. A welded aggregate might scatter seismic waves and still maintain high Q. Whatever the mechanism is determined to be,

it will provide important evidence on the origin and evolution of the lunar interior.

If the outer region of the Moon is composed of blocks of varying dimensions, seismic waves traveling through this material would be intensely scattered. In the case of extreme scattering, the seismic wave energy may be considered to diffuse through the medium in a manner analogous to the flow of heat through a solid or the movement of gas molecules through a gaseous medium. Such propagation is governed by the laws of diffusion in which, as applied to the present case, the seismic energy "flow" is proportional to the gradient of energy density. The applicable equations are given by Latham et al. (ref. 3-3). This hypothesis can be tested by comparing the observed variation of seismic signal energy from the LM impact with the signal predicted from diffusion theory.

Figure 3-11 is the smoothed envelope of the observed seismic signal from the LM impact plotted on an arbitrary decibel scale. Two theoretical curves based upon diffusion theory are also shown. One curve assumes that the outward

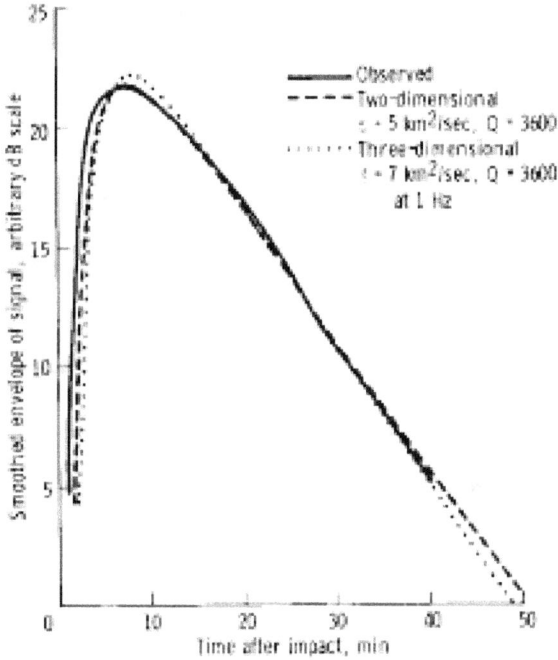

FIGURE 3-11. — Smoothed envelope of the observed wave train from the LM impact compared with the energy distribution with time predicted by diffusion hypothesis.

radiating energy is confined to a near-surface zone (two-dimensional spreading), and the other curve assumes spherical (three-dimensional) spreading. The parameter ξ is the product of a characteristic distance λ, defined as the distance over which one-half of the outward radiating energy is reflected back toward the source, and the average velocity of propagation ν. The symbol Q is the quality factor of the lunar material, as defined previously.

It can be seen from the graph (fig. 3–11) that the diffusion theory accurately predicts the observed signal envelope except for the first few minutes, where the theory predicts somewhat smaller amplitude than is observed. It may be that conventional ray theory provides the best explanation for the early part of the signal. Although this result does not verify the scattering hypothesis, it is sufficiently encouraging to warrant further consideration.

As shown in figure 3–11, the value of ξ that gives the best fit between the theoretical and experimental curves ranges between 5 and 7 km^2/sec. By assuming an average velocity of 3 km/sec and $\xi=6$, the characteristic distance λ is 2 km. This value of λ means that one-half of the propagating seismic energy is reflected back over a distance of 2 km. What this reflection means in terms of the average distribution of discontinuities in the medium is not precisely known. However, it can be assumed that the linear dimension between scattering surfaces must be a fraction of 1 wavelength. The predominant signal frequency is approximately 1 Hz; thus, 1 wavelength is approximately 3 km. Taking 1/10 wavelength to 1 wavelength as limiting values, the inferred separation between discontinuities, or dimensions of blocks, ranges between 300 m and several kilometers. Of course, heterogeneity may exist on a scale outside the range, but would not contribute appreciably to scattering for the observed wavelengths.

The seismic signal detected from the Apollo 12 LM impact (fig. 3–8) demonstrated that prolonged wave trains can be produced on the Moon by relatively small impulsive sources. This result is extremely important to interpretation of the long reverberation as a propagation effect and could be explained by postulating extremely low attenuation (high Q) for the Moon. While the evidence for low attenuation is hard to deny, this phenomenon raises some difficult questions as to its mechanism. Regardless of the explanation of signal duration, the similarity between the impact signal and other prolonged signals suggests that the latter were produced by meteoroid impacts or by near-surface moonquakes at ranges mostly within 100 km of the seismometer.

The seismic energy generated at the point of impact can be calculated from the observed signal amplitude. The calculated energy is on the order of 10^{10} ergs if two-dimensional spreading is assumed and on the order of 10^{13} ergs if three-dimensional spreading is assumed. These values are 10^{-8} and 10^{-5}, respectively, of the kinetic energy of the LM at impact. For these calculations, the signal amplitude at 10 min into the wave train is taken to be 4.2 mm. Simple harmonic motion at a frequency of 1 Hz is assumed. The density of the lunar material is assumed to be 3 g/cm^3, and the effective thickness of the waveguide for the two-dimensional case is taken to be 4 km.

Although the nature of the signals from the LM impact and other L-type events indicates that a considerable amount of scattering of seismic energy has occurred, it is possible that much of the character of the signal can be explained as resulting from propagation through a near-surface waveguide. As mentioned previously, the compressional wave velocity in the regolith near the lunar surface is approximately 100 m/sec. Based upon laboratory measurements on returned lunar rock samples, as discussed in the following paragraphs, this velocity should increase to approximately 6 km/sec at depths of 15 to 20 km, thus forming the lower boundary of the waveguide. Preliminary calculations using ray optics, on simple models consistent with this velocity-depth structure, indicate that the first several minutes of the seismogram can be explained in this manner. Work is continuing on this approach, and at this writing it appears that a reasonable velocity model can be found that will match the major aspects of the records throughout the wave train. Most of the seismic energy from a near-surface source would be trapped in this waveguide. Detailed comparisons between the LM impact signal and other L-signals should allow determination of which of these two possibilities

— scattering or the characteristic of the velocity-depth function — is the more important mechanism in producing the observed wave trains.

To estimate the near-surface properties of the Moon from the available data, lunar seismic velocity models have been constructed on the basis of the measurements of the physical properties of returned lunar samples. (See ref. 3–6.)

Model I assumes the same variation in elastic parameters with pressure (or depth) as measured in the laboratory on a breccia sample. Model III uses the measured properties of a homogeneous igneous rock. Model II is an attempt to combine the properties of the igneous rock and breccia samples to produce a model that will have the elastic parameters of a highly fractured igneous material. For model II, the bulk density is assumed to be that of the igneous sample (approximately 3.1 g/cm³), and the compressibility is assumed to be that of the breccia sample. Poisson's ratio is assumed to be 0.30 for model IIa and 0.27 for model IIb.

The travel time curves for these models are shown in figure 3–12. The corresponding maximum depths of seismic ray penetration are given in table 3–II.

The observed travel times for the first arrival from the LM impact and a later arrival are also plotted, with a range of times indicating the possible reading error. The first arrival is assumed to be the direct P-wave. The wave type of the second arrival, labeled "S(?)," is uncertain. The travel times for these phases fall in between those predicted for the homogeneous igneous model (III) and the fractured igneous model (II). Thus, if it is assumed that the observed phases correspond to direct P- and S-wave propagation, the correct model for the upper 20 km of lunar material in the vicinity of the Apollo 12 landing site must have been a velocity-depth function that falls between those assumed for models II and III but is closer to the homogeneous igneous rock case of model III. This result must be taken to be very tentative since sample-to-sample variations may be large enough that a single predominant rock type may also be found to fit the observations.

The average rates of L-events detected during the Apollo 11 and 12 recording periods are 4 per day and 0.8 per day, respectively. These rates are

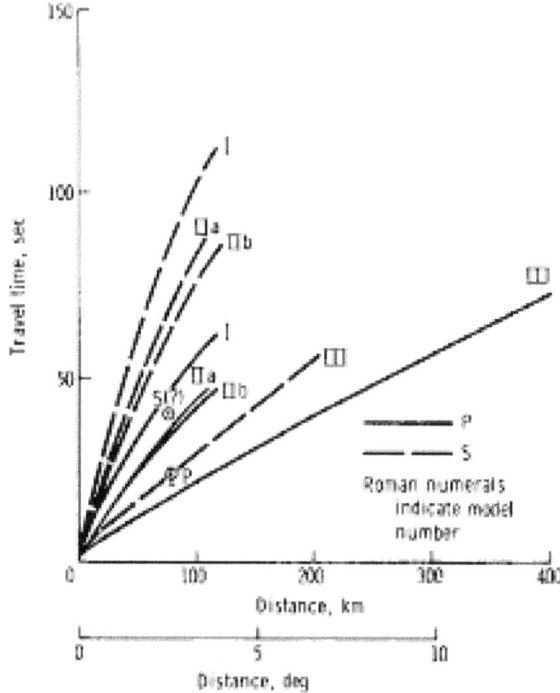

FIGURE 3–12. — First and second arrivals from the LM impact plotted on the travel-time-distance curves based on Apollo 11 lunar sample data from reference 3–6.

consistent with the numbers of detectable meteoroid impacts predicted (refs. 3–7 and 3–8) for a high-Q Moon. Thus, it is possible that all of the observed L-signals were produced by meteoroid impacts. The difference in the observed rates of these events on the two missions may result from failure of the Apollo 12 SPZ seismometer to respond to small signals, as discussed previously. Most of the Apollo 11 L-signals were detected by the SPZ seismometer because the predominant frequencies of these signals are higher than the high-frequency limit (2 Hz) of the LP seismometers.

TABLE 3–II. *Depth of Ray Penetration of Rays That Emerge at a 75.9-km Range*

Model	P-wave depth, km	S-wave depth, km
I	18	21
IIa	15	15
IIb	17	17
III	11	8

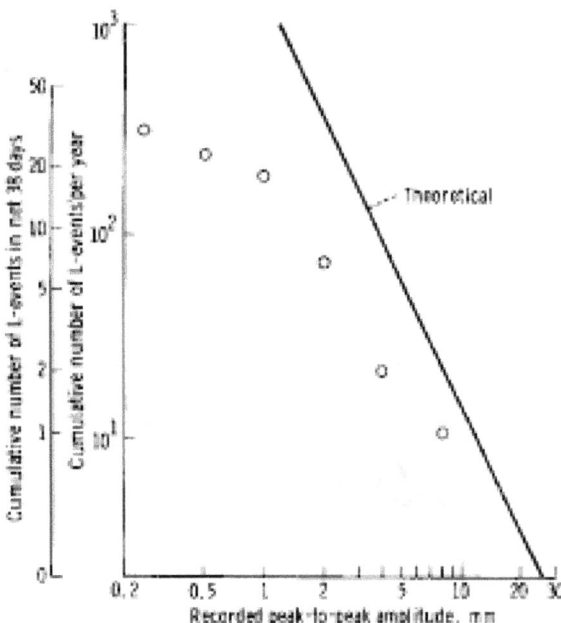

FIGURE 3-13. — Cumulative number of L-events observed to date (denoted by circles). Theoretical curve based on the estimate of meteoroid flux from reference 3-9 is shown by a solid line.

In figure 3-13, the cumulative number of Apollo 12 L-events detected to date by the LP seismometers is compared with the predicted number of detectable meteoroid impacts. The predicted number of impacts assumes (1) the meteoroid flux estimates given by Hawkins (ref. 3-9), (2) attenuation of seismic signals with distance as given by the diffusion model described previously, and (3) that the fraction of impact kinetic energy that is converted to seismic wave energy is the same as that calculated for the LM impact. The conversion efficiency for meteoroids with trajectories more nearly perpendicular to the surface than that of the LM may well be greater than the LM impact efficiency.

In view of the order-of-magnitude estimates involved in these calculations and the short recording time available, the agreement between predicted meteoroid impact signals and the observed data is remarkable. The lack of agreement for small-amplitude signals is probably explained by the inability of the SPZ seismometer to detect small signals. It is concluded from this comparison that there are sufficient numbers of meteoroid impacts to explain all of the observed L-signals as being of meteoroid impact origin, although the presence of some events with other source mechanisms (i.e., moonquakes) is certainly not precluded. By extrapolating the results given in figure 3-13 to large amplitudes, signals with amplitudes equal to that of the LM impact signal are expected to occur at an average rate of three per year.

Few, if any, of the observed signals have patterns normally observed in recordings of seismic activity on Earth (with the possible exceptions of volcanic tremors, microseisms, and landslide signals; i.e., signals from sources that have extended source mechanisms). Phases corresponding to the various familiar types of body and surface waves are either indistinct or absent in the lunar signals, and most wave trains are of long duration with little phase correlation between components.

The fact that no seismic signals with characteristics similar to those typically recorded on the Earth were observed during the combined recording period for Apollo 11 and 12 (63 days at this writing) is a major scientific result. The high sensitivity at which the lunar instruments were operated would have resulted in the detection of many such signals if the Moon were as seismically active as the Earth and had the same transmission characteristics as the Earth. Thus, the data obtained indicate that either seismic energy release is far less for the Moon than for the Earth or the deep interior of the Moon is highly attenuating for seismic waves. Although the material of the outer region of the Moon (to depths of at least 20 km) appears to exhibit very low attenuation in the regions studied, the possibility of the existence of high attenuation at greater depths cannot presently be excluded. The absence of significant seismic activity within the Moon, if verified by future data, would imply the absence of tectonic processes similar to those associated with major crustal movements on the Earth and would imply lower specific thermal energy in the lunar interior than is present in the interior of the Earth.

Interpretation of present data is not concerned with the structure of the deep interior of the Moon since most of the recorded events appear to have occurred at relatively short ranges. The relatively thick zone of self-compaction in which elastic wave velocity increases strongly with depth would perhaps be 20 km thick. The large increase in velocity with depth results in a surface sound channel that may carry the seismic energy of these events. The impact of the Saturn IVB (SIVB) stage in April 1970, at a range of 200 km from the Apollo 12 ALSEP, should enable extension of present interpretation to depths approaching 50 km into the Moon.

It is also important to remember that all results obtained thus far pertain to mare regions. Quite different results may be obtained for nonmare regions in which the structure may differ radically from that of the mare.

With data from one or two lunar seismic stations, construction of a picture of the lunar interior with detail approaching that of Earth models cannot be expected. Seismic experiments have, nevertheless, already revealed some unexpected phenomena, the understanding of which will eventually answer some important questions concerning the structure and dynamics of the Moon with significant implications for lunar history. As a result of the reduced level of detectable lunar seismic activity relative to Earth, results will come more slowly than had been hoped, and there will be greater dependence upon the establishment of a network of stations and upon use of artificial sources such as impacts of the SIVB stage and the LM ascent stage.

Meteoroid impacts are a major factor in shaping the lunar surface. Determination of the size and frequency distribution of meteoroid impacts is necessary to estimate quantitatively the rates of crater formation and erosion. The lunar seismic experiments will provide data that will be uniquely suited to the study of this problem.

References

3-1. LATHAM, G.; EWING, M.; PRESS, F.; and SUTTON, G.: The Apollo Passive Seismic Experiment. Science, vol. 165, no. 3890, July 18, 1969, pp. 241-250.

3-2. LATHAM, G. V.; EWING, MAURICE; et al.: Passive Seismic Experiment. Sec. 6 of Apollo 11 Preliminary Science Report, NASA SP-214, 1969, pp. 143-161.

3-3. LATHAM, G. V.; et al.: Apollo 11 Passive Seismic Experiment. Science, vol. 167, no. 3918, Jan. 30, 1970, pp. 455-457.

3-4. CHRISTENSEN, E. M.; BATTERSON, S. A.; et al.: Lunar Surface Mechanical Properties at the landing Site of Surveyor 3. J. Geophys. Res., vol. 73, no. 12, June 15, 1968, pp. 4081-4094.

3-5. CHOATE, R.; BATTERSON, S. A.; et al.: Lunar Surface Mechanical Properties. Sec. 4 of Surveyor Program Results, NASA SP-184, 1969, pp. 129-169.

3-6. SCHREIBER, E.; et al.: Sound Velocity and Compressibility for Lunar Rocks 17 and 46 and for Glass Spheres From the Lunar Soil. Science, vol. 167, no. 3918, Jan. 30, 1970, pp. 732-734.

3-7. LASTER, S. J.; and PRESS, F.: A New Estimate of Lunar Seismicity Due to Meteorite Impact. Phys. Earth and Planet Interiors, vol. 1, 1968, pp. 151-154.

3-8. MCGARR, A.; LATHAM, G. V.; and GAULT, D. E.: Meteoroid Impacts as Sources of Seismicity on the Moon. J. Geophys. Res., vol. 74, no. 25, Nov. 15, 1969, pp. 5981-5994.

3-9. HAWKINS, GERALD S.: The Meteor Population, Research Report No. 3. NASA CR-51365, Aug. 1963.

4. Lunar Surface Magnetometer Experiment

P. Dyal,[a] C. W. Parkin,[a] and C. P. Sonett[a,†]

The electromagnetic properties of a planetary body yield valuable information concerning the present physical state and the past evolutionary history of the body in the solar system. Magnetic-field measurements have proved to be one of the most useful tools for determining the electromagnetic properties of the Earth's interior and its solar-wind and ionospheric environments.

An extension of these magnetic-field measurement techniques was used on the Apollo 12 mission to investigate the electromagnetic properties of the Moon. The Apollo 12 magnetometer detected a steady magnetic field of approximately 36 gammas superimposed upon the geomagnetic tail, transition region, and interplanetary fields through which the Moon passes during each orbit. The magnetometer also measured a definite lunar magnetic response to the time-varying solar-wind field during both the lunar day and night. The magnetometer was deployed in the Ocean of Storms, a mare-type sea of smooth, uniform surface features that made the site ideal for first measurements of the field of the entire Moon. Large mountains or craters would be more likely to possess local electromagnetic inhomogeneities that would have hindered making whole-body lunar-field measurements with a single instrument.

Magnetometer measurements are used to determine the lunar response to fluctuations in the interplanetary magnetic field and to measure the time-invariant field associated with the whole Moon and with local sources. Measurements of the induced lunar magnetic field permit the electrical conductivity of the lunar interior to be calculated. Because electrical conductivity is a function of temperature, these magnetic-field measurements provide information on the thermal state of the lunar interior. Such information on the interior temperature of the Moon is crucial for differentiation between existing theories of lunar formation and history.

Purpose and Theory of the Experiment

Purpose of the Experiment

The purpose of the magnetometer experiment is to measure the magnetic field on the lunar surface and to determine from these measurements some of the deep-interior electrical properties of the Moon. This experiment will also help to elucidate the interaction between the solar plasma and the lunar surface.

Past Experimental Measurements

Recent experiments have made considerable progress in the determination of the electrodynamic properties of the Moon. Magnetic-field measurements from lunar orbital satellites such as Luna 10 (ref. 4–1) and Explorer 35 (refs. 4–2 and 4–3) indicate a remanent or permanent magnetic dipole moment for the whole Moon of less than 10^{20} G-cm^3, and it may possibly be 0. Thus, the lunar dipole moment is less than 10^{-5} that of the Earth.

The relative permeability μ of the lunar surface material can be estimated from the magnet experiment results of J. N. deWys (ref. 4–4) from Surveyors 5 and 6. She reported the presence of less than 1 percent by volume of magnetic iron material. By assuming this material to be magnetite and by using the dependence of permeability upon the magnetite content of rocks as given by Keller and Frischknecht (ref. 4–5), a relative permeability μ of 1.04 may be derived

---

[a] NASA Ames Research Center.
[†] Principal investigator.

for lunar surface material. Behannon (ref. 4–6) placed an upper limit of 1.8 on the bulk relative permeability by studying the Explorer 35 magnetometer measurements as the Moon traversed the neutral sheet in the geomagnetic tail.

The relative dielectric constant ϵ of the surface material has been measured by Earth- and satellite-based radar experiments. Hagfors (ref. 4–7), in his review of Earth-based observation of the Moon, showed a relative dielectric constant of 2.6 to be consistent with backscattered measurements at 10^8 Hz. Brown et al. (ref. 4–8) estimated a value for ϵ of 3.5 ± 0.7 at 10^{10} Hz by using data from the radar altimeter and the Doppler velocity sensor of the Surveyor 3 spacecraft. The conductivity σ of the lunar surface material was calculated by Kopal (ref. 4–9) from radiometer and radar data to be $\sigma \approx 10^{-4}$ mhos/m at 3×10^8 Hz. More recently, Strangway (ref. 4–10) estimated the dc conductivity to be 10^{-13} to 10^{-16} mhos/m, based upon investigations of radar scattering from the lunar surface. Preliminary examination of the lunar samples from Apollo 11 (ref. 4–11) indicates that the intensity of remanent magnetization varies from 10^{-5} to 10^{-2} emu/g from the various samples.

Theory of the Experiment

Theoretical analysis of the time-series magnetic-field measurements on the lunar surface can be conveniently divided into two parts. The first part covers field fluctuations with periods longer than 10 days, and the second part covers higher frequency phenomena with periods between 10 days and 0.3 sec.

The long-period measurements are associated either with remanent sources localized near the Apollo 12 landing site or with the entire Moon. The magnetic field measured on the lunar surface is a vector sum of the lunar, terrestrial, and solar magnetic fields. The selenomagnetic field associated with the whole Moon or with a localized portion of the Moon should have small amplitude variations for time periods less than 10 days and can, therefore, be separated from the higher frequency fluctuations by measurements obtained through one complete revolution of the Moon around the Earth. Simultaneous measurements from the Apollo 12 magnetometer and the Explorer 35 magnetometer (in lunar orbit) permit calculation of the size, location in the Moon, and magnetic moment of the source. The orbiting magnetometer (ref. 4–2) has a perilune of 1.48 lunar radii (R_M), an apolune of $5.4\ R_M$, and an orbital period of approximately 11 hr.

The short-period measurements involve transient induction of fields in the Moon by the solar field, and by the Earth field, at different locations in the orbit of the Moon. The properties of the driving medium, that is, of the solar plasma, must also be considered in detail to understand the steady-state and transient magnetic fields on the lunar surface. The solar plasma or solar wind emanates from the solar atmosphere and passes the Moon at a velocity of approximately 400 km/sec. The high electrical conductivity of the solar wind allows the solar magnetic fields to be "frozen in" and carried along at the characteristic speed of the plasma. Such fields, which have been measured as a regular feature of the solar wind, originate with the plasma in the atmosphere of the Sun. Because of the rotation of the Sun, the solar magnetic field spirals out from the solar atmosphere in a pattern that makes an angle of approximately 45° with the direction of solar-wind flow near the orbit of the Earth. The magnitude of this interplanetary field at the Earth is normally 5 gammas, and the density of the plasma is approximately 5 particles/cm^3.

The interplanetary magnetic field carries waves of various types that are associated with the different plasma modes in the solar wind. These magnetic-field waves intercept and electromagnetically excite the Moon in a manner similar to magnetotelluric excitation on Earth. McDonald (ref. 4–12) has applied this method to the Earth by using the secular variations in the geomagnetic field to determine the electrical conductivity of the Earth mantle.

In his analysis of the interaction of the Moon with the solar wind, Gold (ref. 4–13) assumed that the electrodynamic interaction could be characterized by a Cowling time constant. Gold postulated that the time constant would be long compared to the rotation period and that a magnetosphere and accompanying shock wave would be produced. Magnetometers on Explorer 35 (refs. 4–2 and 4–3) did not detect a bow shock.

Colburn, Currie, Mihalov, and Sonett (ref. 4-14) and Sonett (ref. 4-15) extended this analysis to include the case in which a steady-state current system is generated in the Moon and closes in the solar plasma. An electric field

$$E = v \times B \quad (4-1)$$

is produced as the solar magnetic field **B**, frozen in the solar plasma, sweeps by the Moon with a relative velocity v. For this to occur, it is necessary for the plasma and the magnetic field to be decoupled near the lunar surface so that the field can diffuse into the interior. The flow of the current driven by this electric field will distort the diffusion of the magnetic field in the lunar interior and cause the interplanetary field to be distorted. This current system will induce a toroidal magnetic field with a magnitude inversely proportional to the total resistance to current flow through the Moon. (See fig. 4-1(a).)

This model was used by Colburn et al. (ref. 4-14) in analysis of the Explorer 35 magnetic-field measurements to determine an effective electrical conductivity of 10^{-9} mhos/m for the whole Moon. By similar analysis of Explorer 35 magnetic-field measurements, Ness (ref. 4-16) found the upper limit of the conductivity to be 10^{-5} mhos/m. Both Colburn and Ness assumed displacement currents to be negligible, the driving magnetic field to be uniform over lunar dimensions, and the Moon to be homogeneous. Colburn also stated that his results were consistent with a model in which the Moon has a highly conducting core surrounded by an insulating layer. Behannon (ref. 4-6) used the Explorer 35 magnetometer measurements made during the traversal of the magnetic tail of the Earth to put an upper limit of 1.8 on the relative magnetic permeability of the Moon. In the tail region, the plasma flow velocity is subsonic with respect to the Moon.

Sill and Blank (ref. 4-17) and Schwartz and Schubert (ref. 4-18) have considered the time-dependent response of the Moon to this dependent driving field. By assuming the proper boundary conditions, it is possible to solve the vector Helmholtz equation for the time-dependent lunar field induced by the time-varying magnetic field and the motional electric field in the solar wind. The magnetometer on Explorer 35 provides a measurement of the external driving field, and the Apollo 12 magnetometer measures the result-

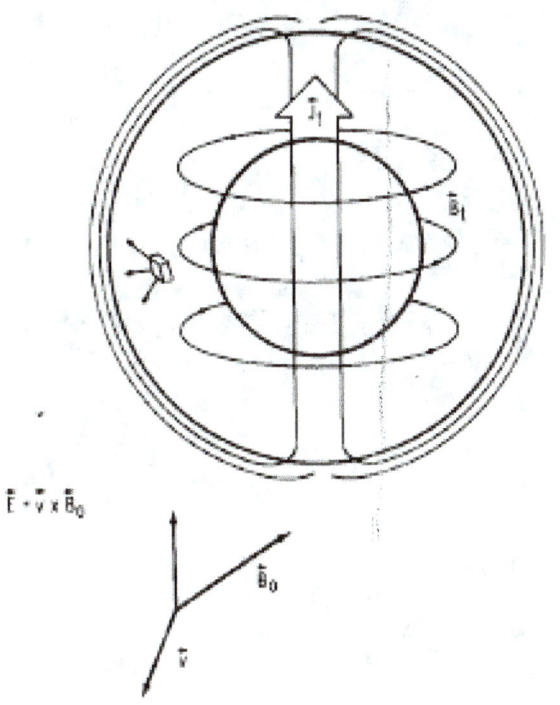

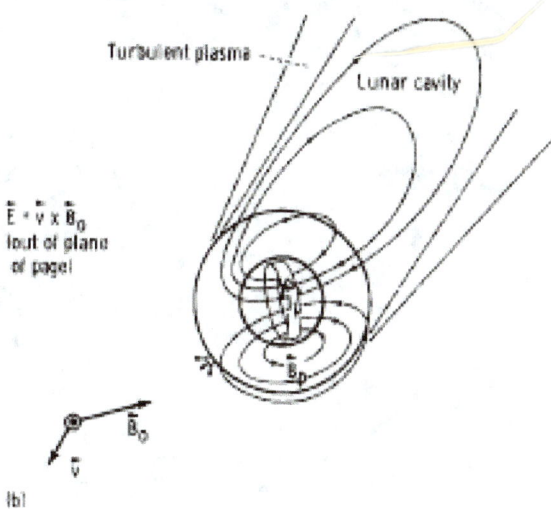

FIGURE 4-1.—Induced magnetic field. (a) Toroidal induced field B_t. View shows the side of the Moon facing the Sun, with north pole at top. (b) Poloidal induced field B_p. View is down upon the lunar northern hemisphere. The equatorial plane B_0 and v are in the plane of the paper. The resultant E is perpendicular to the plane of the paper.

ing induced lunar fields. The eddy currents induced by the rotating sector structure in the solar wind produce a poloidal field that opposes the driving field (fig. 4–1(b)). A conductivity σ can be determined as a function of depth in the lunar interior if it is assumed that displacement currents are negligible. After the radial conductivity distribution has been calculated, a material distribution can be assumed, and the internal temperature T can be calculated by using the method of Rikitake (ref. 4–19); that is,

$$\sigma = \sum_{i=1}^{3} \sigma_i \exp\left(\frac{-E_i}{kT}\right) \quad (4\text{–}2)$$

where E_i is the activation energy of the material, σ_i is the conductivity at infinite temperature, and k is Boltzmann's constant. The three summation terms of equation (4–2) denote impurity, intrinsic, and ion semiconduction that dominate, respectively, at increasing temperatures. England, Simmons, and Strangway (ref. 4–20) have calculated an internal conductivity distribution based upon thermal models from Phinney and Anderson (ref. 4–21) for a young, cold, poorly conducting Moon and for an old, hot, highly conducting Moon. The hot lunar model has a factor of 10^3 greater interior electrical conductivity than the cold model; therefore, determination of the lunar interior temperature profile should allow differentiation between hot and cold models of the Moon.

Experimental Technique

Instrument Requirements

The magnetometer, designed and built to measure the magnetic fields at the Apollo 12 site, is one of the five instruments of the Apollo lunar surface experiments package (ALSEP). Astronauts Conrad and Bean deployed the instrument at 14:00 G.m.t. on November 19, 1969. Figure 4–2 shows the magnetometer immediately following deployment and just prior to instrument activation at 14:39 G.m.t.

In conducting a surface magnetometer experiment, it is important that the geographic site,

FIGURE 4-2. — Apollo 12 magnetometer deployed on the Moon.

instrument sensitivity, and frequency range be selected with care. Time variations of the field are required for sounding measurements, and dc fields are required for a measurement of the unipolar interaction and of the permanent lunar magnetic field. A further requirement is that the magnetometer be able to operate over the lunar day and night, because the mapping of the field lines to determine the interior configuration requires measurements during a complete lunation. The instrument must provide its own internal calibrations on a periodic basis to determine instrument drifts during temperature excursions and over long time periods.

For Earth-based magnetic observations, it is important that a site survey be conducted at the lunar magnetic observatory location. Local deposits of nickel-iron or stony-iron meteoric material may be perined or may produce induction fields that will produce anomalous readings for measurements of the global field response of the Moon. Self-siting of the magnetometer is accomplished by measuring the magnetic field between the three sensor heads and calculating the gradient in the plane of the sensors. These gradient measurements provide information on the locations of sources and will also permit one component of $\nabla \times \mathbf{B}$ to be calculated.

The accuracy of an individual measurement will be fixed because of telemetry bit-rate limitations; therefore, the instrument must have a variable range, and any set range must be capable of being biased by a known amount. The instrument must also measure its position with respect to the lunar coordinate system. This is accomplished by the use of level sensor readings of two angles and a shadowgraph reading taken by an astronaut to determine the azimuthal alinement of the magnetometer. The preceding requirements have resulted in the development of an instrument that has the properties listed in table 4–1 and the configuration illustrated in figures 4–3 and 4–4.

Fluxgate Sensor

The Ames fluxgate sensor is the most important element in the magnetometer and is shown schematically in figure 4–5. The sensor consists of a flattened toroidal core of permalloy that is driven to saturation by a sine wave at a frequency f_0

TABLE 4–I. *Apollo 12 Magnetometer Characteristics*

Parameter	Value
Range	0 to ± 400 gammas
	0 to ± 200 gammas
	0 to ± 100 gammas
Resolution	± 0.2 gamma
Frequency response	dc to 2 Hz
Angular response	Proportional to cosine of angle between magnetic-field vector and sensor axis
Sensor geometry	Three orthogonal sensors at the end of 100-cm booms
	Orientation determination to within 1° in lunar coordinates
Commands	8 ground and 1 spacecraft
Internal calibration and sensor flip	180° flip
	0, ± 25, ± 50, and ± 75 percent of full scale
Field bias offset capability	0, ± 25, ± 50, and ± 75 percent of full scale
Modes of operation	Orthogonal field measurements
	Gradient measurement
	Internal calibration
Power	3.5 W average in daytime
	7.5 W average in nighttime
Weight	8.9 kg
Size	25 by 28 by 63 cm
Operating temperature	$-50°$ to $+85°$ C

of 6000 Hz. The output signal from the sense winding contains a second harmonic signal at $2f_0$, with an amplitude proportional to the input magnetic field and a phase relative to the drive waveform determined by the polarity of the magnetic field. The sensor electronics amplifies and filters the $2f_0$ sense-winding signal and synchronously demodulates it to derive a voltage proportional to the magnetic field.

Electronics

The magnetometer electronics is self-contained, except for ALSEP-supplied power, telemetry timing, and command lines. It weighs 6 lb, occupies a volume of 300 in.³, and contains 1300 active components, 1800 passive components, and 3300 memory core locations. The operation of the electronics can be categorized as illustrated in the functional block diagram (fig. 4–6).

The magnetic-field measurements are made by the fluxgate sensors that are operated by the

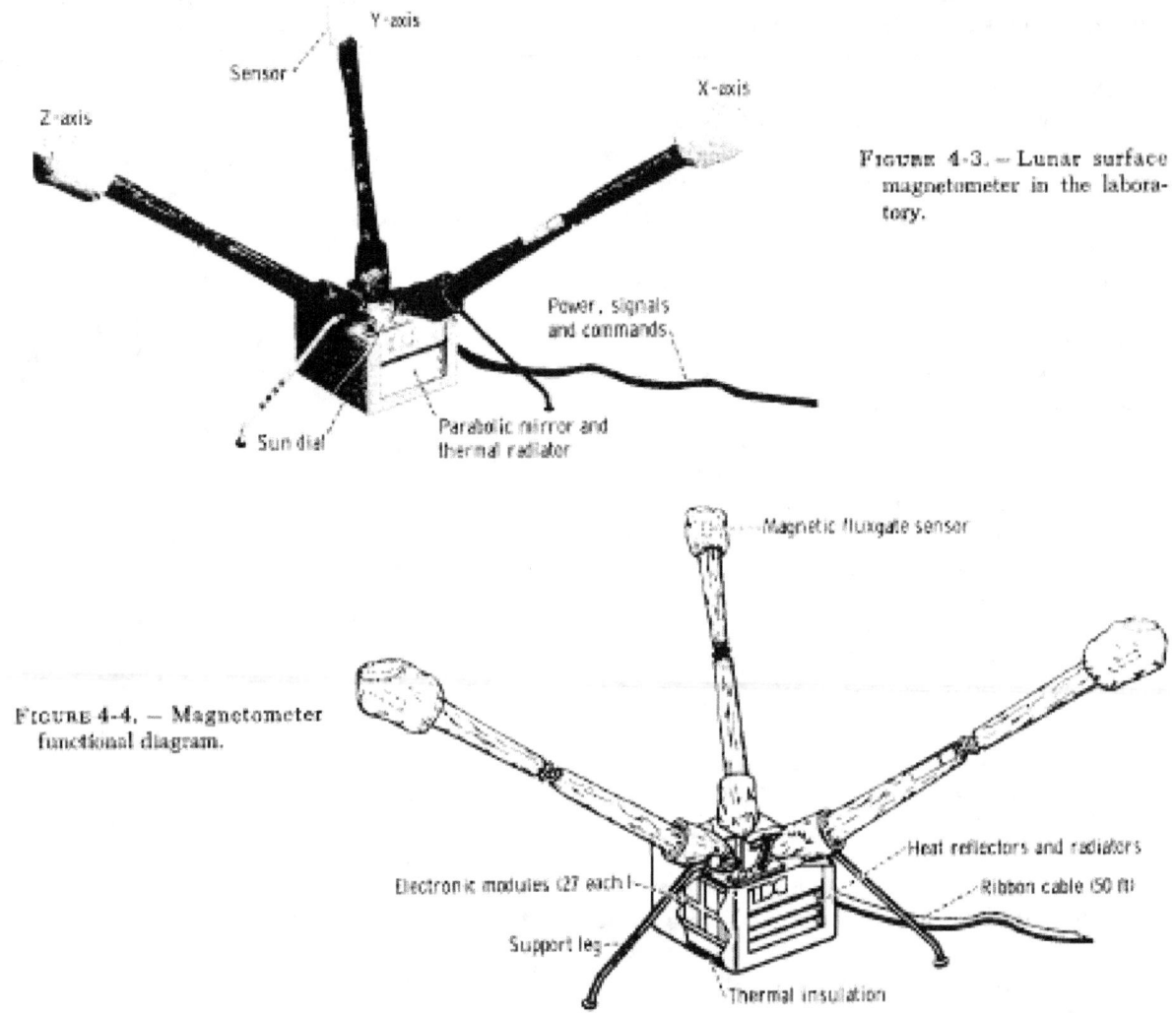

FIGURE 4-3.— Lunar surface magnetometer in the laboratory.

FIGURE 4-4.— Magnetometer functional diagram.

sensor electronics to provide to the analog-to-digital (A/D) converter an output voltage that is proportional to the field. This signal is processed through a low-pass three-pole Butterworth filter required to reduce aliasing errors at the A/D converter sample rate of 26.5 samples per second. This filter has 3-dB attenuation at 2 Hz and 64-dB attenuation at the sample rate of 26.5 Hz. The sensor electronics has three ranges, ±100 gammas, ±200 gammas, or ±400 gammas, that can be selected by ground command. Calibration and dc offset fields are generated by causing precisely known current to flow through the sensor feedback winding.

Internal data processing of the magnetic-field measurements utilizes the major portion of the magnetometer electronics. The analog data are converted to digital form by the A/D converter, filtered by the digital filter, and transferred to the output data buffer. The digital filter has a bandwidth of 0 to 0.36 Hz and will allow magnetic-field measurements to be made over an entire lunar cycle without alias errors being incorporated into the data. The filter is a hard-wired digital computer that accepts the 10-bit word from the A/D converter, stores it in a 3300-bit core-storage memory unit, performs the arithmetic operations for the filter routine, and transfers the data to the output data buffer. The filter can be bypassed by ground command in order to pass higher frequency information. Frequency response for the entire system is shown in figure 4-7.

Mechanical and Thermal Subsystems

The main function of the mechanical subsystem on the magnetometer is to orient the fluxgate sensors in three mutually orthogonal directions and to reorient these sensors automatically for calibration and site-survey operations. The sensors are spread approximately 150 cm apart and are 70 cm above the lunar surface. A functional schematic of the motor and cables necessary to flip and gimbal the fluxgate sensor is shown in figure 4-8.

The thermal subsystem on the magnetometer is designed to allow the instrument to operate over the complete lunar day-night cycle. This thermal system is designed to keep the magnetometer between $-30°$ and $+65°$ C during the night and day excursions. Because of some lunar dust on the thermal-control surfaces, the daytime temperatures were $10°$ C higher than expected from prelaunch tests. This is within the operating temperature limits ($-50°$ to $+85°$ C) of the instrument.

Thermal control is accomplished by a combination of insulation, control surfaces, and heaters. A covering of highly efficient insulation blankets is used to thermally isolate the instrument from its surroundings so that a set of parabolic reflector arrays (fig. 4-9) can control the heat transfer between the instrument and the lunar environment.

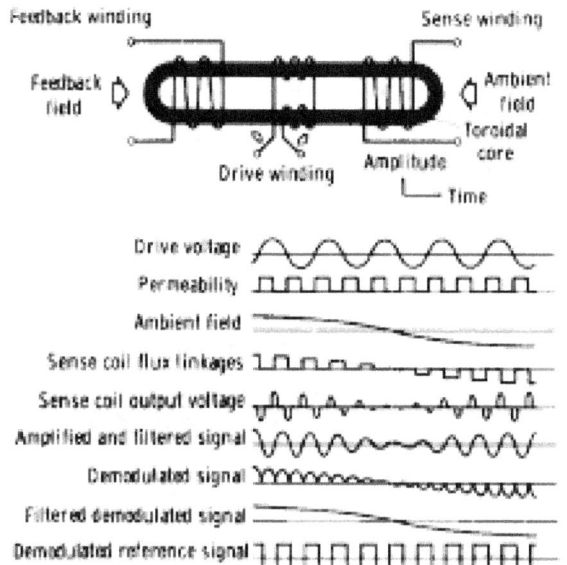

FIGURE 4-5. — Ames fluxgate sensor.

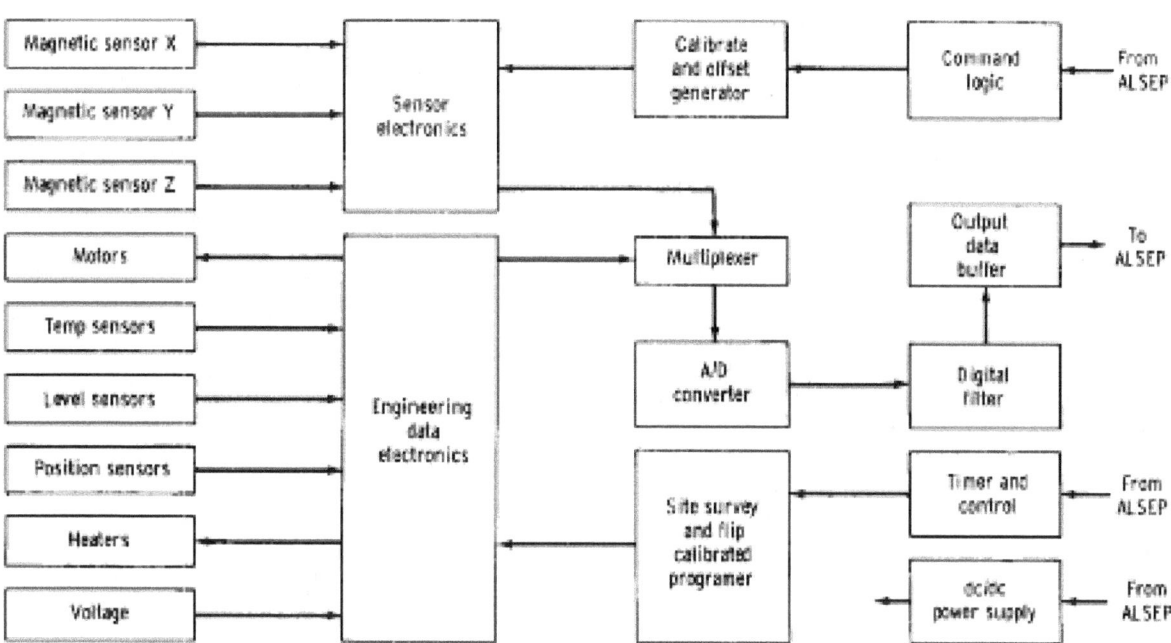

FIGURE 4-6. — Functional block diagram.

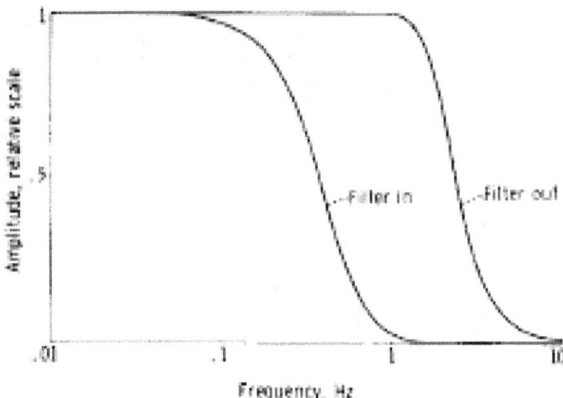

FIGURE 4-7. — Frequency response.

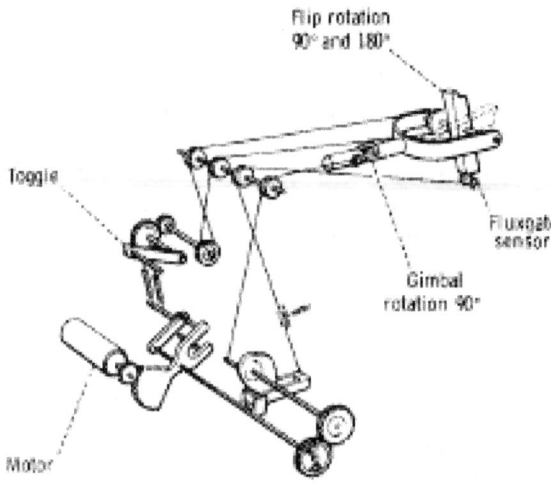

FIGURE 4-8. — Mechanical gimbal and sensor flip unit.

Mission Operation and Data Flow

The Apollo 12 magnetometer experiment is controlled from the NASA Manned Spacecraft Center (MSC), Houston, Tex., by commands transmitted to the ALSEP site from remote tracking stations. These commands are chosen on the basis of the analysis of data received in real time at MSC and are transmitted to the ALSEP by the Manned Space Flight Network (MSFN). Figure 4-10 is a simplified block diagram of the telemetry command and data flow. Ten ground commands and one ALSEP-timer-initiated command control the magnetometer. Table 4-II is a list of these commands.

Science and engineering data are transmitted

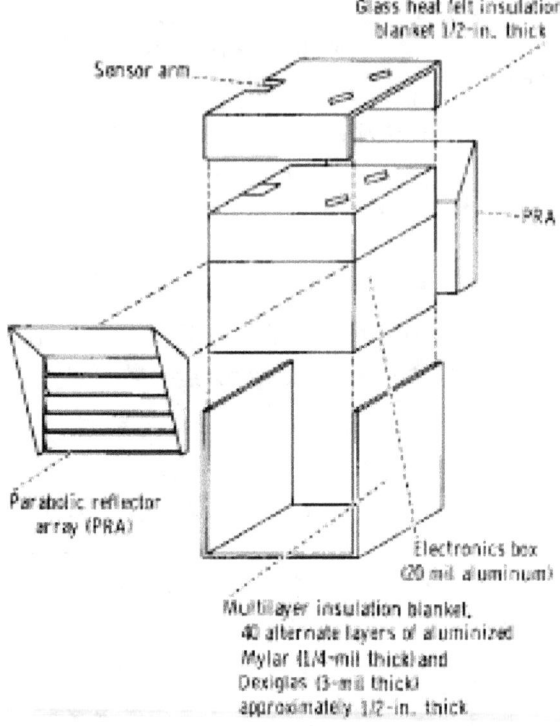

FIGURE 4-9. — Thermal control for electronics box.

back to Earth from the magnetometer experiment. Each analog sensor output is converted to a digital word and transmitted to Earth by means of the telemetry format shown in table 4-III. The digital data are recorded on magnetic tape at the remote sites and sent to MSC for processing. These processed tapes are then sent to Ames Research Center for detailed scientific postmission analysis.

In addition to being recorded on magnetic tape for later use, the data are sent directly to

TABLE 4-II. *Command List*

Command	Function
042	Operational power ON
045	Power standby
123	Range select
124	Field offset
125	Offset address
127	Flip-calibrate inhibit
131	Flip-calibrate initiate (Also initiated every 12 hr by the ALSEP timer.)
132	Digital filter bypass
133	Site survey
134	Thermal control

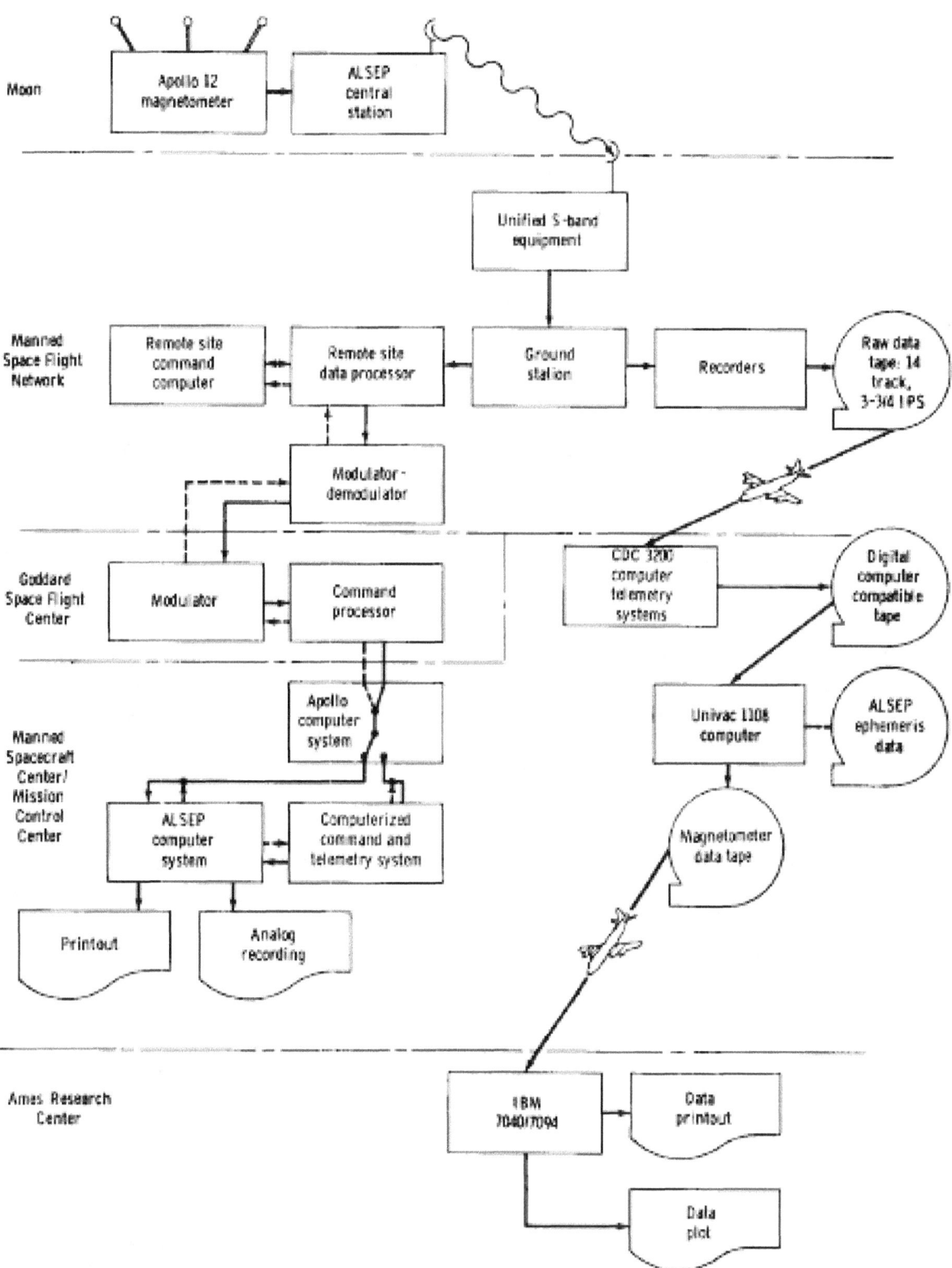

FIGURE 4-10. — Apollo 12 magnetometer telemetry command/data flow.

TABLE 4–III. *Telemetry Data*

ALSEP word	Subcommutated channel (bit 10)	Engineering analog 7 bit (bits 9 to 3)	Status bits	
			1	2
5	1	X-sensor temperature	X-flip position	X-flip
	2	Y-sensor temperature	Y-flip	Y-flip
	3	Z-sensor temperature	Z-flip	Z-flip
	4	Gimbal flip unit temperature	X-gimbal	Y-gimbal
	5	Electronics temperature	Z-gimbal	Temperature control
	6	Level detector 1	Spare	Heater state
	7	Level detector 2	Range	Range
	8	Reference voltage	Spare	Spare
	9	Same as 1	X-offset	X-offset
	10	Same as 2	X-offset	Y-offset
	11	Same as 3	Y-offset	Y-offset
	12	Same as 4	Z-offset	Z-offset
	13	Same as 5	Z-offset	Z-offset
	14	Same as 6	Address	Address
	15	Same as 7	Filter	Inhibit
	16	Same as 8	Spare	Spare
17	10 bit	X-axis magnetic field		
19	10 bit	Y-axis magnetic field		
21	10 bit	Z-axis magnetic field		
49	10 bit	X-axis magnetic field		
51	10 bit	Y-axis magnetic field		
53	10 bit	Z-axis magnetic field		

MSC for real-time analysis to allow proper control of the experiment. Real-time control is necessary to establish the proper range, offset, frequency response, thermal control, etc. One of the most important requirements for real-time operation is to control the one time-irreversible sequence of events that performs the gradient measurements during a site survey. It is necessary to assess both the magnetic-field amplitude stability and the instrument performance before sending the command to start the site-survey internal sequences. Data analysis for the real-time operations is presented continuously on a brush recorder and is presented intermittently upon command on a high-speed printer.

Results

The Apollo 12 mission provided the first magnetic-field measurements from the lunar surface. The sequence of operations during the first part of the mission is given in table 4–IV.

The magnetometer location on the lunar surface is 23.35° W longitude and 2.97° S latitude. Robert Sutton of the U.S. Geological Survey determined these coordinates by matching lunar features on panoramic surface photographs with the corresponding Lunar Orbiter photographs on a Mercator projection prepared by the Army Map Service, Corps of Engineers, for NASA.

The experiment is deployed so that each sensor is directed at an angle approximately 35°

TABLE 4–IV. *Apollo 12 Magnetometer Experiment Operations*

Date	Time (G.m.t.) hr:min	Operation
Nov. 14, 1969	16:22	Liftoff from Kennedy Space Center
Nov. 18, 1969	14:32	Lunar landing
Nov. 19, 1969	12:55	ALSEP removed from lunar module
Nov. 19, 1969	13:41	Magnetometer removed from ALSEP
Nov. 19, 1969	14:01	Magnetometer deployed by Conrad and Bean
Nov. 19, 1969	14:02	Magnetometer photographed
Nov. 19, 1969	14:39	Magnetometer turned ON
Nov. 19, 1969	14:45	Magnetometer range selected
Nov. 20, 1969	11:09	First flip-calibrate sequence completed
Nov. 20, 1969	14:26	Lunar module ascent
Nov. 22, 1969	22:50	Site-survey sequence started

above the horizontal. The Z-sensor is pointed toward the east and the X-sensor toward the northwest. The Y-sensor completes a right-handed orthogonal system. A photograph of the deployed instrument taken by Astronaut Bean is shown in figure 4–2. This photograph was taken after he had leveled and azimuthally alined the instrument along the ALSEP-to-Sun line by moving the instrument around until the bubble level and shadowgraph read within marked preset values. A subsequent photograph (fig. 4–11), also taken by Astronaut Bean, shows the shadowgraph and visual bubble level after instrument alinement. The shadowgraph reading was transmitted over the voice telemetry link and indicated that the instrument was alined azimuthally to within 0.5° of the instrument shadowgraph-to-Sun line. Two orthogonal electronic level sensors are monitored every 5 sec as part of the magnetometer engineering data. The angular readings are shown in figure 4–12 and indicate that the instrument changed its orientation by about 2° during the first lunar day.

During each orbit around the Earth, the Moon is embedded in each of the different magnetic-field regions shown schematically in figures 4–13 and 4–14. The magnetic-field environment is dominated by the solar wind in interplanetary space, by the interaction of the solar wind and the Earth magnetic field in the bow shock and transition region, and by the Earth intrinsic field in the geomagnetic-tail region. A detailed measurement of the magnetic field as a function of time on the lunar surface and in the immediate vicinity of the Moon will permit the electrical conductivity σ, permeability μ, and dielectric constant ϵ to be calculated. These electrical parameters that are calculated for field measurements at one point on the lunar surface may be associated with either the whole Moon or with that part of the lunar body in the vicinity of the Apollo 12 landing site. A magnetometer network is required to measure unambiguously the whole-body inductive response of the Moon to time-varying magnetic fields.

Magnetic-field data were received immediately

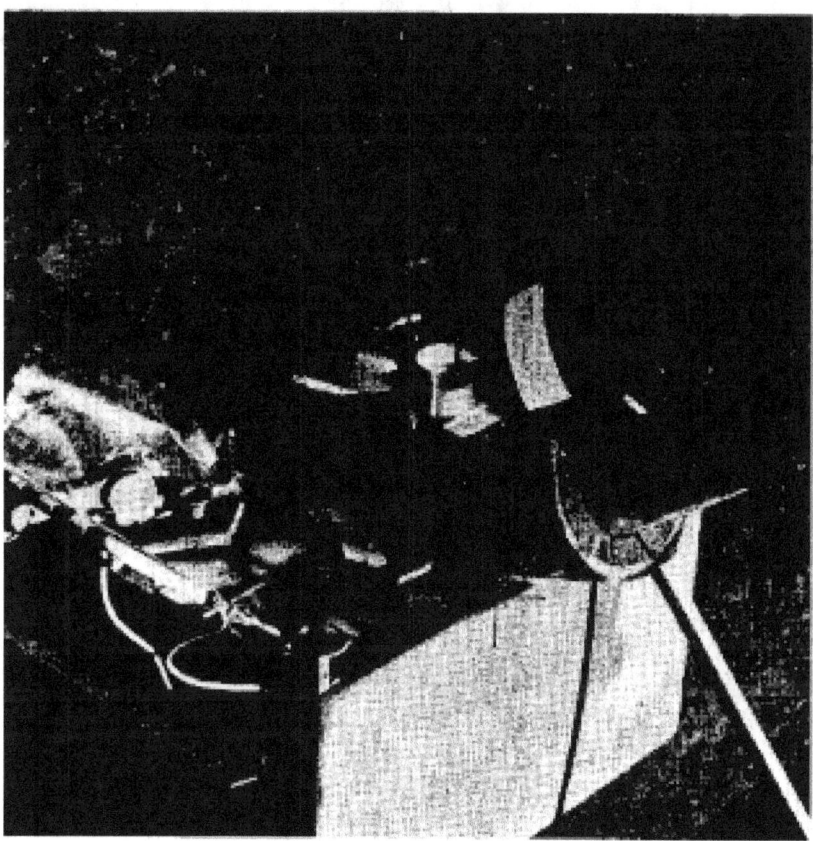

FIGURE 4–11.—Astronaut photograph of shadowgraph.

after instrument turn-on, and ground commands were sent to establish the proper range, field offset, and operational mode for the instrument.

Low-frequency magnetic-field data were averaged over the first 20-day period. A preliminary examination indicates that the average field magnitude is 36±5 gammas and that the field is directed downward and toward the southeast, as shown in figure 4-15. A more detailed analysis over a period of several lunations will be required to reduce the uncertainty in this measurement.

A site survey was performed during a 2-hr period starting at 22:50 G.m.t. on November 22, 1969. The magnitudes for the vector measurements at each of the sensor-head locations were as follows:

Sensor:	B
X	32.7±0.2 gammas
Y	32.7±0.2 gammas
Z	32.8±0.2 gammas

The scalar gradient calculated in the plane determined by the three sensors is

$$\frac{\partial |\mathbf{B}|}{\partial r} \leq 4 \times 10^{-3} \text{ gammas/cm} \quad (4\text{-}3)$$

Higher frequency data, which will be used to measure the inductive response of the Moon, are shown in figures 4-16 to 4-21. These figures show typical field measurements obtained during

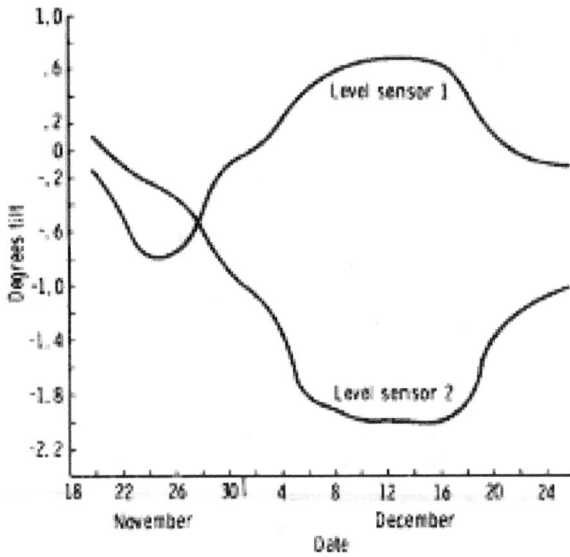

FIGURE 4-12. — Level sensor data.

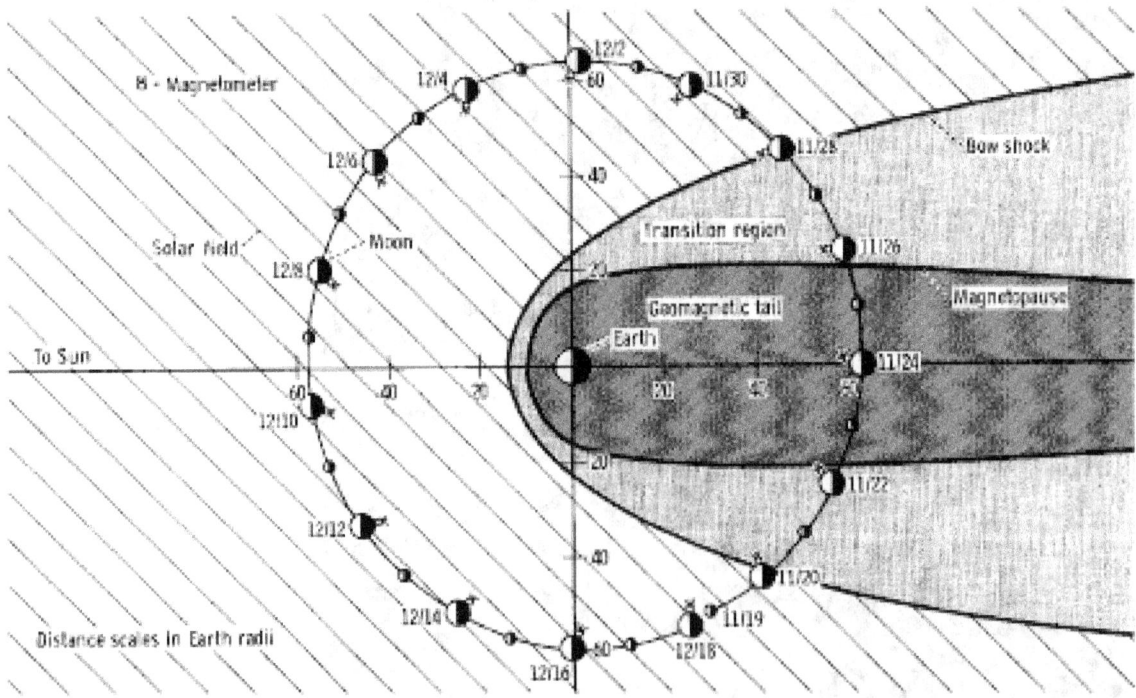

FIGURE 4-13. — Lunar orbit in the solar ecliptic plane.

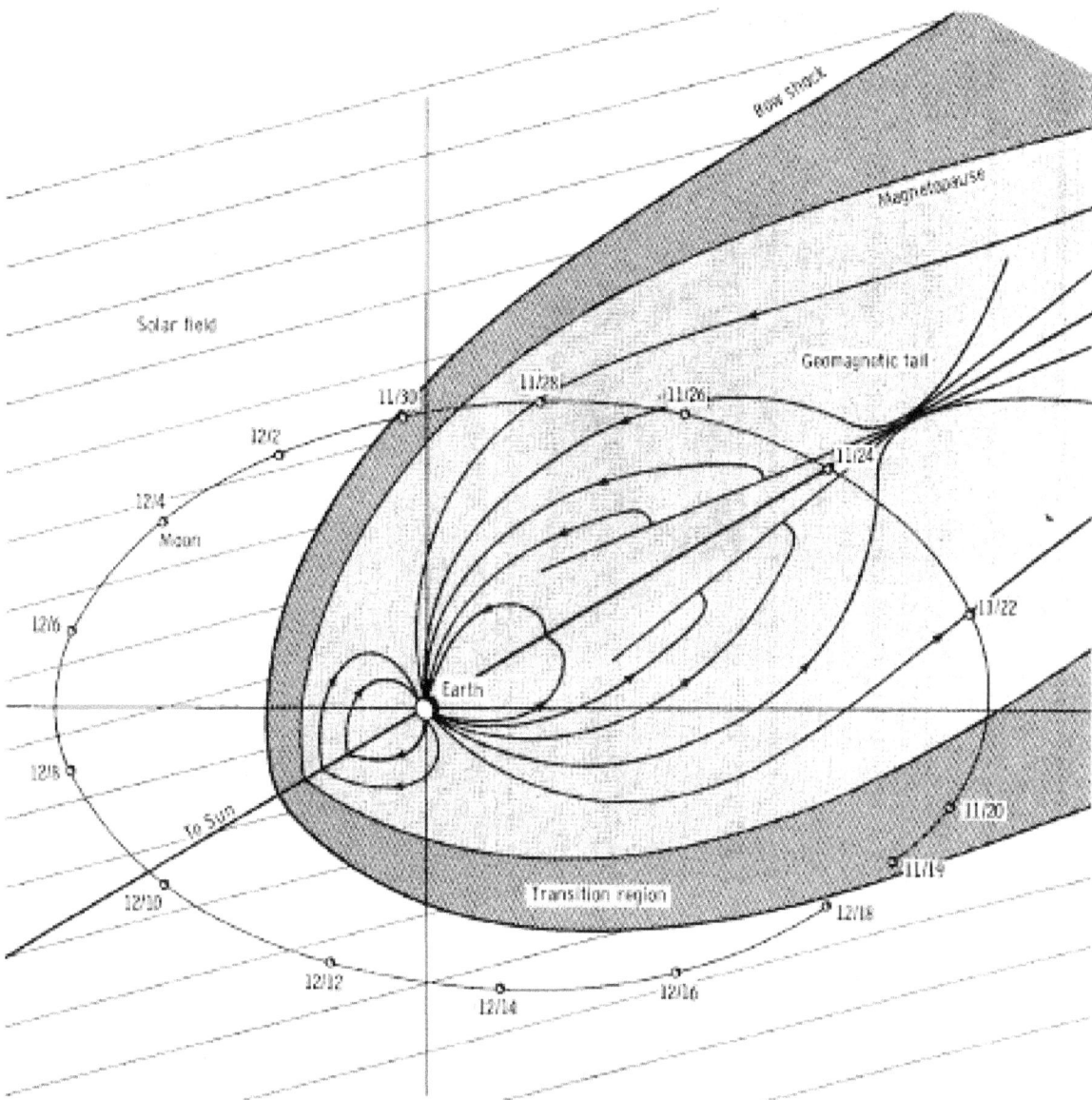

FIGURE 4-14. — Three-dimensional view of the lunar orbit.

a 6-min period in each of the three regions shown figure 4-13. The figures are reproductions of real-time records and do not have all the instrumental parameters removed from the data. The steady 36±5 gamma field is readily observed in all the real-time records.

Figure 4-16 is a time-series plot of the three vector components of the magnetic field in the instrument coordinate system. During this measurement period, the Moon was in interplanetary space, and the instrument was on the sunlit side of the Moon (fig. 4-13). The field variations are caused by the fluctuating solar field that is transported to the lunar surface by the solar plasma. These measurements correlate in time with data from the solar-wind spectrometer.

Figure 4-17 is a plot of the three vector components during a time period when the Moon was in interplanetary space and the magnetometer was on the dark side of the Moon. It can be seen that the resultant lunar surface field lacks the short-period fluctuations appearing in

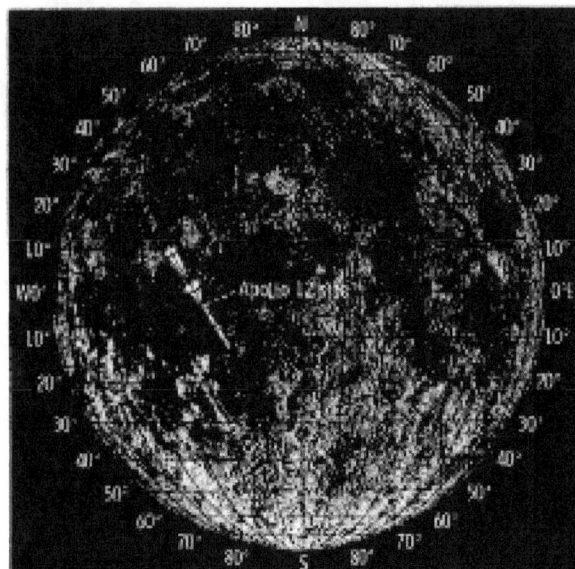

FIGURE 4-15. — Orientation of the average magnetic-field vector at the Apollo 12 site.

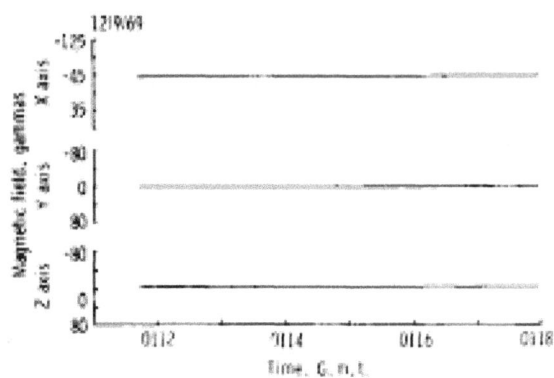

FIGURE 4-17. — Interplanetary field region on the dark side of the Moon.

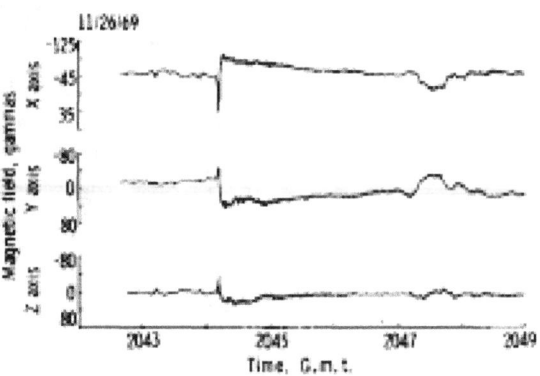

FIGURE 4-18. — Shock crossing.

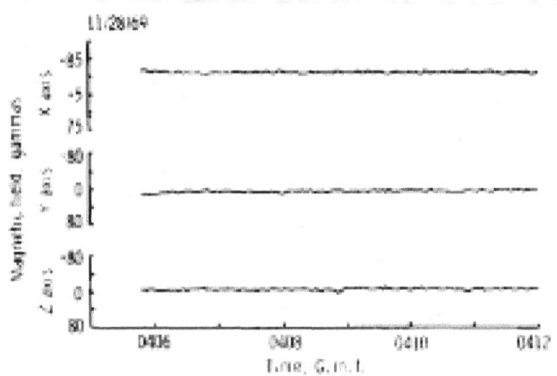

FIGURE 4-16. — Interplanetary field region on the sunlit side of the Moon.

data taken on the sunlit side. This indicates that the Moon has an inductive response to magnetic-field fluctuations in the solar wind.

Figure 4-18 shows the vector components of the surface magnetic field during a time when the Moon was in the vicinity of the Earth magnetohydrodynamic bow shock. This is a large-amplitude shock, and it contains frequency components throughout the bandpass of the instrument.

Figure 4-19 shows typical measurements obtained in the transition region between the bow shock and the magnetopause. In this region, the field fluctuations are of greater amplitude and contain higher frequencies than in the interplanetary solar-field region. These measurements also correlate with data from the solar-wind spectrometer.

Figure 4-20 shows a measurement taken in the geomagnetic-tail field region. The field has low amplitude and frequency fluctuations as a function of time.

Figure 4-21 is a time-correlated plot of Apollo 12 magnetometer data with Explorer 35 magnetometer data transformed into the same coordinate system. The data were obtained when the Moon was in the transition region. A typical orbit plot of Explorer 35 is shown in figure 4-22. It can be seen from the relative amplitude scales

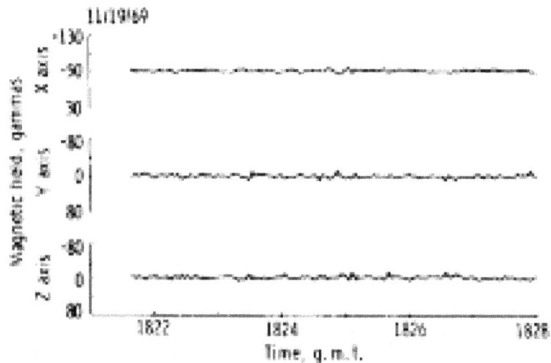

FIGURE 4-19. — Transition region between the magnetopause and the Earth bow shock.

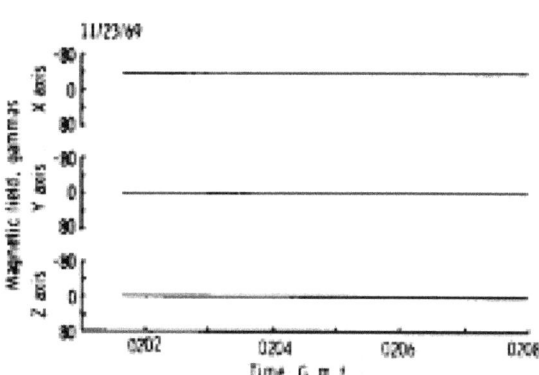

FIGURE 4-20. — Geomagnetic-tail region.

in figure 4-21 that the horizontal (Y and Z) components of the magnetic field at the Apollo 12 site are amplified when compared to the Explorer 35 measurements.

Temperatures are measured at five different locations in the instrument and transmitted to Earth every 5 sec. A plot of the temperatures during the first half of the lunar day is shown in figure 4-23. The temperatures are approximately 20° C higher than expected from prelaunch tests because of an accumulation of lunar dust on the thermal control surfaces during instrument deployment.

A reference voltage is measured and transmitted with other engineering information to the ground stations. This voltage remained constant to within 0.2 percent during the first 3-week period on the Moon.

Sixteen different instrument status functions are transmitted to Earth every 10 sec. The status information has been analyzed and found to be normal for all modes of operation. A total of nine different ground commands are utilized to establish the proper operating modes to optimize gain, sensor orientation, gradient measurement, calibration, etc. More than 200 separate commands have been transmitted to the magnetometer, and all have been properly executed.

The magnetometer is calibrated periodically both by internally generated ALSEP-timer commands and by ground command from MSC. The calibration sequence consists of four sets of amplitude steps and one mechanical 180° rotation of each of the three fluxgate sensors. The entire sequence takes approximately 5 min, and an example is shown in figure 4-24. To date, 130 calibrations have been performed with normal instrument response.

There have been two anomalies in the operation of the magnetometer since its deployment on November 19, 1969. One of the three filters in the data processing electronics was bypassed by ground command at 03:57 G.m.t. on November 22, 1969, after a malfunction was discovered. This digital filter is a small computer that low-pass filters the data and has a hardwired routine with the characteristic of a four-pole Bessel filter. The anomaly indicated that a subroutine in the digital computer was erroneously multiplying the data by a zero. After the electronics temperature decreased to 50° C from a high of 75° C during the lunar day, the filter was commanded back into the data link. The filter characteristics were measured, and it was determined to be functioning properly. Preliminary analysis indicates that a welded connection may have parted at high temperature. Normal data were received during the period in which the digital filter was bypassed. The bandpass of the instrument is increased from 0.3 to 3 Hz by bypassing the filters, so that the data may be aliased if the field contains high-amplitude information at frequencies greater than the Nyquist frequency of 1.5 Hz. Recorded data during this period indicated a very low probability of aliased information. This anomaly did not recur, how-

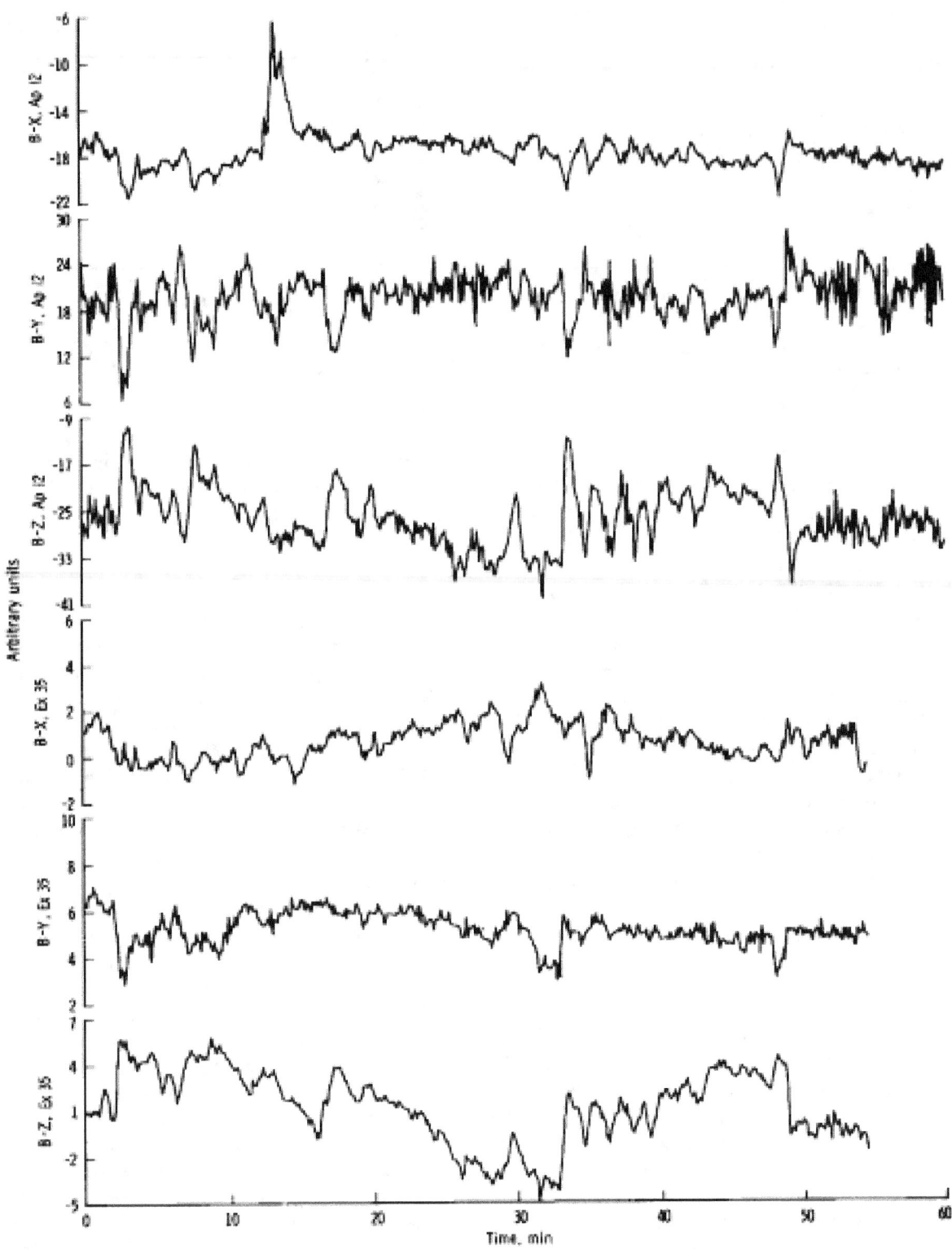

FIGURE 4-21.—Comparison between Explorer 35 and Apollo 12 field measurements.

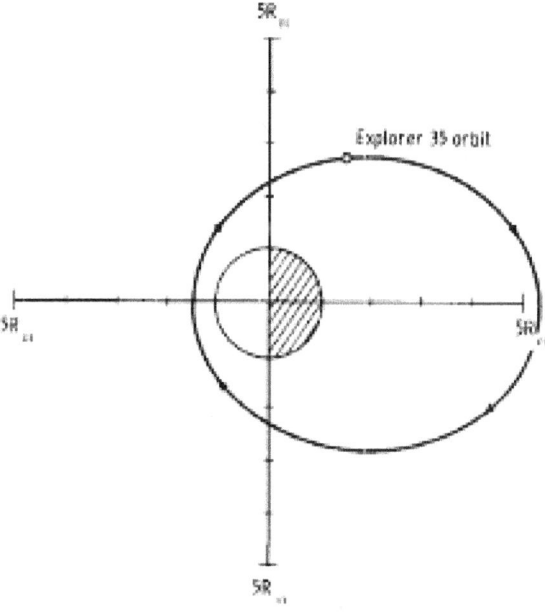

FIGURE 4-22. — Representative Explorer 35 lunar orbit.

ever, during the second lunar day when the electronics again reached 75° C.

The second anomaly occurred at 20:18 G.m.t. on December 11, 1969, when the three vector-component measurements dropped off scale. Subsequent commands permitted the X-component measurement to be brought back on scale but not the Y- and Z-sensor outputs. All subsystems were operating normally except for the sensor electronics. All three sensors resumed on-scale operation shortly following lunar sunrise on December 18, 1969, and the sensor electronics were calibrated and determined to be functioning normally. These two anomalies do not place a constraint on following missions.

Analysis

The magnetic field measured on the lunar surface is a vector sum of the lunar, terrestrial, and solar magnetic fields. The selenomagnetic field

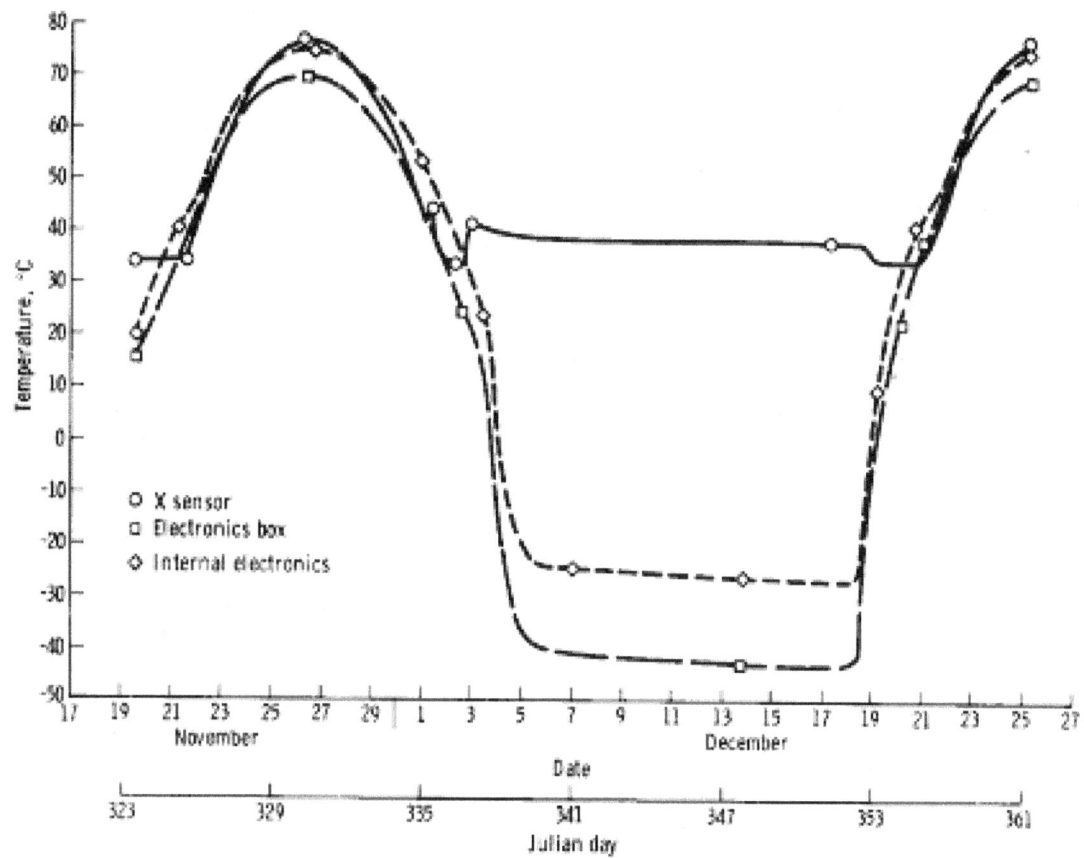

FIGURE 4-23. — Instrument temperatures versus time.

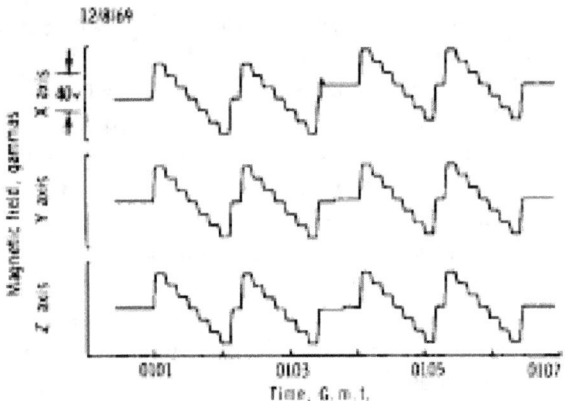

FIGURE 4-24. — Magnetometer calibration sequence.

associated with a local portion of the Moon should have small-amplitude variations for time periods on the order of days and can, therefore, be separated from the higher frequency transients by measurements taken during a period of one complete revolution around the Earth. A preliminary analysis of the field measured during one-half a lunar orbit shows that the field is 36 ± 5 gammas in magnitude and is directed downward approximately $50°$ from the vertical toward the southeast. The magnetic-field gradient was measured to be less than 4×10^{-3} gamma/cm in the plane tangent to the lunar surface.

If it is assumed that the source of this field is a dipole with magnetic moment m and that it is positioned a distance r away from the magnetometer in the plane of the sensors on the lunar surface, then the magnetic field B at the magnetometer is

$$\mathbf{B} = \frac{\mu_0 m}{4\pi r^3} (2 \cos \theta \, \mathbf{r} + \sin \theta \, \boldsymbol{\theta}) \quad (4-4)$$

where μ_0 is the permeability of free space, and θ is the angle between m and B. The total magnetic-field magnitude is then

$$|\mathbf{B}| = \frac{\mu_0 m}{4\pi r^3} (4 \cos^2 \theta + \sin^2 \theta)^{1/2} \quad (4-5)$$

The site survey or gradient measurement at the Apollo 12 site showed the gradient in the plane of the sensors to be less than 4×10^{-3} gammas/cm. (See eq. (4–3).) The gradient calculated from equation (4–5) is

$$\frac{\partial |\mathbf{B}|}{\partial r} = \frac{-3\mu_0 m}{4\pi r^4} (4 \cos^2 \theta + \sin^2 \theta)^{1/2} \quad (4-6)$$

From equations (4–5) and (4–6), a minimum distance from the dipole source to the magnetometer card can then be calculated.

$$\frac{\partial |\mathbf{B}|/\partial r}{|\mathbf{B}|} \leq \frac{-3}{r} \quad (4-7)$$

$$r_{min} = \frac{-3 |\mathbf{B}|}{\partial |\mathbf{B}|/\partial r} \simeq 2 \times 10^4 \text{ cm} \quad (4-8)$$

Therefore, the source must be located at least 0.2 km from the magnetometer if the source lies on the surface of the Moon.

If the measured 36 ± 5 gamma field were the result of a localized dipole source m located at the center of the Moon, which can be approximated by a point dipole, then the component form of equation (4–4) could be used to calculate m and θ. In component form

$$B_r = \frac{2\mu_0 m}{4\pi r^3} \cos \theta \quad (4-9)$$

$$B_\theta = \frac{\mu_0 m}{4\pi r^3} \sin \theta \quad (4-10)$$

Solving for m and θ with $r = 1.738 \times 10^6$ m yields $m = 1.4 \times 10^{21}$ G-cm^3 and $\theta = 72°$ in a plane containing m and a radius vector to the magnetometer site.

However, magnetic-field measurements by Sonett et al. (ref. 4–2) and Ness et al. (ref. 4–3) on the lunar orbital spacecraft Explorer 35 indicated that the dipole moment is less than 10^{20} G-cm^3. Therefore, to be consistent with Explorer 35 measurements, the 36-gamma field must be the result of a localized source near the Apollo 12 landing site rather than a uniform dipole moment associated with the whole Moon.

The foregoing preliminary analysis of the real-time data indicates that a portion of the Moon near the Apollo 12 landing site is magnetized. Further examination of the data from the Apollo 12 magnetometer during the lunar day and night and during large field transients as well as correlation with the Explorer 35 magnetometer indicates a strong lunar inductive response to external magnetic fields. This inductive response may be the result of the whole Moon or a part of the lunar material near the Apollo 12 landing site that is chemically and/or electrically differentiated from the whole Moon. One possible explanation of the remanent magnetic-field source is that a region of lunar material may have cooled below its Curie point after a meteorite impact or a whole-Moon cooling period, thereby causing the ambient magnetic field to be frozen in the material.

Discovery of both the remanent and induced lunar magnetic-field sources places a strong requirement for a network of magnetometers to make independent multiple measurements and ultimately to calculate the physical state of the lunar interior. Data analysis in the near future will be directed toward understanding the inductive lunar response and toward calculating an electrical-conductivity distribution of the lunar interior.

References

4-1. DOLGINOV, SH. SH.; EROSHENKO, E. G.; ZHUZGOV, L. N.; and PUSHKOV, N. V.: Measurements of the Magnetic Field in the Vicinity of the Moon by the Artificial Satellite Luna 10. Akad. Nauk USSR, Doklady, vol. 170, Sept. 21, 1966, pp. 574-577.

4-2. SONETT, C. P.; COLBURN, D. S.; and CURRIE, R. G.: The Intrinsic Magnetic Field of the Moon. J. Geophys. Res., vol. 72, no. 21, Nov. 1, 1967, pp. 5503-5507.

4-3. NESS, N. F.; BEHANNON, K. W.; SCEARCE, C. S.; and CANTARANO, S. C.: Early Results From the Magnetic Field Experiment on Lunar Explorer 35. J. Geophys. Res., vol. 72, no. 23, Dec. 1, 1967, pp. 5769-5778.

4-4. DEWYS, J. N.: Results and Implications of Magnet Experiments on Surveyor 5, 6, and 7 Spacecrafts (abs.). Trans. Am. Geophys. Union, vol. 49, no. 1, Mar. 1968, p. 249.

4-5. KELLER, G. V.; and FRISCHKNECHT, F. C.: Electrical Methods in Geophysical Prospecting. Pergamon Press, Inc., 1966, p. 58.

4-6. BEHANNON, K. W.: Intrinsic Magnetic Properties of the Lunar Body. J. Geophys. Res., vol. 73, no. 23, Dec. 1, 1968, pp. 7257-7268.

4-7. HAGFORS, T.: Review of Radar Observations of the Moon. The Nature of the Lunar Surface, W. N. Hess, D. H. Menzel, and J. A. O'Keefe, eds., Johns Hopkins Press, 1966, pp. 229-239.

4-8. BROWN, W. E., JR.; DIBOS, R. A.; GIBSON, C. B.; MUHLEMAN, D. O.; PEAKE, W. H.; and PEOPLES, T. V.: Scientific Results. Ch. 6 of Surveyor III Mission Report, Part II. TR32-1177, JPL, Calif. Inst. Tech., June 1, 1967, pp. 189-194.

4-9. KOPAL, Z.: An Introduction to the Study of the Moon. D. Reidel Pub. Co. (Dordrecht-Holland), 1966, p. 369. (Available from Gordon & Breach Sci. Pub.)

4-10. STRANGWAY, D. W.: Moon: Electrical Properties of the Uppermost Layers. Science, vol. 165, no. 3897, Sept. 5, 1969, pp. 1012-1013.

4-11. ANON.: Apollo 11 Preliminary Science Report. NASA SP-214, 1969.

4-12. MCDONALD, K. L.: Penetration of the Geomagnetic Secular Field Through a Mantle With Variable Conductivity. J. Geophys. Res., vol. 62, no. 1, Mar. 1957, pp. 117-141.

4-13. GOLD, T.: The Magnetosphere of the Moon. The Solar Wind, R. J. Mackin, Jr., and M. Neugebauer, eds., Pergamon Press, 1966, pp. 381-389. (Also available as JPL Tech. Rept. no. 32-630.)

4-14. COLBURN, D. S.; CURRIE, R. G.; MIHALOV, J. D.; and SONETT, C. P.: Diamagnetic Solar-Wind Cavity Discovered Behind Moon. Science, vol. 158, no. 3804, Nov. 24, 1967, pp. 1040-1042.

4-15. SONETT, C. P.: Principle of Planetary Unipolar Generators. Planetary Electrodynamics. Vol. II, S. C. Coroniti and J. Hughes, eds., Gordon & Breach Sci. Pub., 1969.

4-16. NESS, N.F.: Electrical Conductivity of the Moon (abs.). Trans. Am. Geophys. Union, vol. 49, no. 1, Mar. 1968, p. 242.

4-17. SILL, W. R.; and BLANK, J. L.: Method for Estimating the Electrical Conductivity of the Lunar Interior. Tech. Rept. no. TR-69-103-7-4, Bellcomm, Inc., Sept. 22, 1969.

4-18. SCHWARTZ, KENNETH; and SCHUBERT, GEORGE: Time-Dependent Lunar Electric and Magnetic Fields Induced by a Spatially Varying Interplanetary Magnetic Field. J. Geophys. Res., vol. 74, no. 19, Sept. 1, 1969, pp. 4777-4781.

4-19. RIKITAKE, T.: Electromagnetism and Earth's Interior. Elsevier Pub. Co., Amsterdam, 1966.

4-20. ENGLAND, A. W.; SIMMONS, G.; and STRANGWAY, D.: Electrical Conductivity of the Moon. J. Geophys. Res., vol. 73, no. 10, May 15, 1968, pp. 3219-3226.

4-21. PHINNEY, R. A.; and ANDERSON, D. L.: Internal Temperatures of the Moon. Report of the "Tycho" Study Group, Elec. Eng. Dept., Univ. of Minnesota, Aug. 1965.

5. The Solar-Wind Spectrometer Experiment

Conway W. Snyder,[a][1] *Douglas R. Clay,*[a] *and Marcia Neugebauer* [a]

The solar-wind spectrometer experiment was designed for the Apollo lunar surface experiments package (ALSEP) with the objective of detecting whatever solar plasma might strike the surface of the Moon. Thus, the spectrometer was required to be sensitive enough to detect the normal solar-wind flux at any angle in the lunar sky and to measure enough of the properties of the bulk solar wind to establish the nature of its interaction with the Moon. In 1966, at the time the experiment was proposed, nothing was known about this interaction. Subsequently, measurements of the solar wind and the magnetic field were made near the Moon by the lunar orbiter Explorer 35 (refs. 5-1 to 5-6). No evidence of a plasma shock ahead of the Moon was discovered, and the distortion of the solar-wind magnetic field by the Moon was observed to be very small. Both of these results imply that the solar wind probably strikes the surface directly, but they do not rule out the possibility of some more complex type of interaction very near the lunar surface and especially near the terminator plane. The Apollo 11 solar-wind composition experiment (ref. 5-7) detected rare gas atoms deposited in an aluminum foil on the lunar surface and demonstrated that the magnitude and direction of their incident velocity was very roughly as would be expected for the undisturbed solar wind.

The scientific objectives of the solar-wind spectrometer experiment are as follows:

(1) *The existence of the solar-wind plasma on the Moon* — to compare solar-wind properties measured at the lunar surface with those measured in space near the Moon to determine whether the Moon has any effect on the solar plasma other than simply absorbing it

(2) *The properties of the lunar surface and interior* — to determine whether there are any subtle effects of the Moon on the solar-wind properties and to relate these effects to properties of the Moon such as its magnetic field, its electrical conductivity, the possibility of the Moon retaining an atmosphere, or the possible effect of solar corpuscular radiation on the lunar surface layer by the mechanism of sputtering or electrical charging

(3) *General solar-wind properties* — to study the motion of waves or discontinuities in the solar wind by measuring the time intervals between the observations of changes in plasma properties at the Moon and at the Earth

(4) *The magnetospheric tail of the Earth* — to make inferences as to the length, breadth, and structure of the magnetospheric tail of the Earth from continuous measurements made for 4 or 5 days around the time of full Moon

Instrument Description

The basic sensor in the solar-wind spectrometer is a Faraday cup that measures the charged-particle flux entering the cup. By collecting these ions and using a sensitive current amplifier, the resultant current flow is determined. Energy spectra of positively and negatively charged particles are obtained by applying fixed sequences of square-wave ac retarding potentials to a modular grid and measuring the resulting changes in current. Similar detectors have been flown on a variety of space probes. Such detectors are described in reference 5-8.

To be sensitive to solar-wind plasma from any direction (above the horizon of the Moon) and to ascertain the solar-wind angular distribution,

---

[a] Jet Propulsion Laboratory, California Institute of Technology.
[1] Principal investigator.

the solar-wind spectrometer has an array of seven cups. Since the cups are identical, an isotropic particle flux would produce equal currents in each cup. If the flux is not isotropic but appears in more than one cup, analysis of the relative amounts of current in the collectors can provide information on the direction of plasma flow and its anisotropy. The central cup faces vertically, and the remaining six cups symmetrically surround the central cup. Each of the six cups faces 60° off vertical. The combined acceptance cones of all cups cover most of the upward hemisphere. Each cup has a circular opening, five circular grids, and a circular collector (fig. 5-1). The functions of the grid structures are to apply an ac modulating field to incoming particles and to screen the modulating field from the inputs to the sensitive preamplifiers. The entrance apertures of the cups were protected from damage or dust by covers that remained in place until after the departure of the lunar module. The angular dependence of the Faraday-cup sensor has been measured by laboratory plasma calibrations. The result, averaged over all seven cups, is shown in figure 5-2 and agrees quite well with the measured optical transparency.

The electronics for the solar-wind spectrometer is in a temperature-controlled container that hangs below the sensor assembly. The electronics includes power supplies, a digital programer that controls the voltages in the sensors as required, current-measuring circuitry, and data-conditioning circuits.

The solar-wind spectrometer operates in an invariable sequence in which a complete set of

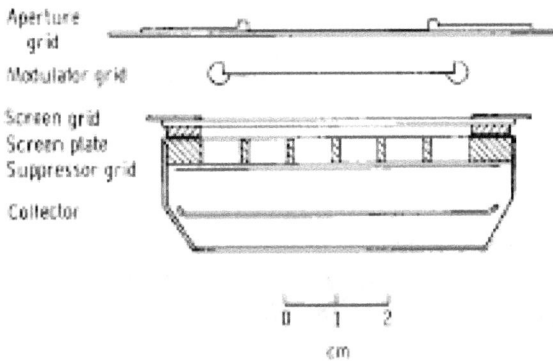

FIGURE 5-1. — Faraday-cup sensor.

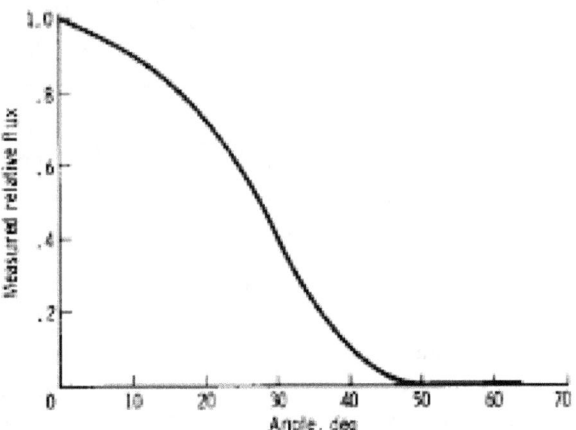

FIGURE 5-2. — Angular response of the Faraday cup.

plasma measurements is made every 28.1 sec. The sequence consists of 14 energy steps spaced a factor of $\sqrt{2}$ apart for positive ions and seven energy steps spaced a factor of 2 apart for electrons. A large number of internal calibrations are provided, and every critical voltage is read out at intervals of 7.5 min or less.

On the Moon, the solar-wind spectrometer is hung from a pair of knife edges so that it is free to swing about an east-west horizontal axis and, hence, is self-leveling in one dimension. Leveling about the north-south axis is indicated by a Sun sensor that peaks at the time that the Sun is 30° east of the axis of the central cup.

Instrument Deployment

The solar-wind spectrometer was deployed without difficulty by Astronaut Charles Conrad, Jr. Figure 5-3 shows the spectrometer on the lunar surface. From the pattern of the shadow on the radiator (left side of the instrument), it has been determined that the east-west axis of the instrument is actually alined approximately 2.8° north of east. The data from the Sun sensor indicate that the axis is off level by approximately 2.5°, with the west edge low. Both these values are well within specified limits. The level about the east-west axis cannot be determined from the photograph but should be insured by the self-leveling suspension.

Shortly after deployment, the spectrometer was turned on to provide background data with the sensor covers in place. Approximately 1 hr

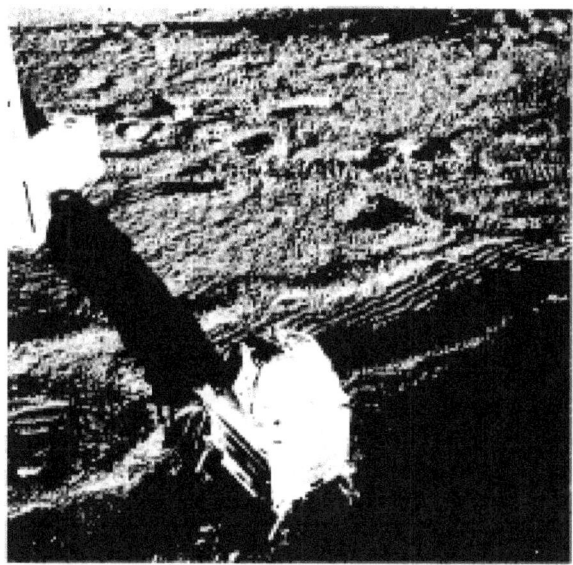

FIGURE 5-3.—Solar-wind spectrometer experiment deployed on the Moon.

after lunar module ascent, the covers were removed by command from Earth, and detection of solar plasma began.

Instrument Performance

During the first month, the solar-wind spectrometer has operated as expected. All functions have performed properly. The thermal control has proved adequate with temperatures ranging between $-16°$ and $+63°$ C inside the electronics package and between $-134°$ and $+63°$ C inside the sensor assembly. Comfortable margins for possible future degradation exist, because thermal vacuum tests have demonstrated no adverse effects of electronics package temperatures from $+108°$ and $+100°$ C or of sensor temperatures from $-147°$ to $+111°$ C.

Types of Spectra Observed

The data discussed in this report have been obtained from the high-speed printer in the ALSEP control room at the NASA Manned Spacecraft Center. The data consist of intermittent samples of data from 1 to 8 min in duration. The positive-ion spectra observed have been of the following general types:

(1) The peak current is in an energy window corresponding to a proton velocity of 400 to 550 km/sec. The bulk of the remaining currents is in the two adjacent energy windows, and a small (4 to 10 percent) current is in the second higher energy window. Figure 5-4 is a histogram of a typical spectrum of this type. This type of spectrum would be expected for the unperturbed (normal interplanetary) solar wind.

(2) The second type of spectrum has a peak current in a window corresponding to a proton velocity of 250 to 450 km/sec and has significant currents in several adjacent windows on either side of the peak. Such spectra typically have positive-ion densities of approximately 5 particles/cm^3, and adjacent spectra (spaced 28.1 sec apart) often show large velocity changes. Double-peaked spectra occasionally appear, probably indicating rapid velocity fluctuations. Figure 5-5 is a histogram of a typical spectrum of the second spectrum type. Such spectra are typical of the transition region between the Earth bow shock and the geomagnetosphere. However, other sources of perturbation, such as solar conditions at the time the solar-wind particles left the Sun or lunar interaction with the solar wind, cannot be ruled out until further information is available.

(3) At times, detectable currents appear in only one energy step or in two adjacent steps so that certain plasma properties, such as velocity distribution, are not calculable for these positive-ion spectra.

(4) During most of the lunar night, there is no detectable flux of solar-wind particles within the 100- to 900-km/sec range of the solar-wind spectrometer.

A further type of spectrum for positive ions was observed early on November 27, 1969, but for only a few hours. This special type of spectrum had characteristics that were highly variable in time. Many such spectra were of the perturbed solar-wind type except for increased fluxes in all windows from 180 to 900 km/sec (and perhaps beyond, since the spectrometer was in low modulation gain, and higher velocities were not measured) and except for significant fluxes detected in other cups facing as far as 60° from the Sun angle. One of these spectra is shown in figure 5-6. These spectra are not typical and do not seem to fit easily into the general types listed previously.

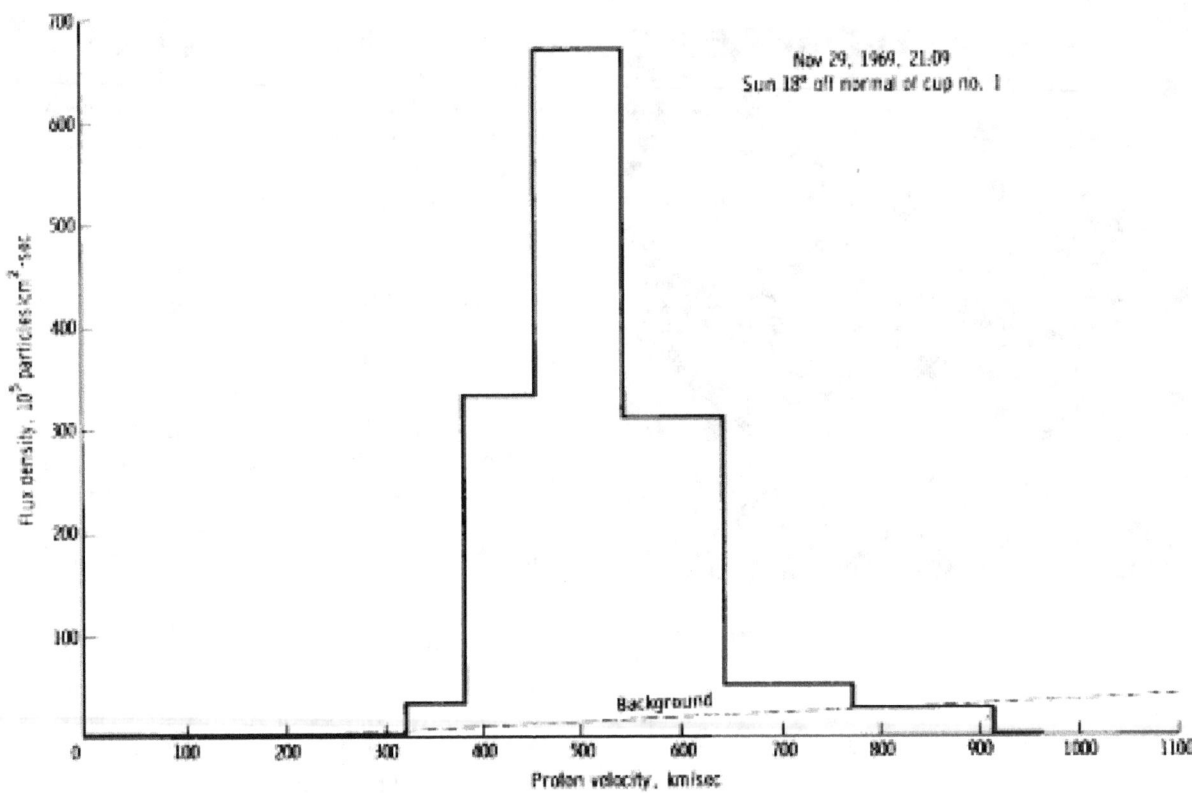

FIGURE 5-4.— Typical positive-ion spectrum obtained in the interplanetary solar wind. The background reading of each cup is slightly dependent upon modulation voltage, which is caused primarily by pickup in the electrometer inputs. Until all the data become available and are analyzed, this background is only estimated, giving an uncertainty in readings in the higher velocity windows. The magnitude of this source of uncertainty is indicated in each spectrum as a dashed line labeled "background."

The electron component of the solar wind has also been detected. However, since this type of plasma probe typically does not completely distinguish between plasma electrons and photoelectrons, interpretation of the data is difficult. Analysis of the complete data from magnetic tape will be required before conclusions can be drawn from the electron data. The times of appearance and disappearance of photoelectrons gave clear indications of sunset and sunrise.

Observations

At the time of solar-wind spectrometer dust-cover removal (15:25:30 G.m.t. on November 20, 1969), the positive-ion spectra observed were of the second type (perturbed solar wind), and the lunar surface magnetometer indicated that the Moon was behind the plasma bow shock of the Earth. This type of solar-wind data continued, with large fluctuations in bulk velocity and density, until approximately 03:00 G.m.t. on November 21, 1969, when a region of no plasma was observed. For the next 5 days, there were only occasional sampling periods of spectrometer printout that indicated plasma was present (always of the second type). The magnetometer indicated that the Moon was in the magnetospheric tail of the Earth (fig. 5–7).

Commencing at approximately 10:00 G.m.t. on November 26, 1969, the solar-wind spectrometer entered a region wherein the majority of spectra were of the second type (perturbed). From approximately 12:00 G.m.t. on November 28, 1969 (when the magnetometer indicated passage out through the bow shock into interplanetary space), until sunset on December 3, 1969,

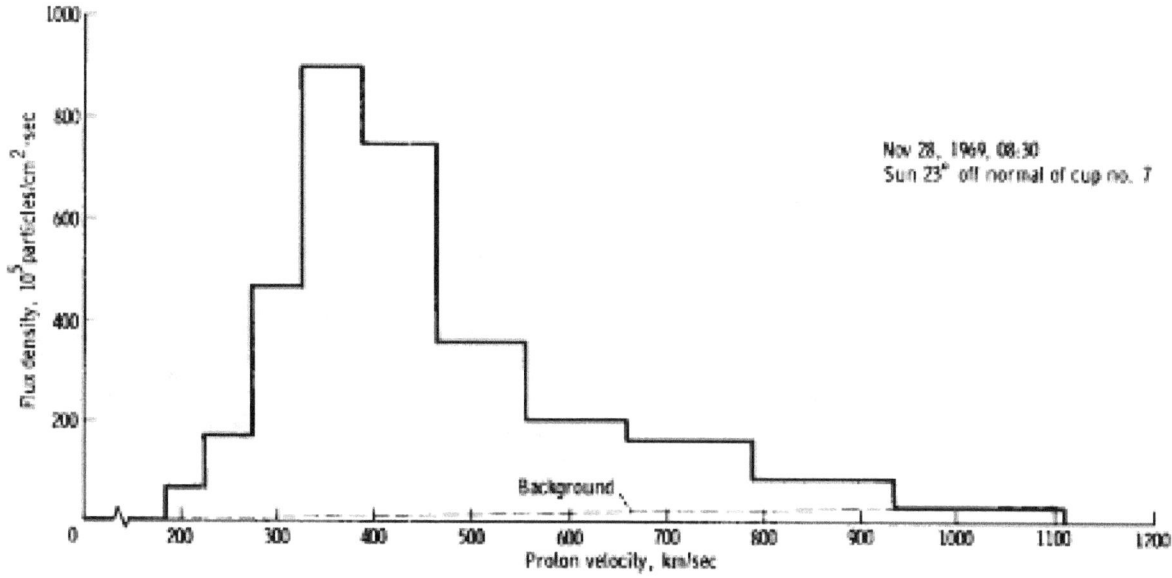

FIGURE 5-5. — Typical positive-ion spectrum obtained in the magnetospheric tail (perturbed solar wind) of the Earth.

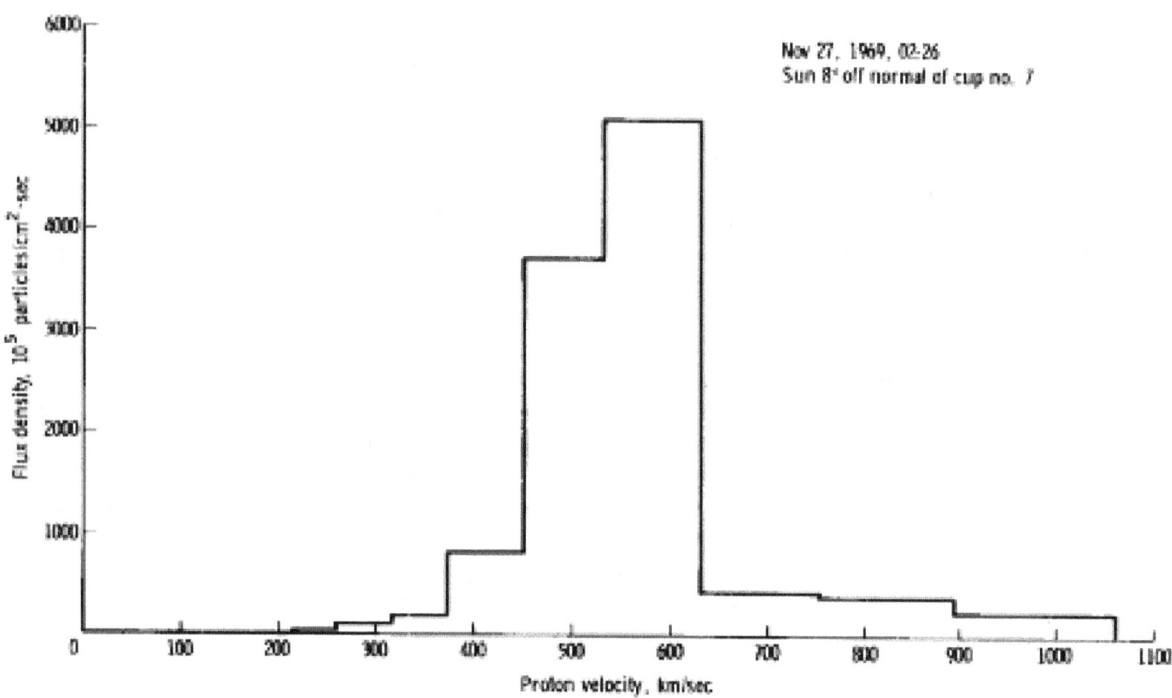

FIGURE 5-6. — One of many unusual spectra obtained on November 27, 1969.

the spectra were mainly of the first type (interplanetary solar wind); although fairly frequent, the flux density became low enough to make classification according to spectrum type difficult. Later during November 28, 1969, several short periods of type (2) spectra were tentatively identified.

As indicated by the cessation of photoelectrons detected in the cup nearest the solar direction, the spectrometer was in darkness beginning at

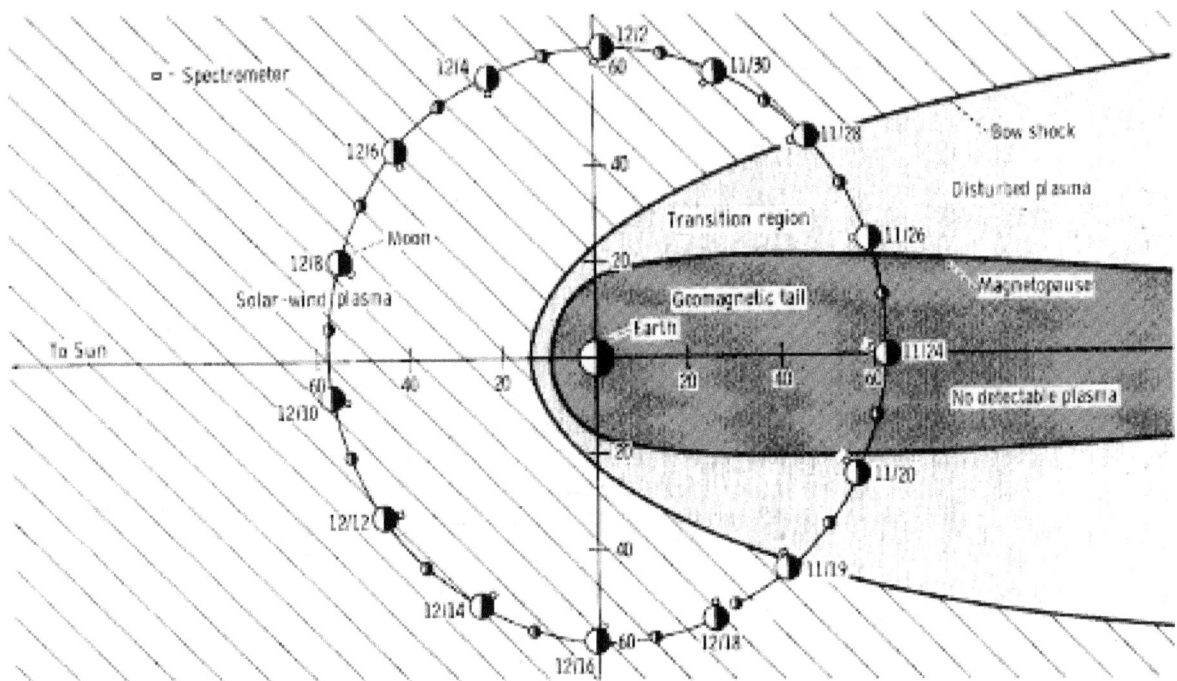

FIGURE 5-7. — Solar-wind plasma regions.

15:22:20 G.m.t. on December 3, 1969. For several hours preceding sunset, the density of the solar-wind plasma had appeared to decrease steadily, but this apparent behavior may be partially caused by reduced detector sensitivity at angles near the horizon. The plasma signal continued to decrease until between 19:00 and 20:00 G.m.t. on December 3, 1969, when the instrument threshold of sensitivity was reached. In the following 14 days of lunar darkness, no times of plasma detection have been observed in the limited data scanned.

The first photoelectrons of sunrise were detected at 11:38 G.m.t. on December 18, 1969, and within 20 min the first telemetry indication of warming of the sensor assembly was received. However, by sunrise, protons of the solar wind had been observed for nearly 12 hr. Several large fluctuations in proton flux occurred, varying from a high flux approximately 18 min before sunrise to no detectable plasma as late as 15 min after sunrise. These fluctuations may be caused by changes in the solar-wind flux or flow direction or by a lunar interaction effect. Comparison with interplanetary solar-wind data will be required to resolve this uncertainty. Data samples for December 22 to 24, 1969, also show no detectable plasma, probably because the instrument was once more in the magnetospheric tail of the Earth.

Typically, the preponderance of flux entered only one cup — the one most nearly facing the Sun. However, during the time when the solar-wind spectrometer was most sensitive to angular variations, from 12:00 G.m.t. on November 28, 1969, to 8:00 G.m.t. on November 29, 1969, there were indications that the direction of bulk velocity varied as much as $\pm 10°$ from the mean angle 2° to 4° east of the optical direction of the Sun. This is the only time span for which angular information is now available.

Summary and Results

This report is based upon examination of the printer output of the data for the first 35 days after deployment of the solar-wind spectrometer. These data are only a small percentage of the data that will ultimately be available for this period.

The solar plasma at the lunar surface is superficially indistinguishable from that at a distance from the Moon, both when the Moon is ahead of and when the Moon is behind the plasma bow shock of the Earth. No detectable plasma appears to exist in the magnetospheric tail of the Earth or in the shadow of the Moon.

Times of passage through the bow shock or through the magnetospheric-tail boundary, as indicated by the solar-wind spectrometer and by the lunar surface magnetometer, are in agreement when comparison of data has been possible. These times are given more accurately in the lunar surface magnetometer experiment section of this document.

Generally, observations have been in accordance with expectations, but the highly variable spectra observed on November 27, 1969, and at sunrise may prove to involve unexpected phenomena. Complete data and detailed comparison with other solar-wind measurements will be required before firm or quantitative conclusions can be drawn.

References

5-1. COLBURN, D. S.; CURRIE, R. G.; MIHALOV, J. D.; and SONETT, C. P.: Diamagnetic Solar-Wind Cavity Discovered Behind Moon. Science, vol. 158, no. 3804, Nov. 24, 1967, pp. 1040-1042.

5-2. SONETT, C. P.; COLBURN, D. S.; and CURRIE, R. G.: The Intrinsic Magnetic Field of the Moon. J. Geophys. Res., vol. 72, no. 21, Nov. 1, 1967, pp. 5503-5507.

5-3. NESS, N. F.; BEHANNON, K. W.; TAYLOR, H. E.; and WHANG, Y. C.: Pertubations of the Interplanetary Magnetic Field by the Lunar Wake. J. Geophys. Res., vol. 73, no. 11, June 1, 1968, pp. 3421-3440.

5-4. NESS, N. F.; BEHANNON, K. W.; SCEARCE, C. S.; and CANTARNO, S. C.: Early Results from the Magnetic Field Experiment on Lunar Explorer 35. J. Geophys. Res., vol. 72, no. 23, Dec. 1, 1967, pp. 5769-5778.

5-5. LYON, E. F.; BRIDGE, H. S.; and BINSACK, J. H.: Explorer 35 Plasma Measurements in the Vicinity of the Moon. J. Geophys. Res., vol. 72, no. 23, Dec. 1, 1967, pp. 6113-6117.

5-6. SISCOE, G. L.; LYON, E. F.; BINSACK, J. H.; and BRIDGE, H. S.: Experimental Evidence for a Detached Lunar Compression Wave. J. Geophys. Res., vol. 74, no. 1, Jan. 1, 1969, pp. 59-69.

5-7. BUEHLER, F.; EBERHARDT, P.; GEISS, J.; MEISTER, J.; and SIGNER, P.: Apollo 11 Solar Wind Composition Experiment: First results. Science, vol. 166, no. 3912, Dec. 19, 1969, pp. 1502-1503.

5-8. HUNDHAUSEN, A. J.: Direct Observations of Solar Wind Particles. Space Sci. Rev., vol. 8, no. 5/6, 1968, pp. 690-749.

ACKNOWLEDGMENTS

In the 51 months between the proposal of the solar-wind spectrometer for ALSEP and its successful deployment on the Moon, a rather large number of people contributed significantly to the design, development, and testing of the instrument. The authors regret that it is not practicable to acknowledge all of them. Among the engineers at the Jet Propulsion Laboratory whose deep and prolonged involvement in the project was clearly indispensable were David D. Norris, Gary L. Reisdorf, James W. Rotta, Jr., Gary J. Walker, and Robert H. White. Albert J. Fender, the quality assurance representative, should be added to this list.

The initial design and fabrication of the electronics package was done by Electro-Optical Systems, Inc. (EOS) of Pasadena, Calif. Among EOS employees who made major contributions, the authors particularly wish to express appreciation to Bill F. Lane, Thomas D. MacArthur, and Richard D. McKeehan. The instrument integration team at Bendix Aerospace Division, Ann Arbor, Mich., under Charles J. Weatherred was always responsive to suggestions, and the authors are especially indebted to Albert D. Robinson, who was the instrument engineer. Recognition is also due to the NASA Manned Spacecraft Center engineers, Richard A. Moke, Carl O. McClenny, Ansley B. Carraway, and William P. LeCroix, who had management cognizance of the instrument at various times.

This report presents the results of one phase of research carried out at the Jet Propulsion Laboratory, California Institute of Technology, under NASA Contract NAS 7-100, sponsored by the National Aeronautics and Space Administration.

6. Suprathermal Ion Detector Experiment (Lunar Ionosphere Detector)

J. W. Freeman, Jr.,[a][†] H. Balsiger,[a] and H. K. Hills[a]

The suprathermal ion detector experiment (SIDE), a part of the Apollo lunar surface experiments package (ALSEP), is designed to achieve the following experimental objectives:

(1) Provide information on the energy and mass spectra of the positive ions close to the lunar surface that result from solar-ultraviolet or solar-wind ionization of gases from any of the following sources: residual primordial atmosphere of heavy gases, sporadic outgassing such as volcanic activity, evaporation of solar-wind gases accreted on the lunar surface, and exhaust gases from the lunar module descent and ascent motors and the astronauts' portable life-support equipment

(2) Measure the flux and energy spectrum of positive ions in the Earth's magnetotail and magnetosheath during those periods when the Moon passes through the magnetic tail of the Earth

(3) Provide data on the plasma interaction between the solar wind and the Moon

(4) Determine a preliminary value for the electric potential of the lunar surface

The Instrument

The suprathermal ion detector experiment consists of two positive ion detectors. The first of these, the mass analyzer, is provided with a crossed electric- and magnetic-field (or Wien) velocity filter and a curved-plate electrostatic energy-per-unit-charge filter in tandem in the ion flightpath. The requirement that the detected ion must pass through both filters allows a determination of its mass per unit charge. The ion sensor itself is a channel electron multiplier operated as an ion counter that yields saturated pulses for each input ion. The second detector, the total ion detector, employs only a curved-plate electrostatic energy-per-unit-charge filter. Again, the ion sensor itself is a channel electron multiplier operated as an ion counter. Both channel electron multipliers are biased with their input ends at -3.5 kV, thereby providing a post-analysis acceleration to boost the positive ion energies in order to yield high detection efficiencies. Figure 6–1 illustrates the general detector concept, and figure 6–2 is a cutaway drawing of the suprathermal ion detector experiment which illustrates the location of the filter elements and the channel electron multipliers.

A primary objective of the experiment is to provide a measurement of the approximate mass-per-unit-charge spectrum of the positive ions near the lunar surface as a function of energy for ions from approximately 50 eV down to near-thermal energies. Therefore, the mass analyzer measures mass spectra at six energy levels: 48.6, 16.2, 5.4, 1.8, 0.6, and 0.2 eV. However, for the Apollo 12 instrument, dependable laboratory calibrations were achieved only at the two highest energy levels. The total ion detector measures the differential positive ion energy spectrum from 3500 eV down to 10 eV in 20 energy steps. For the Apollo 12 mass analyzer, the range of the mass spectrum covered is approximately 10 to 1000 atomic mass units (amu). Twenty mass channels span this range. The relative width for each mass channel $\Delta M/M$ is approximately 0.2 near the lower masses. In principle, the flux of ions with masses less than 10 amu per unit charge can be obtained by subtracting the integrated mass spectrum flux obtained with the mass analyzer from the total

[a] Rice University.
[†] Principal investigator.

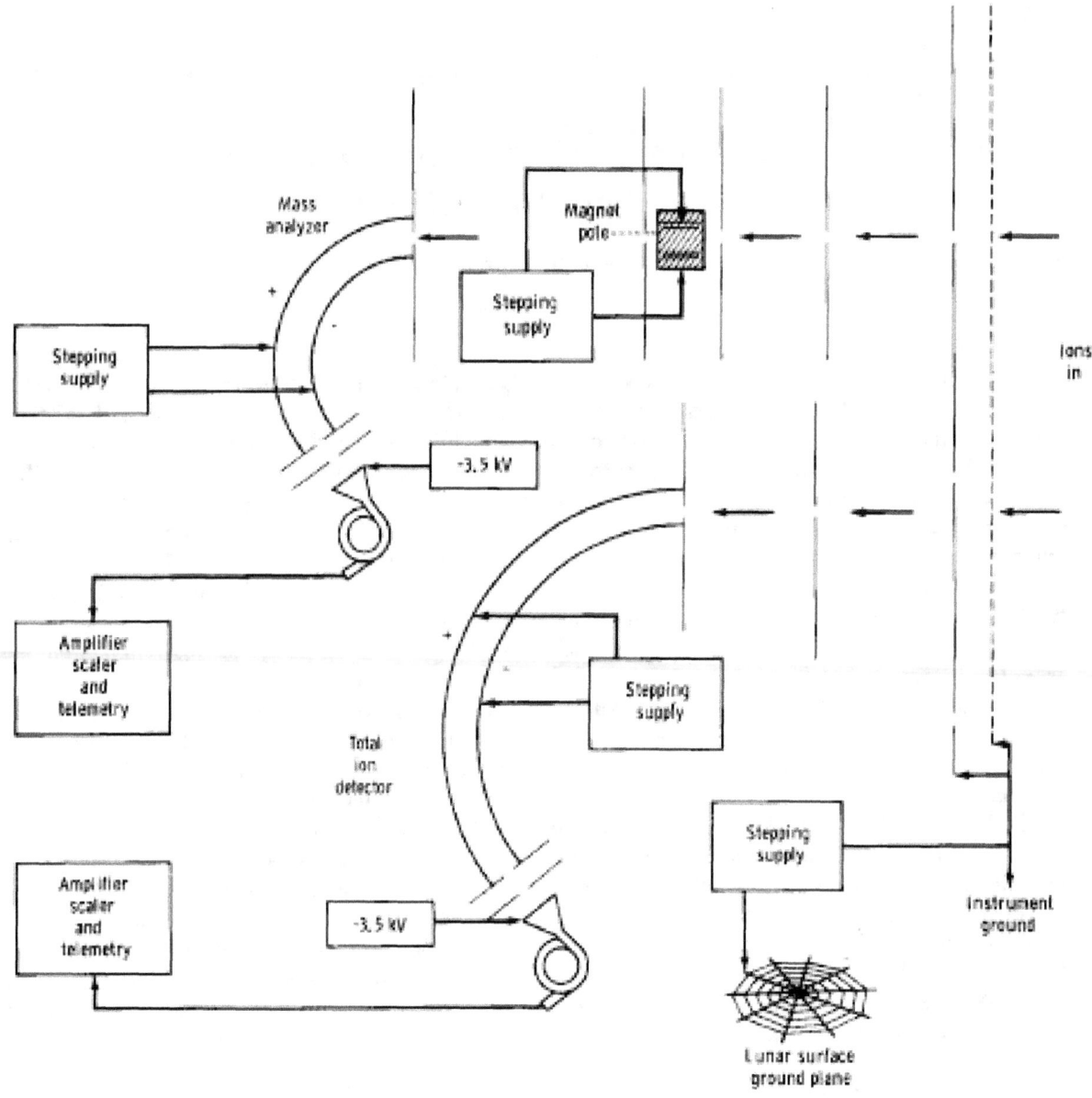

FIGURE 6-1. — A schematic diagram of the suprathermal ion detector experiment.

ion flux, at the same energy, obtained with the total ion detector.

To compensate for a possibly large (tens of volts) lunar surface electric potential, a wire screen is deployed on the lunar surface beneath the suprathermal ion detector. This screen is connected to one side of a stepped voltage supply, the other side of which is connected to the internal ground of the detector and to a grounded grid mounted immediately above the instrument and in front of the ion entrance apertures (fig. 6-1). The stepped voltage is advanced only after a complete energy and mass scan of the mass analyzer (i.e., every 2.58 min). The voltage supply is programed to step through the following voltages: 0, 0.6, 1.2, 1.8, 2.4, 3.6, 5.4, 7.8, 10.2, 16.2, 19.8, 27.6, 0, −0.6, −1.2, −1.8, −2.4, −3.6, −5.4, −7.8, −10.2, −16.2, −19.8, and −27.6. This stepped supply and its ground screen may function in either of two

SUPRATHERMAL ION DETECTOR EXPERIMENT

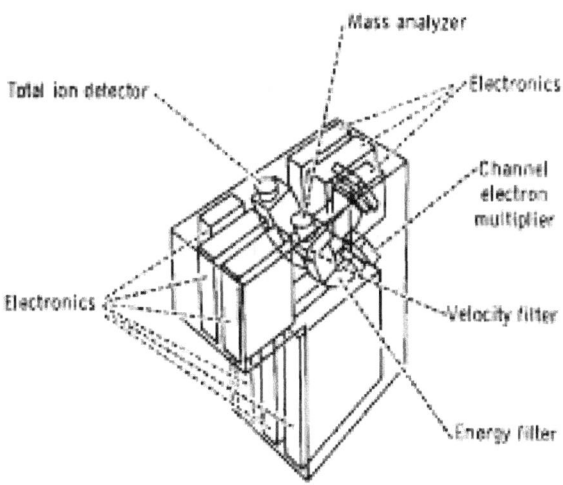

FIGURE 6-2. — A cutaway drawing showing the interior of the suprathermal ion detector experiment.

ways. If the lunar surface potential is large and positive, the stepped supply, when on the appropriate step, may counteract the effect of the lunar surface potential and, thereby, allow the low-energy ions to reach the instrument with their intrinsic energies. However, if the lunar surface potential is near zero, then on those voltage steps that match or nearly match the energy levels of the mass analyzer or the total ion detector (1.2, 5.4, etc.), thermal ions may be accelerated into the suprathermal ion detector at energies optimum for detection. The success of this method depends on the Debye length and on the extent to which the ground-screen potential approximates that of the lunar surface. It is not yet possible to assess either of these two factors.

Figure 6-3 shows the SIDE deployed on the lunar surface. The experiment is deployed approximately 50 ft from the ALSEP central station in a southwesterly direction. The top surface stands 20 in. above the lunar surface. The sensor look directions include the ecliptic plane, and the look axes are canted 15° from the local vertical and to the west. Figure 6-4 shows the look directions in an Earth-Sun coordinate system at various points along the lunar orbit. The field of view of each sensor is roughly a square solid angle, 6° on a side. The sensitivities of the total ion detector and mass analyzer are approximately 5×10^{17} and 10^{17} counts/sec/A of entering ion flux, respectively.

In addition to detecting ions directly, the suprathermal ion detector is also sensitive to the ambient neutral gas pressure through the back-

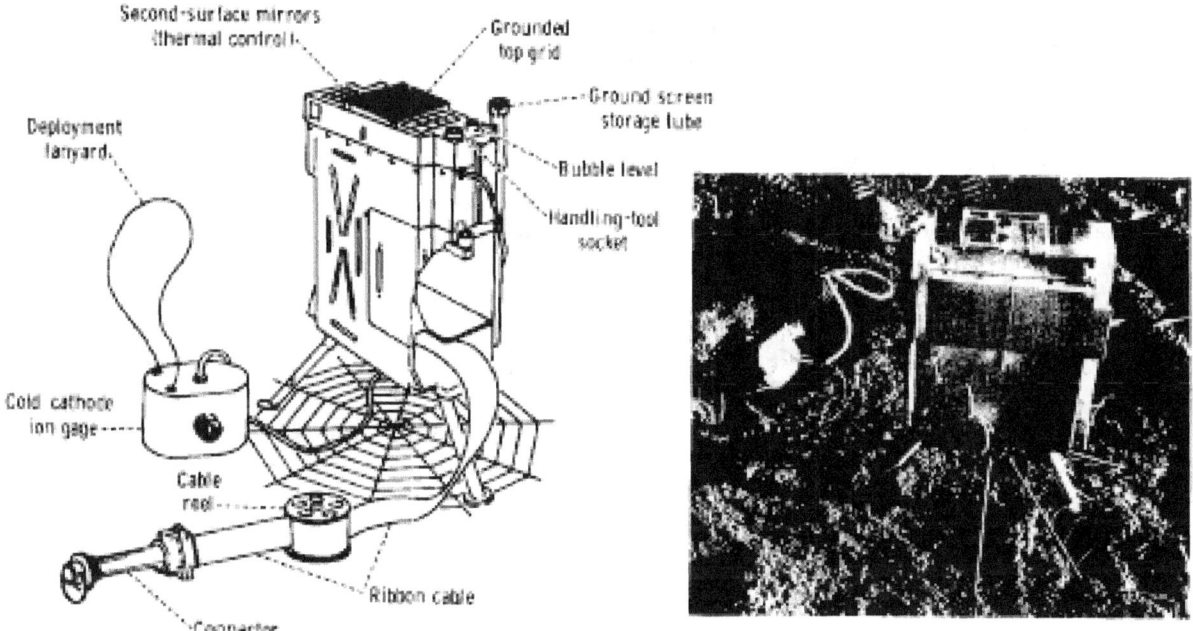

FIGURE 6-3. — The suprathermal ion detector experiment as deployed on the Moon.

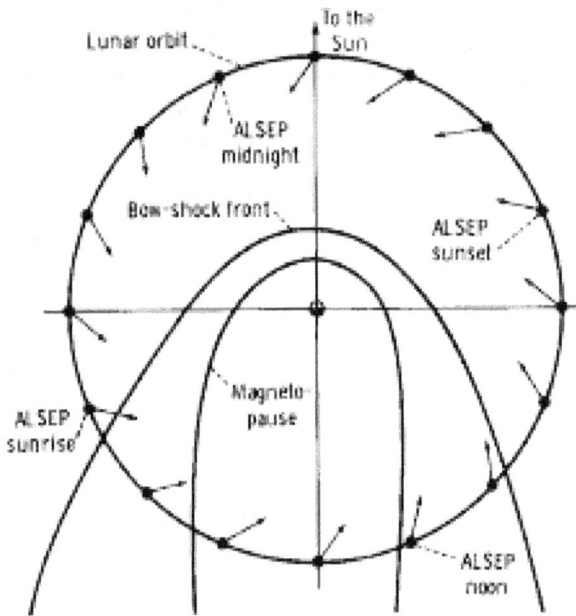

FIGURE 6-4. — The look directions of the suprathermal ion detector at various points along the lunar orbit. The diameter of the Earth is not drawn to scale.

ground counting rate of the channel electron multipliers. The background counting rate is defined as those counts present when the energy and velocity filter voltages are turned off, direct entrance of charged particles being, therefore, impossible. This property can provide a rough measure of the instrument outgassing.

Most of the data used for the preliminary analysis presented in this report consist of quick-look analog strip-chart recordings for which only counts of more than four per frame are detectable. This limitation prohibits the observation of low-intensity events in the quick-look data. Furthermore, no long-time averaging has yet been attempted. For these reasons, only the readily discernible features of the data will be discussed.

Performance of the Suprathermal Ion Detector

At the time of preparation of this report, 45 days after deployment, the operation of the suprathermal ion detector continues to be excellent. All temperatures and voltages have been nominal. Only two operational anomalies have been noted. First, outgassing associated with the high temperatures at lunar noon has caused the channel-electron-multiplier high voltage to be commanded off for several days on either side of lunar noon. The situation at the second lunar noon showed an improved situation over that experienced on the first lunar noon, and it is expected that 100-percent duty-cycle operation will be possible within the next 3 lunar days. Second, on a few occasions, a calibration signal for the total ion detector has been intermittent for a short time. The calibration signal is used as a diagnostic check on the digital logic and is not essential to the operation of the instrument as long as the digital logic functions properly.

General Results

Low-Energy Events

Shortly after turn-on of the SIDE, several low-energy events were detected in the total ion detector. These ions appear to come in clouds that remain for approximately 10 min. Several clouds were accompanied at the outset by higher energy ions (500 to 750 eV). One arrival of such a cloud was coincident with a magnetometer variation that indicated the passage of a current sheet nearby.

At the same time these low-energy events were seen by the total ion detector, the mass analyzer detected ions in the 48.6-eV range (frames 0 to 19). Figure 6-5 shows the total ion detector counts and the mass analyzer counts of a typical event. In figure 6-6, an average for five mass spectra is shown. The only mass spectra identified in the preliminary analysis originate in the period shortly after turn-on, and all these spectra show a shape similar to the one presented in figure 6-6, that is, with a peak between frames 2 and 6. The number of spectra detected to date is rather low because within a few hours after turn-on the background rate had increased with increasing temperature so that additional similar events may be hidden. Such low-magnitude events can be evaluated only after a detailed background study. Furthermore, the investigation was restricted to events that are simultaneously apparent in both detectors (the total ion detector and the mass analyzer) at about the same energy step. When this criterion was used,

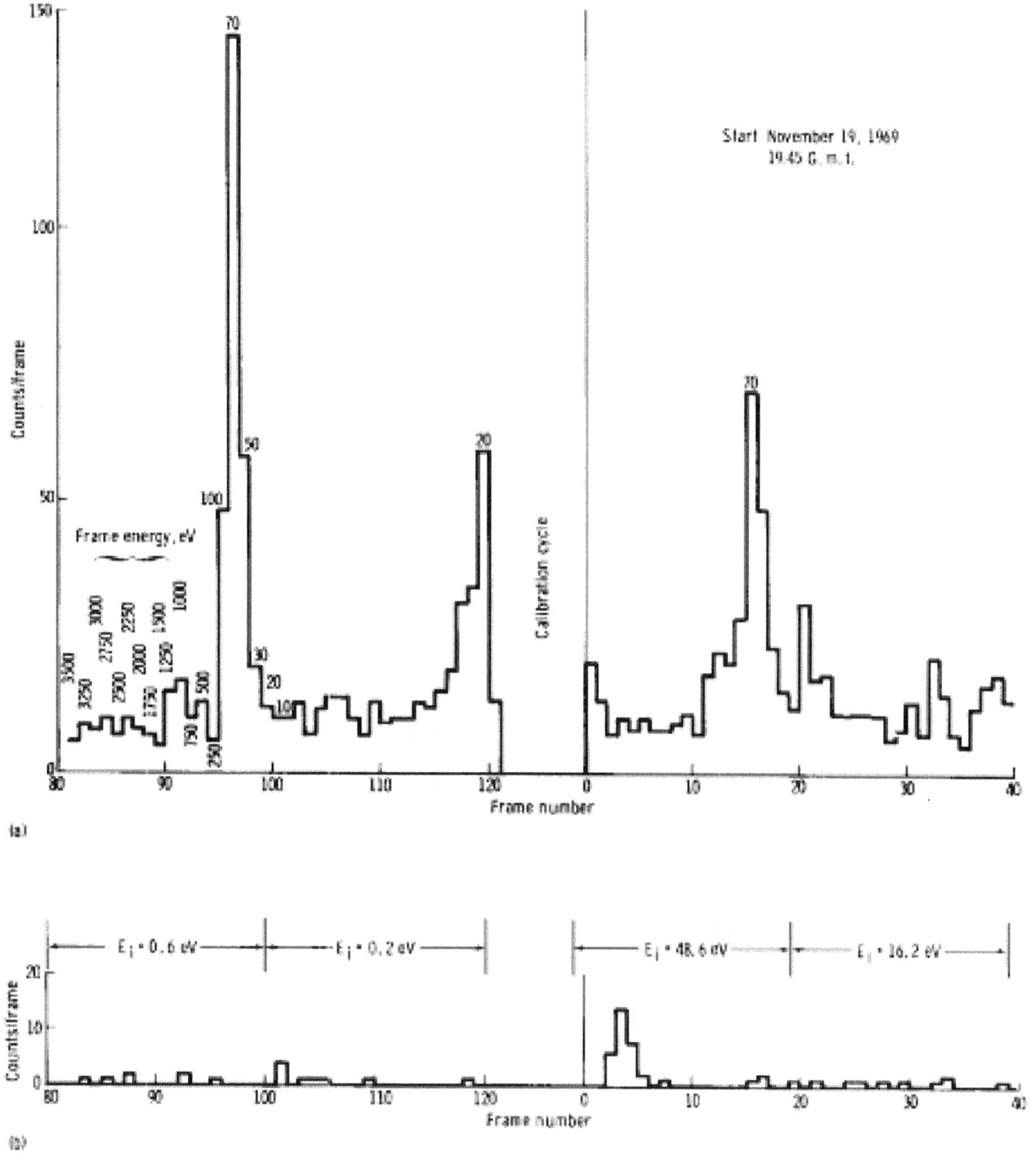

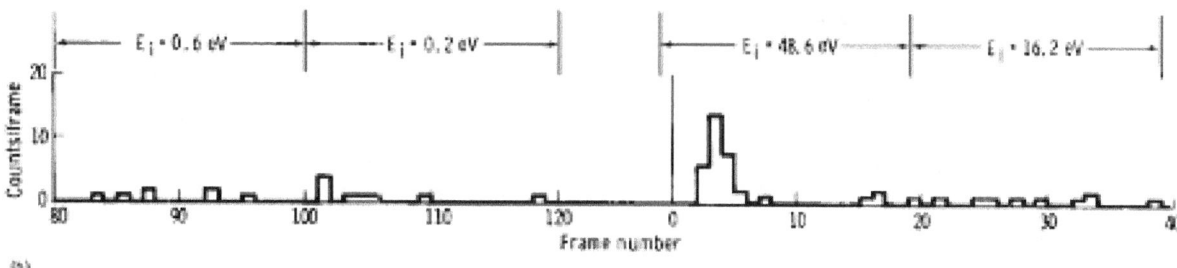

FIGURE 6-5.—Samples of the simultaneous total ion detector and mass analyzer data from November 19, 1969. Each frame is 1.2 sec long, and the counts are accumulated in that time interval. The total ion detector energy spectrum is repeated every 20 frames, except for a calibration cycle from frames 121 to 0. The mass analyzer sweeps through a mass spectrum at each energy in 20 frames; therefore, a complete spectrum at all six energies is obtained every 2.58 min (including the calibration cycle). Note the repeated peaks in the total ion detector data in the 20- to 100-eV energy range. Note also the peak in frames 2 to 6 in the mass analyzer data. (a) Total ion detector data. (b) Mass analyzer data.

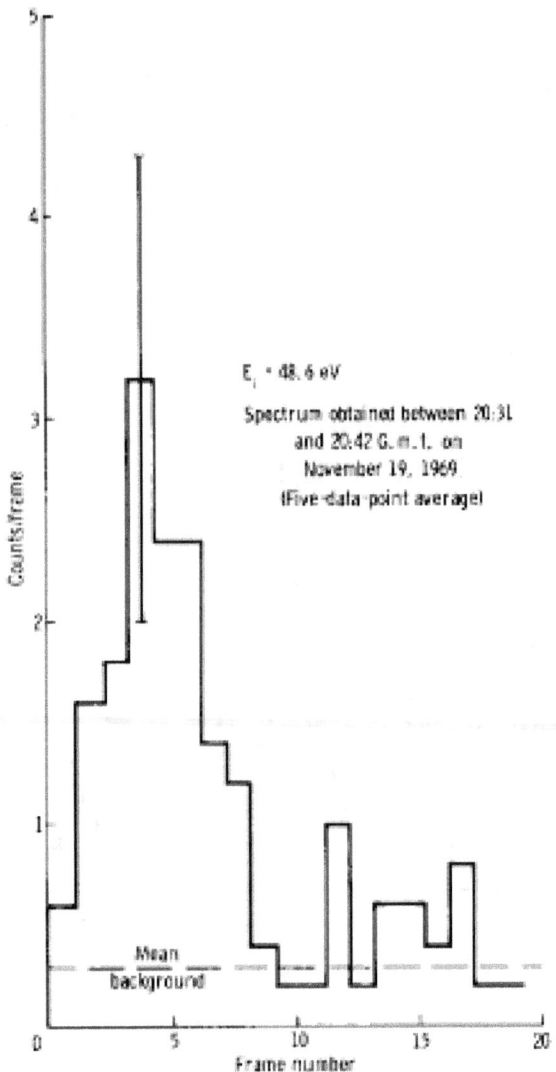

FIGURE 6-6. — A five-data-point average of the mass analyzer 48.6-eV range on November 19, 1969.

no events were found in the lower energy steps of the mass analyzer.

As mentioned previously, the only mass spectra found to date peak in frames 2 to 6, corresponding to a mass-per-unit-charge range of 18 to 50 amu/q. Evaluations of these events are not complete enough at this time to discuss what gases these events represent. However, it seems reasonable that ions would be found in this medium mass range since possible sources such as lunar module rocket exhaust products and thermalized or sputtered solar-wind ions may exist. It should be kept in mind that the light gases H, H_2, and He cannot be detected by the mass analyzer. They can be detected only by the total ion detector.

In addition to the previously described events that were detected in both the total ion detector and the mass analyzer, low-energy (10 to 250 eV) ions have often been seen in the total ion detector only. The total ion detector yields six complete energy spectra in the time required for a complete mass-energy spectrum from the mass analyzer. The probability is high, therefore, that the mass analyzer will not be at the appropriate energy level to simultaneously detect the ions seen by the total ion detector. These ions are often quite monoenergetic. For example, several consecutive spectra may occur with high counts seen in only one energy channel. Figure 6-7 is an example of such a spectrum. In other events, the peaks are wider, covering two to three energy channels, or monoenergetic peaks of different energies are mixed in several consecutive spectra.

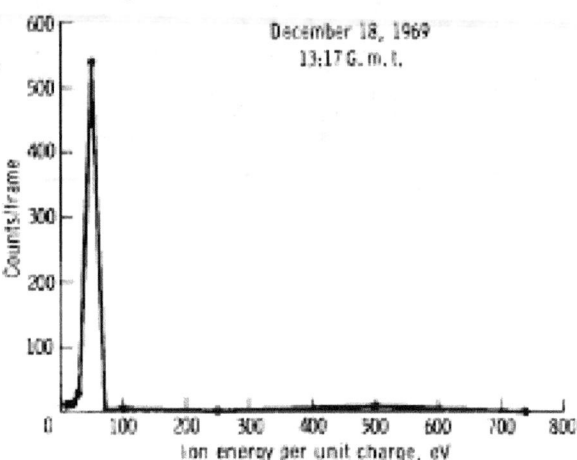

FIGURE 6-7. — The energy spectrum for a typical 50-eV monoenergetic event.

It is interesting to speculate on the possibility that the energization process for these ions is the $E \times B$ drift acceleration by the solar wind. The frequent appearance of such suprathermal ions suggests a general acceleration mechanism.

Higher Energy Phenomena

A variety of interesting higher energy spectra have been observed with the total ion detector. The majority of these have not yet been examined in detail; however, two categories of spectra have been singled out for illustration.

The first category is characterized by large counts near the upper end of the energy range of the total ion detector, that is, between 1 and 3 keV. Figure 6-8 is an example of several spectra of this category. Data of this type come predominantly from that portion of the orbit near, and up to 4 days following, sunset at the ALSEP site. These ions appear sporadically, often lasting for tens of minutes and then disappearing for hours. These ions have been tentatively identified as protons escaping from the bow shock front of the Earth and moving generally along the interplanetary magnetic field lines at the "garden hose" angle. Such effects have been reported at the much closer Earth orbit of the Vela satellites (ref. 6-1).

The second category of spectra is characterized by an energy peak near or slightly below solar-wind energies. The interesting feature of this type of spectra is that it has also been seen during the lunar night. Figure 6-9 shows examples of spectra taken approximately 3 days before the sunrise terminator crossing. At this time, the detector look direction is only 25° from the antisolar direction, and the ALSEP is nearly 40° from the sunrise terminator. The maximum flux found to date for these ions is approximately 10^6 ions/cm^2-sec-sr. The solar wind appears to be deflected in some way around the limb of the Moon for a substantial distance.

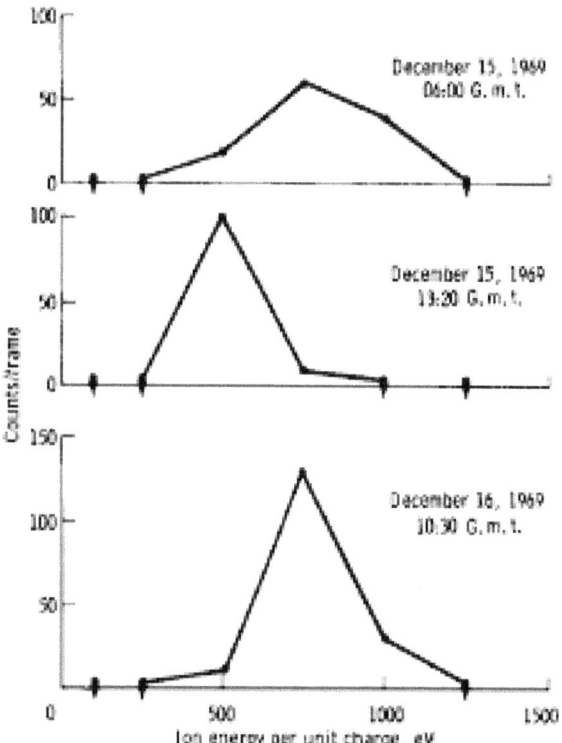

FIGURE 6-9.—Typical spectra showing the solar-wind energy ions seen before sunrise.

Special Events

Lunar Module Liftoff

At the time of liftoff of the lunar module ascent stage, there was a slight increase in the counting rates of the total ion detector in lower energy channels. No significant change was seen in the mass analyzer data, but the mass analyzer

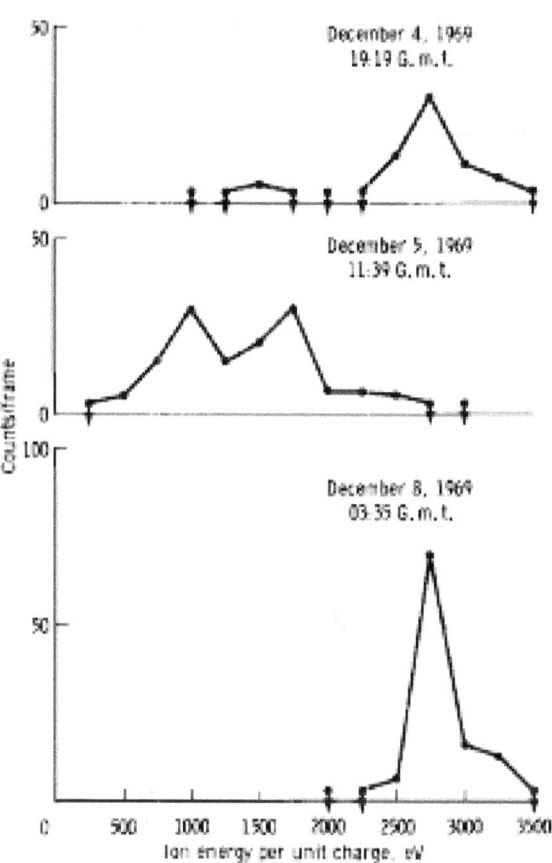

FIGURE 6-8.—Typical spectra showing the 1- to 3-keV ion seen after sunset.

was looking at 5.4-eV and lower energy ions at the time. Four consecutive spectra taken during this event are shown in figure 6-10. The first spectrum is typical of several consecutive spectra prior to it. The next two spectra show the increased counting rates in the channels covering the broad energy range of 10 to 500 eV. By the time of the fourth spectrum, the counting rates had returned to background rate. The instrument was heating up during this period of time, and the consequent outgassing caused the fairly high background counting rates observed.

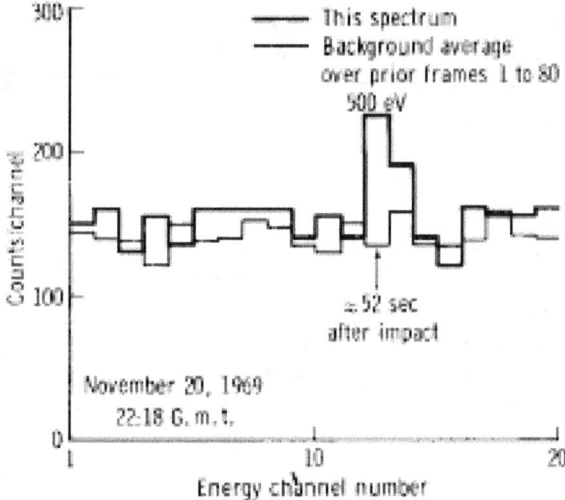

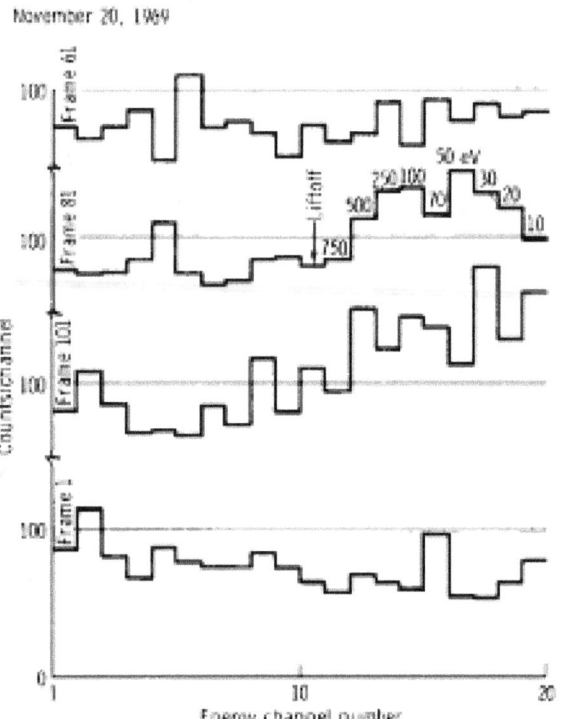

FIGURE 6-10. — Four successive total ion detector spectra before, during, and after lunar module ascent. An eight-frame calibration cycle occurs between the spectrum that starts with frame 101 and the spectrum that starts with frame 1.

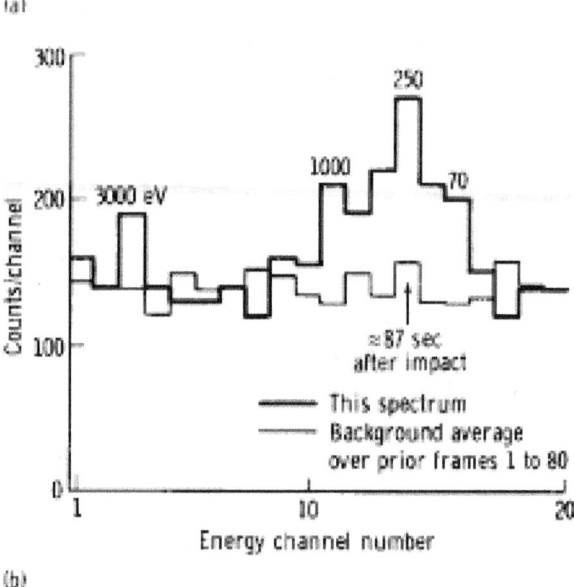

FIGURE 6-11. — Total ion detector spectra following lunar module impact. (a) First spectrum after ascent stage impact in which increased counting rate was observed. (b) Spectrum following instrument calibration frames.

Lunar Module Impact

At 22:17:17 G.m.t. on November 20, 1969, the lunar module ascent stage impacted the Moon 74 km east-southeast of the ALSEP. At this time, the total ion detector had been counting at an average rate of 142 counts per frame with no significant variations in any frame. Figure 6-11(a) is the first spectrum after the impact in which an increased counting rate was observed. A significant ion flux was detected in the 500- and 250-eV channels. These channels were sampled 52 sec after impact. After this spectrum, there were eight frames of instrument calibration followed by the spectrum shown in figure 6-11(b). The peak at 3000 eV is approximately 4σ above the background rate and is, therefore, certainly real, but the feature of main interest is the high flux of ions in the 70- to 1000-eV channels. However, these high fluxes vanish within the next 12 sec. The following spectra show no counting rates that are significantly greater than the background rate.

On the basis of the foregoing, the inference exists that the gases liberated on lunar module impact with the lunar surface may have either triggered a temporary perturbation to the solar wind that rendered the solar wind partially detectable to the suprathermal ion detector or, more probably, these gases themselves may have been ionized as the spherical gas shell moved outward from the impact point. The ionized impact-generated gases and lunar module consumable gases may have then executed $E \times B$ drift motion brought on by the solar wind and, hence, gained access to the suprathermal ion detector along the general direction of solar-wind flow. In this connection, it is noted that the ALSEP was 68 km down solar wind of the impact point and that the time of arrival of the first gas burst at the suprathermal ion detector would correspond to a velocity of approximately 1 km/sec; for thermal expansion of the cloud, the requisite temperature is several thousand degrees Kelvin. The efficiency for such a mechanism is difficult to estimate because of the tortuous paths followed by the ions after pickup by the solar wind and because of the large gyroradii of the ions. However, a conservative order-of-magnitude estimate yields a flux consistent with the observed flux of approximately 10^7 ions/cm^2-sec-sr-keV.

High Background Rates Observed During the Second Lunar Day

Three hours following the first sunrise, the mass-analyzer background rate became very high and erratic. These high counts came in bursts, slowly at first but with gradually increasing frequency. They continued throughout the lunar day with slowly decreasing intensity and disappeared abruptly within 12 hr after sunset. Figure 6-12 shows the general intervals during which the counts were seen.

These high counts are clearly not the result of the low-energy positive ions the instrument was designed to detect since they are seen during background measurement frames and since neither the velocity-filter nor the energy-filter

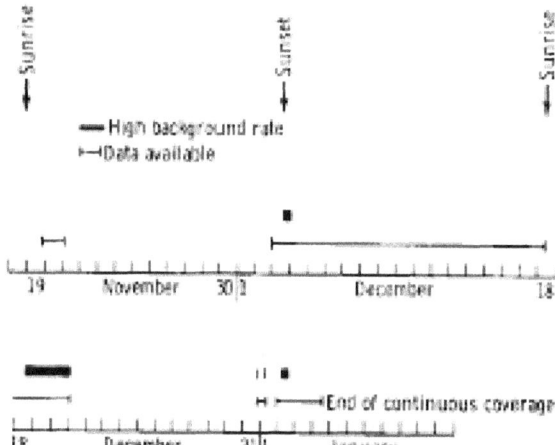

FIGURE 6-12. — A chart showing the time intervals during which the high, erratic background rates were observed in the mass analyzer data.

voltages appear to have an effect on the counting rates. Possible sources for the phenomena include a high ambient pressure of neutral gas or a temperature-sensitive malfunction of the channel electron multiplier. Channel electron multipliers are known to often exhibit high background rates in the event of cracks in the glass walls. Generally, however, such rates are relatively steady rather than highly erratic, as is the case observed for the Apollo 12 detector. Also, such rates have not been known to disappear once a crack or break has developed, whereas the background rate observed during the second lunar night was very low and perfectly normal.

Concerning the likelihood that these counts represent an enhanced lunar surface pressure, it is noted that approximately 1200 lb of propellant remained in the lunar module descent-stage tanks. These tanks are opened by the astronauts prior to departure. It is suspected, however, that the vents for the tanks may freeze and periodically inhibit the escape of the sublimating fuel (Aerozine 50) and oxidizer (N_2O_4). On the Apollo 11 mission, the passive seismic experiment saw seismic activity attributed to "venting or circulating gases or liquids or both" (ref. 6-2). This activity occurred for approximately 192 hr on the first lunar day after deployment. It seems possible that the high and erratic background rates of the mass analyzer indicate the periodic and impulsive escape of these gases. It is believed that a detailed comparison of the data with those from the Apollo 12 passive seismic experiment and a study of some apparent periodicities and decay rates of the bursts may resolve this question.

Summary

The performance of the suprathermal ion detector has been good. The preliminary data analysis yields the following features:

(1) Mass spectra of 50-eV ions are available from early in the experiment life. These spectra show a concentration of ions in the 18- to 50-amu/q mass-per-unit-charge range.

(2) Ions appear frequently in the ten- to several-hundred electron-volt range. This is highly suggestive of solar-wind acceleration of ambient ions of either natural or lunar-module-associated origin.

(3) One- to 3-keV ions are present sporadically early in the lunar night. These ions are thought to be energetic protons escaping upstream from the bow shock front of the Earth.

(4) Solar-wind energy ions are present on the nightside of the Moon approximately 4 days before lunar sunrise at the ALSEP site.

(5) Energetic ion fluxes are seen in good time correlation with the impact of the lunar module ascent stage onto the lunar surface. Again, there is a strong suggestion that the impact-released gases have been ionized and accelerated by the solar wind.

(6) High background count rates seen during the second lunar day may be indicative of large quantities of gas escaping impulsively from the lunar module descent-stage tanks.

Numerous other phenomena are apparent in the data. These have not been examined in sufficient detail to allow categorization or discussion at this time. Furthermore, no effort has been made to search for low-level phenomena that require averaging of the data.

References

6-1. ASBRIDGE, J. R.; BAME, S. J.; and STRONG, I. B.: Outward Flow of Protons From the Earth's Bow Shock. J. Geophys. Res., vol. 73, no. 17, Sept. 1, 1968, pp. 5777-5782.

6-2. LATHAM, G. V.; et al.: Passive Seismic Experiment. Sec. 6 of Apollo 11 Preliminary Science Report, NASA SP-214, 1969.

ACKNOWLEDGMENTS

The authors gratefully acknowledge the dedicated support of many who have contributed to the success of the suprathermal ion detector experiment. In particular, thanks are extended to Wayne Andrew Smith, Rice University Program Manager; Paul Bailey, Rice University Assistant Program Manager; James Ballentyne, Rice University Reliability and Quality Assurance Engineer; Robert Shane and David Young, space science graduate students who participated in the design and calibration of the instrument; Alex Frosch, computer programer; and the staff of the Rice University Space Science Facilities. Thanks are also due to William Sandstrom, Dean Aalami, and the staff at Time Zero Corporation (formerly Marshall Laboratories), the subcontractor for design and fabrication of the suprathermal ion detector, and to James Boswell, Defense Contract Audit Service Representative. Derek Perkins of the Bendix Aerospace Systems Division and Jim Sanders and Ernie Weeks of the NASA Manned Spacecraft Center provided invaluable support to the project. Martha Fenner, space science graduate student, assisted with the data analysis.

This research has been supported by NASA contract NAS9-5911. H. Balsiger is a European Space Research Organization/National Aeronautics and Space Administration International Postdoctoral Fellow.

7. Cold Cathode Gage (Lunar Atmosphere Detector)

F. S. Johnson,[a][1] D. E. Evans,[b] and J. M. Carroll[a]

Purpose of the Experiment

Although the lunar atmosphere is known to be extremely tenuous, its existence cannot be doubted. At the very least, the solar wind striking the lunar surface constitutes a source mechanism. The expected atmospheric concentration depends upon the equilibrium between source and loss mechanisms. The observations of the lunar atmosphere will be of greatest significance if the dominant source mechanism for the atmosphere is internal (i.e., geochemical) rather than external (i.e., the solar wind).

The dominant loss mechanisms for lunar gases are expected to be thermal escape, for particles lighter than neon, and removal through ionization, for particles heavier than neon. At the temperatures encountered on the lunar surface, thermal velocities for the lighter gas particles are such that a significant fraction of the particles has greater than escape velocity. The average lifetime before escape for particles on the warmest portion of the Moon is approximately 10^4 sec for helium and 10^7 sec for neon. Heavier particles require much longer to escape by thermal motion. However, all particles exposed to solar ultraviolet radiation become ionized in approximately 10^7 sec. Once the particles are ionized, they are accelerated by the electric field associated with the motion of the solar wind. The initial acceleration is at right angles to the direction of both the solar wind and the imbedded magnetic field. The direction of motion is then deviated by the magnetic field such that the particle acquires a velocity equal to the component of the solar-wind velocity that is perpendicular to the imbedded magnetic field. The time required for this second acceleration is approximately equal to the ion gyro period in the imbedded magnetic field, and the radii of gyration for most ions are comparable to the lunar radius or greater. Consequently, most particles in the lunar atmosphere are swept away into space within a few hundred seconds (the ion gyro period) after becoming ionized. Thus, the time required for ionization regulates the loss process and results in lifetimes of the order of 10^7 sec.

The cold cathode gage gives indications of the amount of gas present but not of the composition of the gas. The measured amount of gas can be compared with the amount expected from the solar-wind source to indicate whether or not other sources are present. Contamination from the vehicle system, of course, constitutes an additional source mechanism, but such a source should decrease with time in an identifiable way. Eventually, however, measurements of actual composition should be made with a mass spectrometer to examine constituents of particularly great interest and to discriminate against known contaminants from the vehicle system.

The Instrument

The basic sensing element of the cold cathode gage consists of a coaxial electrode arrangement as depicted in figure 7-1. The cathode consists of a spool surrounded by a cylindrical anode. A magnetic field of approximately 900 G is applied in the direction of the axis, and a voltage of +4500 V is applied to the anode. A self-sustained Townsend discharge develops in the gage. In this discharge, trapped electrons in the magnetic field have enough energy to ionize any gas particles they strike. The current of ions

---

[a] University of Texas at Dallas.
[b] NASA Manned Spacecraft Center.
[1] Principal investigator.

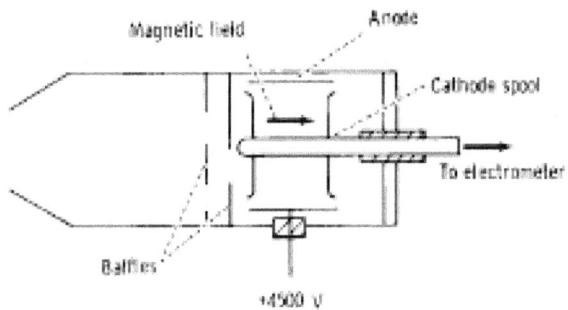

FIGURE 7-1. — Diagram of cold cathode gage sensor.

collected at the cathode is a measure of the gas density in the gage.

Figure 7-2 shows the response of the cold cathode gage in terms of cathode current as a function of pressure. It is usual to express the response in terms of pressure, although the gage is actually sensitive to gas density. If the gage temperature differs significantly from approximately 21.1° C, the reading should be corrected for the temperature difference. The gage response is also slightly dependent on gas composition. As long as the gas composition remains unknown, a fundamental uncertainty remains in the interpretation of the data. It is usual to express the results in terms of the nitrogen pressure that would produce the observed response. The true pressure will vary from this nitrogen pressure by a factor usually smaller than 2.

A temperature sensor was mounted on the gage to determine the gage temperature. There was no temperature control; therefore, the expected range was approximately 100° to 400° K.

The gage was closed (sealed) with a dust cover. This cover did not provide a vacuum seal. The cover was removed on command by utilizing a squib motor to release the cover, which was then pulled aside by a spring. Because the gage was not evacuated and sealed, adsorbed gases could produce an elevated level of response when the gage was first turned on. The baking of the gage on the lunar surface at 400° K for more than a week during the lunar day was expected to drive the adsorbed gases out of the gage.

The Electrometer

An auto-ranging, auto-zero electrometer monitors current outputs from the sensor or from the calibration current generators in the 10^{-13} to 10^{-6} A range. The output of the electrometer ranges from -15 mV to -15 V. The output of the electrometer is fed to the analog-to-digital (A/D) converters. The electrometer consists of a high-gain, low-leakage differential amplifier with switched high-impedance feedback resistors and an auto-zero network.

The electrometer operates in three automatically selected, overlapping ranges: range number 1, most sensitive; range number 2, midrange; and range number 3, least sensitive. Range number 1 senses current from approximately 10^{-13} to 9.3×10^{-11} A. Range number 2 senses currents from approximately 3.3×10^{-12} to 3.2×10^{-9} A. Range number 3 senses currents from approximately 10^{-9} to 9.3×10^{-7} A. The electrometer transfer function is shown in figure 7-3 in A/D readout counts as a function of input current (theoretical and measured curves) for a typical system.

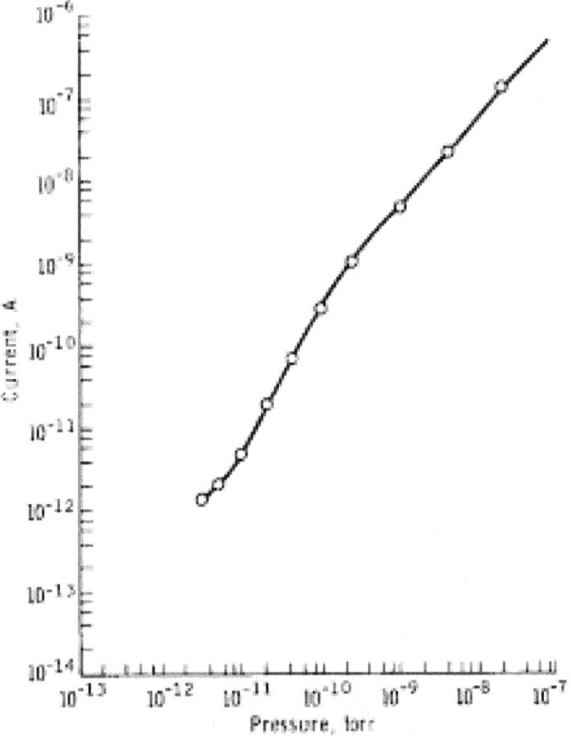

FIGURE 7-2. — Cold cathode gage response.

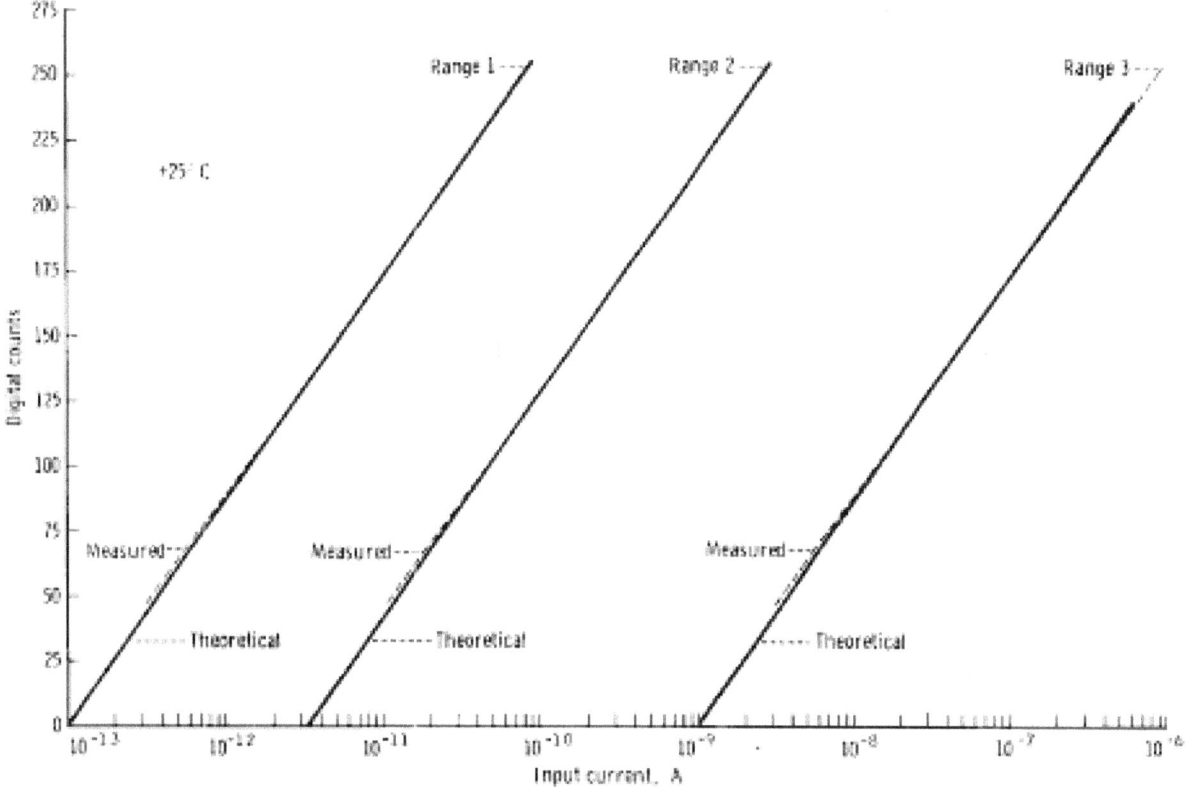

FIGURE 7-3. — Electrometer transfer function.

The 4500-V Power Supply

Basically, the power supply consists of a regulator, a converter, a voltage-multiplier network, and the associated feedback network of the low-voltage power supply. The regulator furnishes approximately 24 V for conversion to a 5-kHz square wave to be applied to the converter transformer. The output of the converter transformer is applied to a voltage-multiplier network (stacked standard doublers) and then is filtered and applied to the gage anode.

Deployment

The electronics for the cold cathode gage were contained in the suprathermal ion detector experiment (SIDE). The command and data-handling systems of the SIDE also served the cold cathode gage. The gage was physically separable from the SIDE package and was connected to it by a cable approximately 1 m long.

When the SIDE was deployed, the cold cathode gage was removed from its storage position in the SIDE. It was intended that the gage opening would be oriented horizontally and would face the pole, generally away from the descent stage of the lunar module (LM). The cable proved to be cold and stiff, and in the lunar gravity, even the relatively heavy gage was not adequate to hold the extended cable straight. Consequently, the gage tipped to face in a generally upward direction.

Results

The cold cathode gage was turned on at approximately 19:18 G.m.t. on November 19, 1969. At first, a full-scale response was obtained because of gases trapped within the gage. The time history is shown in figure 7-4. After approximately 1 hr, the response changed perceptibly from the full-scale reading. After 7 hr, the indication was approximately 1.2×10^{-8} torr. When the LM was depressurized, prior to the second extravehicular activity (EVA) period, the re-

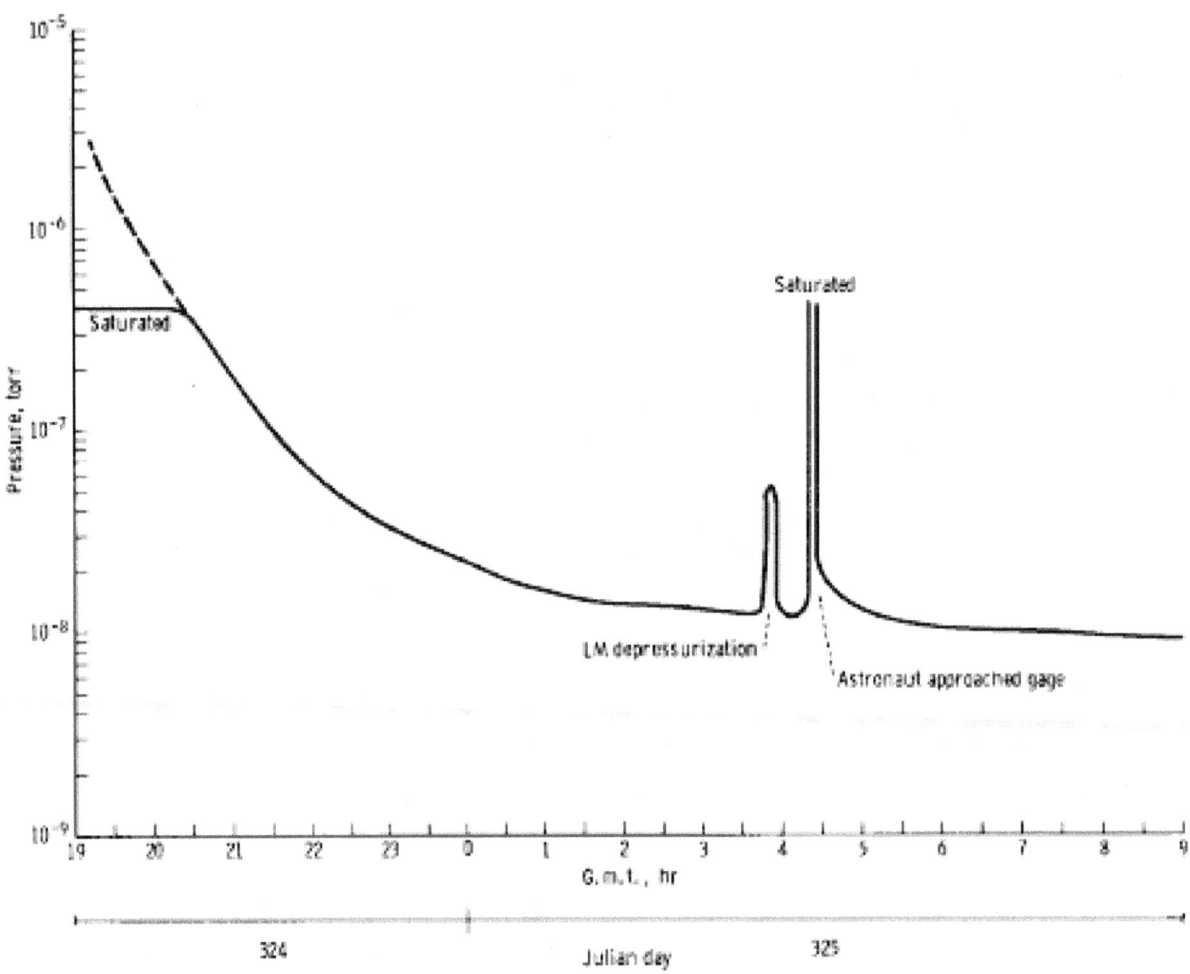

FIGURE 7-4. — Cold cathode gage response history.

sponse rose to at least 5×10^{-8} torr. The exact value is in doubt, because a calibration cycle obscured the data at the time of maximum pressure. The increase in pressure at the gage as a result of the release of gas from the LM is in reasonable agreement with expectations, based upon the amount of gas released. The release of 3 kg of gas in 10^2 sec should have produced a directed or dynamic pressure of 2×10^{-7} torr at a distance of 200 m. However, the obscuration of data near the peak of the pressure pulse probably eliminates any prospect of making meaningful diffusion studies based on the data.

During the second EVA period the gage response went off scale when an astronaut approached because of gases released from his portable life-support system. This response is also in agreement with expectations, but no close comparison with the predicted response can be made because of the lack of quantitative information on the separation between the astronaut and the gage.

An apparent catastrophic failure occurred after approximately 14 hr of operation when the 4500-V power supply shut off. Two possibilities exist. Either there was a failure, such as a short circuit, in the high-voltage supply, or the toggle command failed. (Its failure mode was such as to turn off the high voltage.) There appears to be no way to distinguish between these two possibilities, but the latter appears to be the more likely. In test and development, no failures were encountered with the high-voltage power supply. However, logic failures did occur that were brought about as a result of arcing when the package was tested under inadequate vacuum. The failure may have been brought about by arcing, which was associated with gassing in the electronics package as it heated up on the lunar surface.

In summary, the results show that the ambient lunar atmospheric pressure is less than 8×10^{-9} torr. Contamination of the experiment site by the landing operations does not produce a local atmosphere in excess of 8×10^{-9} torr after approximately 20 hr. The gas cloud around an astronaut on the lunar surface exceeds the upper range of the gage (approximately 10^{-6} torr) for a distance of several meters from the astronaut; however, no perceptible residual contamination at the 10^{-8} torr level remains around the gage for longer than a few minutes after his departure.

8. The Solar-Wind Composition Experiment

J. Geiss,[a][1] P. Eberhardt,[a] P. Signer,[b] F. Buehler,[a] and J. Meister[a]

The abundances of H^+ and He^{2+} ions in the solar wind have been monitored for several years (refs. 8-1 to 8-3). The He/H ratio is highly variable with time and ranges from less than 0.01 to 0.25 (ref. 8-1) with an average of about 0.04 (refs. 8-1 and 8-4 to 8-6). Oxygen ions have been measured during periods of low ion temperature by means of unmanned satellites and space probes; under very favorable conditions, even ^3He has been detected (ref. 8-7). The Apollo program has made it possible to introduce a new approach to the measurement of solar-wind ion composition. Targets can be exposed to the solar wind outside the magnetosphere of the Earth. In this way, solar-wind ions can be collected for laboratory analysis after return of the target. An experiment of this kind, the solar-wind composition experiment (refs. 8-8 and 8-9), was flown for the first time on the Apollo 11 mission. The experiment has yielded absolute solar-wind fluxes of ^4He, ^3He, ^{20}Ne, and ^{22}Ne averaged over a 77-min exposure period.

Prior to the first manned landing on the Moon, Explorer 35 plasma and magnetic field measurements had established that, to a good approximation, the Moon behaves like a passive obstacle to the solar wind, and no evidence for a bow shock had been observed (refs. 8-10 and 8-11). Thus, during the normal lunar day, the solar-wind particles were expected to strike the lunar surface with essentially unchanged energy. This was substantiated by the Apollo 11 solar-wind composition experiment (ref. 8-9). It was shown that solar-wind helium reaches the lunar surface in an unimpeded, highly directional flow.

[a] Physikalisches Institut, University of Bern.
[b] Institut für Kristallographie und Petrographie, Federal Institute of Technology, Zurich.
[1] Principal investigator.

Principle of the Experiment

An aluminum foil 30 cm wide and 140 cm long, with an area of approximately 4000 cm^2, was exposed to the solar wind at the lunar surface by the Apollo 12 crew on November 19, 1969, at 12:35 G.m.t. The foil was positioned perpendicular to the solar rays (fig. 8-1), exposed for 18 hr and 42 min, and brought back to Earth. Laboratory experiments (ref. 8-12) have determined that solar-wind particles, arriving with an energy of approximately 1 keV/nucleon, penetrate approximately 10^{-5} cm into the foil, and a large and calibrated fraction are firmly trapped. In the laboratory, the returned foil is analyzed for trapped solar-wind noble gas atoms. Parts of the foil are melted in ultra-high vacuum systems, and the noble gases of solar-wind origin thus released are analyzed with mass spectrometers for element abundance and isotopic composition. In addition, a search will be conducted for the possible presence in the solar wind of the radioactive isotopes tritium and ^{56}Co.

Instrumentation and Lunar Surface Operation

The experiment hardware was the same as that flown on Apollo 11 (ref. 8-8) and consisted of a metallic telescopic pole approximately 4 cm in diameter and approximately 40 cm in length when collapsed. In the stowed position, the foil was enclosed in the tubing and rolled up on a spring-driven roller. The instrument weighed 430 g. When extended at the lunar surface, the pole was approximately 1.5 m long and a 30- by 130-cm foil area was exposed. Only the foil assembly was recovered at the end of the lunar exposure period; it was rolled on the spring-driven roller and returned to Earth. Figure 8-1 shows the instrument as deployed on the lunar

FIGURE 8-1. — Apollo 12 solar-wind composition experiment as deployed on the lunar surface (NASA photograph AS12-47-6899).

surface. By evaluating a number of Apollo 12 photographs, it was concluded that the foil was reclined by 10° during exposure. The average solar elevation during exposure was 13°, and thus the average direction of incidence of the sunlight on the foil was 3° above the foil normal.

After retrieval, the return unit was placed in a special Teflon bag and returned to Earth in the documented sample return container. In the Manned Spacecraft Center Lunar Receiving Laboratory (LRL), the unit was taken out of the Teflon bag, and the foil was inspected without unrolling it. The upper portion of the foil was found to be tightly and smoothly rolled. The outer windings of the foil were bulky. This, however, does not affect the quality of scientific data obtained from the experiment. Inside the Teflon return bag, a quantity of about a gram of fine lunar soil material was found, including grain sizes of 1 to 2 mm. This lunar material must have entered the bag during return or during postflight handling of the lunar sample return container. First analyses of light noble gases in the foil have shown that this dust contamination can be eliminated sufficiently by ultrasonic treatment. Dust contamination during return could be lowered by using a larger return bag, which can be closed more effectively.

Preliminary Results

The foil had the same dimensions, general makeup, and trapping properties as the Apollo 11 foil, described in detail in reference 8-8. Again, as on Apollo 11, test pieces were incorporated that had been irradiated before flight with a calibrated amount of neon. In addition, an unirradiated test foil was mounted in a position that remained shielded from the solar wind.

In the Apollo 11 solar-wind composition experiment, part of the foil had been sterilized and released from the LRL before termination of the quarantine period for lunar material. No such early release was attempted with the Apollo 12 experiment foil to restrict foil handling and to avoid additional contamination with lunar dust in the LRL quarantine cabinets.

The Apollo 12 foil was received in the laboratory in the middle of January 1970. For the first analysis, three small foil pieces were decontaminated by means of the ultrasonic treatments that had proved their efficiency in the Apollo 11 foil analyses. The results of these first measurements are given in table 8–I. It may be seen that the shielded foil piece had a ^{4}He concentration per unit area that was less than 1 percent of the concentrations found in the foil pieces exposed to the solar wind. The agreement is good between the concentrations and the ^{4}He/^{3}He ratios measured in the two exposed foil pieces.

TABLE 8–I. *First Results From Apollo 12 Solar-Wind Composition Experiment Foil Analyses*

Sample number	Elevation above lunar surface, cm	Area, cm^2	^{4}He concentration per unit area, $\times 10^{14}$ atoms/cm^2	^{4}He/^{3}He ratio
Shielded foil:				
10-2	145	7.7	0.4	—
Exposed foil:				
10-1	139	9.8	45.6	2580
9-1	123	10.9	44.5	2610

The average ^{4}He flux during the Apollo 12 exposure period can be calculated by using the data given in table 8–I. The trapping probabilities of the foil for noble gas ions depend only slightly on energy in the general solar-wind velocity region. For helium with a velocity greater than 300 km/sec, the trapping probability is 89 ± 2 percent for normal incidence and 5 percent less for an incidence angle of 70° to 75°. In the Apollo 11 experiment, the angular distribution of the arriving helium ions has actually been determined (ref. 8–9). The same experiment with the Apollo 12 foil has not yet been completed; therefore, in this paper, the expected angle of incidence on the foil has been estimated. The average angle of incidence of the sunlight on the foil was 87°. The effects of corotation and aberration lower this value to about 84° for the solar wind (refs. 8–13 and 8–14). During the time of the Apollo 12 foil exposure, the Moon had probably already passed into the magnetosheath of the Earth, and the ion flow direction was changed relative to the undisturbed solar wind. The tilt of the magnetosphere (ref. 8–15) lowers the expected direction of incidence by approximately 5° (ref. 8–15), and an additional lowering by a few degrees can be expected as a result of the change of flow direction in the shockfront. Thus, the expected angle of incidence on the foil is 70° to 75°. With this assumption, the average ^{4}He flux during the Apollo 12 exposure period is as given in table 8–II and is compared with the flux observed during the Apollo 11 landing. The two figures are similar and are in good agreement with average fluxes derived from He/H ratios observed with solar-wind energy/charge spectrometers. (Compare with ref. 8–6.) The expected direction of solar-wind incidence is 25° to 30° above the lunar horizon. Even if helium would be heated to 1 to 2 million degrees centigrade in the shock transition, the portion of the helium flux cut off by the horizon would be negligible. Thus, the ^{4}He flux given herein should be directly comparable to fluxes obtained by Earth satellites during the same period.

TABLE 8–II. *Comparison Between the Preliminary Average ^{4}He Flux Obtained from the Apollo 12 Solar-Wind Composition Foil Exposure Period and the Flux Obtained From the Apollo 11 and the Flux Obtained From the Apollo 11 Solar-Wind Composition Experiment*

Mission	Exposure initiated	Exposure duration	Average ^{4}He flux, $\times 10^6$ atoms/cm^2-sec
Apollo 11	July 21, 1969, 03:35 G.m.t.	77 min	6.3 ± 1.2
Apollo 12	Nov. 19, 1969, 12:35 G.m.t.	18 hr 42 min	8.2 ± 1.0

From the first two Apollo 12 foil pieces analyzed, ^{4}He/^{3}He = 2600 ± 200 is obtained for the

Apollo 12 exposure period. This value is higher than the $^4He/^3He$ ratio obtained so far from the analyses of pieces of the Apollo 11 foil. Comparative analyses of pieces of foils from the two flights are being continued to confirm this difference. Actually, time variations in isotopic ratios in the solar wind can be expected (ref. 8-16), and the $^4He/^3He$ ratio has to be determined repeatedly to assess the range of occurring variations before an average for the present-day solar wind can be established. This average is of high astrophysical significance, since it can be compared with ancient $^4He/^3He$ ratios derived from solar-wind gases trapped in the lunar surface (ref. 8-17) or in meteorites. If a secular increase in the solar $^3He/^4He$ ratio should be found, this could be interpreted as a result of mixing inside the Sun or as a result of nuclear reactions at the solar surface.

The results obtained from the analyses of the first small pieces of the Apollo 12 foil indicate that, from larger foil areas, fluxes and isotopic composition can be obtained not only for helium but also for neon and argon.

Reference

8-1. HUNDHAUSEN, A. J.; ASBRIDGE, J. R.; BAME, S. J.; GILBERT, H. E.; and STRONG, I. B.: Vela-3 Satellite Observations of Solar Wind Ions: A Preliminary Report. J. Geophys. Res., vol. 72, no. 1, Jan. 1, 1967, pp. 87-100.

8-2. SNYDER, C. W.; and NEUGEBAUER, M.: Interplanetary Solar Wind Measurements by Mariner II. Space Research IV. Proceedings of the Fourth International Space Science Symposium, vol. IV, P. Muller, ed., North-Holland Pub. Co. (Amsterdam), 1964, pp. 89-113.

8-3. WOLFE, J. H.; SILVA, R. W.; McKIBBIN, D. D.; and MASON, R. H.: The Compositional, Anisotropic, and Nonradial Flow Characteristics of the Solar Wind. J. Geophys. Res., vol. 71, no. 13, July 1, 1966, pp. 3329-3335.

8-4. SNYDER, C. W.; and NEUGEBAUER, M.: The Relation of Mariner-2 Plasma Data to Solar Phenomena. The Solar Wind, R. J. Mackin, Jr., and Marcia Neugebauer, eds., Pergamon Press, 1966, pp. 25-32.

8-5. OGILVIE, K. W.; BURLAGA, L. F.; and WILKERSON, T. D.: Plasma Observations on Explorer 34. J. Geophys. Res., vol. 73, no. 21, Nov. 1, 1968, pp. 6809-6824.

8-6. ROBBINS, D. E.; HUNDHAUSEN, A. J.; and BAME, S. J.: Helium Abundance and Plasma Properties in the Solar Wind. Preprint ST531, Trans. Am. Geophys. Union, vol. 50, no. 4, April 1969, p. 302.

8-7. BAME, S. J.; HUNDHAUSEN, A. J.; ASBRIDGE, J. R.; and STRONG, I. B.: Solar Wind Ion Composition. Phys. Rev. Letters, vol. 20, no. 8, Feb. 19, 1968, pp. 393-395.

8-8. GEISS, J.; EBERHARDT, P.; SIGNER, P.; BUEHLER, F.; and MEISTER, J.: The Solar-Wind Composition Experiment. Apollo 11 Preliminary Science Report, sec. 8, NASA SP-214, 1969, pp. 183-186.

8-9. BUEHLER, F.; EBERHARDT, P.; GEISS, J.; MEISTER, J.; and SIGNER, P.: Apollo 11 Solar Wind Composition Experiment: First Results. Science, vol. 166, no. 3912, Dec. 19, 1969, pp. 1502-1503.

8-10. LYON, E. F.; BRIDGE, H. S.; and BINSACK, J. H.: Explorer 35 Plasma Measurements in the Vicinity of the Moon. J. Geophys. Res., vol. 72, no. 23, Dec. 1, 1967, pp. 6113-6117.

8-11. NESS, N. F.; BEHANNON, K. W.; SCEARCE, C. S.; and CANTARANO, S. C.: Early Results from the Magnetic Field Experiment on Lunar Explorer 35. J. Geophys. Res., vol. 72, no. 23, Dec. 1, 1967, pp. 5769-5778.

8-12. BUEHLER, F.; GEISS, J.; MEISTER, J.; EBERHARDT, P.; HUNEKE, J. C.; and SIGNER, P.: Trapping of the Solar Wind in Solids, Part 1. Trapping Probability of Low Energy He, Ne, and Ar Ions. Earth and Planet. Sci. Letters, vol. 1, no. 5, Sept. 1966, pp. 249-255.

8-13. HUNDHAUSEN, A. J.: Direct Observations of Solar-Wind Particles. Space Sci. Rev., vol. 8, no. 4, Sept. 24, 1968, pp. 690-749.

8-14. AXFORD, W. I.: Observations of the Interplanetary Plasma. Space Sci. Rev., vol. 8, no. 3, Jan. 5, 1968, pp. 331-365.

8-15. HUNDHAUSEN, A. J.; BAME, S. J.; and ASBRIDGE, J. R.: Plasma Flow Pattern in the Earth's Magnetosheath. J. Geophys. Res., vol. 74, no. 11, June 1, 1969, pp. 2799-2806.

8-16. GEISS, J.; HIRT, P.; and LEUTWYLER, H.: On Acceleration and Motion of Ions in Corona and Solar Wind. Solar Physics, 1970 (in press).

8-17. Proceedings of the Lunar Science Conference. Science, vol. 167, no. 3918, Jan. 30, 1970, pp. 449-784.

9. Apollo 12 Multispectral Photography Experiment

A. F. H. Goetz,[a][e] F. C. Billingsley,[b] E. Yost,[c] and T. B. McCord[d]

The lunar multispectral photography experiment was successfully accomplished on Apollo 12. A number of photographs were returned in the blue, green, red, and infrared (IR) portions of the optical spectrum. Preliminary data analysis shows no color boundaries in the frame containing the Fra Mauro formation and the Apollo 13 landing site. Color differences were found in the frame containing Lalande η, establishing the existence of small-scale color differences on the lunar surface.

Purpose of the Experiment

The goal of the lunar multispectral photography experiment was to obtain vertical strip photography in three portions of the optical spectrum — blue, red, and IR — at resolution one to two orders of magnitude higher than is obtainable from Earth. A fourth camera, which had a green filter, was added to the array for operational purposes. However, for the sake of the following discussion, this camera will be considered to be part of the experiment.

The further objectives of this experiment were as follows:

(1) To photograph future Apollo landing sites so that ground-truth information provided by the returned samples may be extrapolated to other points on the lunar surface

(2) To produce photometrically accurate, two- and three-color images by photographic and computer processing methods that will accurately delineate lunar color boundaries and their magnitudes

(3) To evaluate, under closely controlled conditions, the photographic versus the computer image-processing techniques for reduction of lunar multispectral photography

Lunar Color Measurement

Lunar color and its variation across the lunar surface have interested planetary astronomers for many years. The interest has heightened with the growing weight of evidence, obtained from accurate Earth-based photoelectric photometry, that points toward a positive correlation between color and compositional differences (ref. 9–1). Once ground-truth samples have been obtained at several sites of differing color, it may be possible to extrapolate compositional color information to large areas of the Moon that will not be sampled *in situ*.

In the present context, color differences mean relative differences in spectral reflectivity between points on the surface. The general spectral reflectivity curve of the Moon shows a near-linear increase in reflectivity from 400 to 800 nm (ref. 9–2). Areas designated red or blue reflect more energy in their respective wavelength regions than a standard lunar area. In most cases, the greater the separation in wavelength, the greater the color difference obtained. The differences between 400 and 800 nm, among points on the lunar surface, average 4–8 percent (ref. 9–1).

Photographic (ref. 9–3) and photoelectric (ref. 9–1) methods have been used in the past for the measurement of lunar color variations. The best known photographic method used to date is Whitaker's sandwich printing technique (ref. 9–3) in which a negative ultraviolet plate and a positive IR plate are sandwiched together and printed. Color differences then show up as varying shades of gray. The disadvantages of

[a] Bellcomm, Inc., Washington, D.C.
[b] JPL, California Institute of Technology.
[c] Long Island University.
[d] Massachusetts Institute of Technology.
[e] Principal investigator.

this method are that it is not quantitative and that albedo changes can masquerade as color differences if the density-logarithm-exposure ($D \log E$) curves of the two plates are not extremely well matched. The advantage of the method lies in the image form of data display.

Recent advances in photoelectric instrumentation allow ground-based relative color measurements to be made to 0.1 percent accuracy (ref. 9-4). Spectral reflectivity measurements of the Apollo 11 soil samples show an excellent correlation with ground-based telescope data of the Tranquility site (ref. 9-5). However, point-by-point measurement is a time-consuming process and is not suited for image display. Work now in progress with electronic imaging systems incorporates the advantages of both methods.

Computer image processing (ref. 9-5) of photographs combines the advantages of photographic image display and provides quantitative color information for the entire picture, although at accuracies less than are obtainable from photoelectric photometry.

Equipment and Operation

The experiment camera array consisted of four 70-mm Hasselblad cameras with 80-mm lenses. The filter and black-and-white film combinations were as follows: blue 47B filter, type 3401 film; green 58 filter, type 3401 film; red 29 + 0.6ND filter, type 3401 film; and infrared 87C filter, type SO-246 film. Type 3401 is Plus-X aerial film, and type SO-246 is type 5424 infrared aerographic film coated on a 4-mil base. The center wavelength of each filter/film combination is as follows: blue, 430 nm; green, 540 nm; red, 660 nm; and IR, 860 nm.

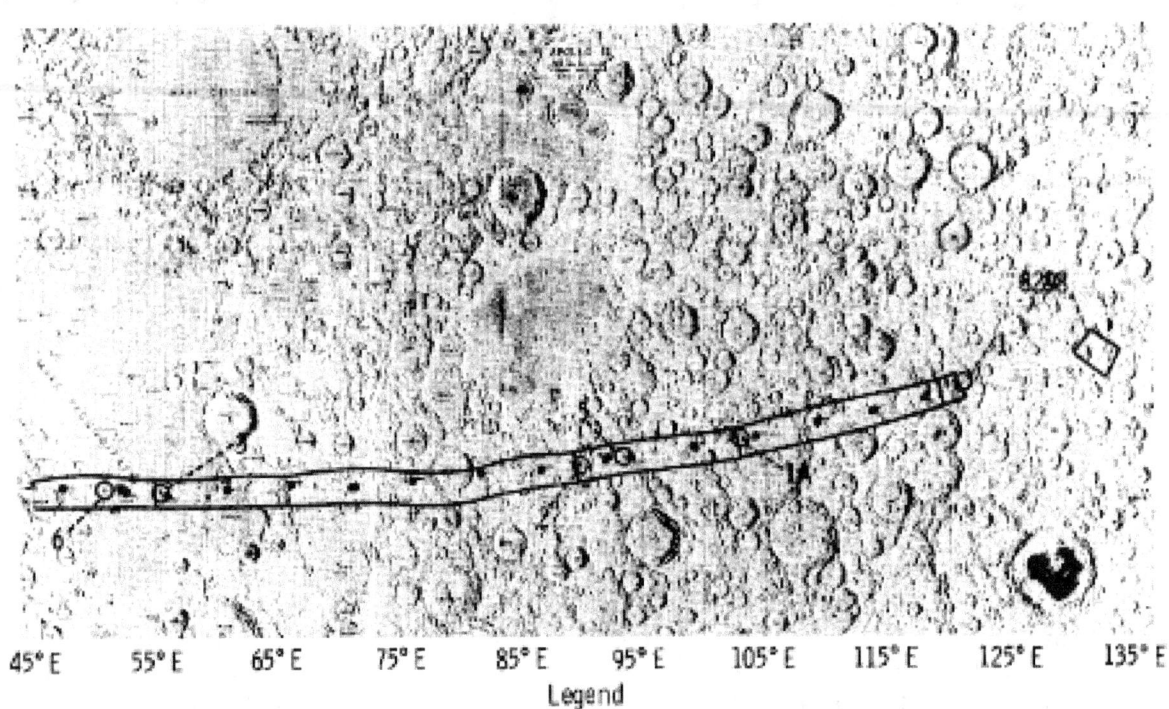

Legend

1. Start blue, green, red stereo sequence (1/60, f/5.6)
1A. Start IR (1/60, f/4.0)
2. All stop
3. Start blue, green, red
5. All start (blue, green, IR, f/8.0) second revolution (red, f/5.6)
6. All stop
• Principal point

FIGURE 9-1. — Apollo 12 multispectral photography ground track from 135° E to 45° E.

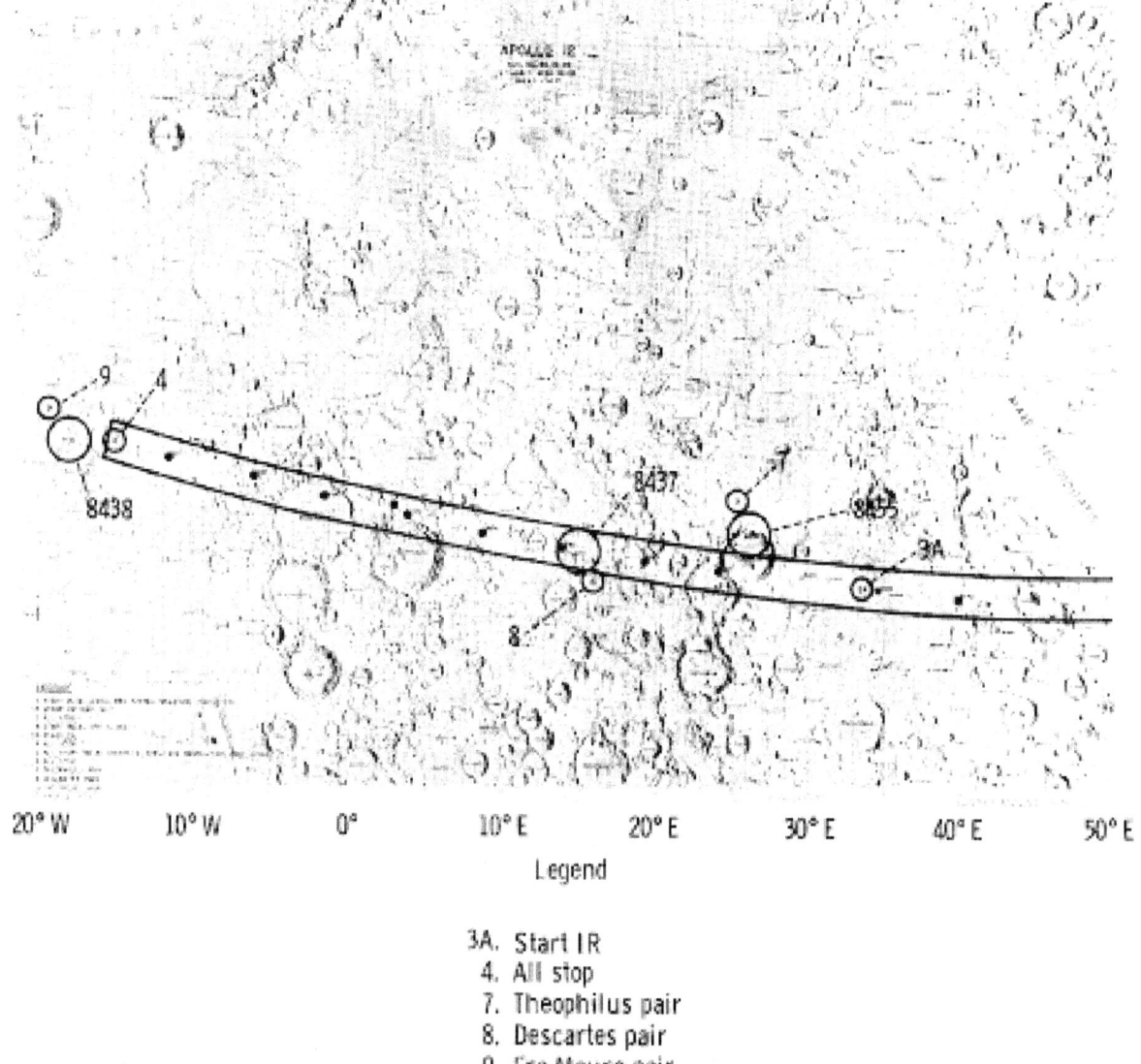

FIGURE 9-2.—Apollo 12 multispectral photography ground track from 50° E to 20° W.

An intervalometer tripped all shutters simultaneously at 20-sec intervals. For operational purposes, the shutter speed on all four cameras was fixed at 1/60 sec. Focus settings were fixed at 44 ft for the IR camera and at infinity for the other three cameras. To facilitate f-stop changes, the vertical strip photography was broken into three segments. During orbital revolution 27, the blue, green, and IR cameras were set at f/5.6, and the red camera was set at f/4.0. Photography was taken between longitudes 120° E and 90° E and between 54° E and 15° W. The minimum Sun angle was approximately 25°. The remainder of the vertical photography was carried out on revolution 28 at f/8/5.6. The off-vertical targets of opportunity were exposed at f/5.6/4.0 for Theophilus and Descartes and at f/2.8/2.8 for Fra Mauro. Figures 9–1 and 9–2 show the photography ground track.

Film Calibration and Processing

All flight-film calibration and processing were accomplished by the NASA Manned Spacecraft Center (MSC) Photographic Technology Laboratory (PTL). Preflight calibration was accomplished by applying a 21-step gray wedge to the film in a 1-B sensitometer. For the experiment, a special step tablet was constructed to provide four 21-step wedges arranged to fit in a 60- by 60-mm format to facilitate film-scanning procedures. In addition, a preflight standard wedge and a postflight special tablet were applied to the leader of the film.

Preflight process controls were established to develop the films to the following gammas: blue, 1.7; green, 1.65; red, 1.6; and IR, 1.5. The different gammas were chosen to compensate for the increasing transmission of the standard wedge toward longer wavelengths. In other words, the absolute gamma should be 1.7 for each film. The relatively high gammas were chosen to give the maximum exposure differentiation commensurate with the required dynamic range on the film. The following gammas were obtained: blue, 1.68; green, 1.48; red, 1.42; and IR, 1.44. The reasons for the discrepancy among control and flight-film gammas are not completely understood but can probably be attributed to radiation fogging and latent image decay. The PTL is investigating this effect.

Film Return

Each of the blue-, green-, and red-filtered cameras returned 142 frames, while the IR camera, in which the film had been rationed, returned 105 frames. The resolution in the returned type 3401 film is approximately 30 m. This limit is approximately the motion resulting from the shutter speed of 1/60 sec. The densities on all frames fell within the approximate straight-line portion of the respective $D \log E$ curves shown in figure 9-3. This requirement was necessary for data reduction by photographic methods. For reasons not understood at this time, all IR frames have a 4-mm-wide underexposed strip at the leading and trailing edges of each frame. The IR frames are not in focus because of an apparent film-magazine malfunction, and they will not be usable for color difference analysis.

Data Reduction

The color or spectral reflectivity differences sought in this study are not detectable by eye or on normal color film. The eye is very sensitive to small color variations under controlled laboratory conditions, in particular when the objects have the same brightness and are juxtaposed. However, the eye is incapable of reliably detecting small differences in spectral reflectivity in conjunction with brightness differences such as in a lunar surface scene. Normal color-additive techniques using color-separation photographs also fail to show up differences, even at higher saturation.

The two methods mentioned previously — the photographic sandwich and computer image processing techniques — are basically techniques for ratioing two pictures. Ratioing is necessary to remove the brightness variations caused by general albedo and slope differences. Because the film density is a function of the logarithm of the exposure, ratios are formed by taking differences in densities between two pictures. Such difference pictures for two colors have been produced by photographic (ref. 9-3) and computer (ref. 9-6) methods. Data reduction for the experiment will be carried out in three colors by using extensions of both methods. Details of these procedures will be the subject of later publications.

Photographic data reduction to date has included construction of three-color difference pictures of five frame sets. The main difficulties in analyzing these composites are anomalous colors introduced by nonlinearities in the $D \log E$ curves (which limit the range of brightnesses for which the color construction is valid) and brightness nonuniformities that are due to camera vignetting. While several frames show color differences, more work must be completed on establishing confidence limits before interpretations can be made.

Extensive computer image processing has been accomplished on only two frames: 8438 and 8392. Only two-color difference pictures have been constructed to date by methods used previously on Earth-based photography (ref. 9-6). The basic method consists of the following steps:

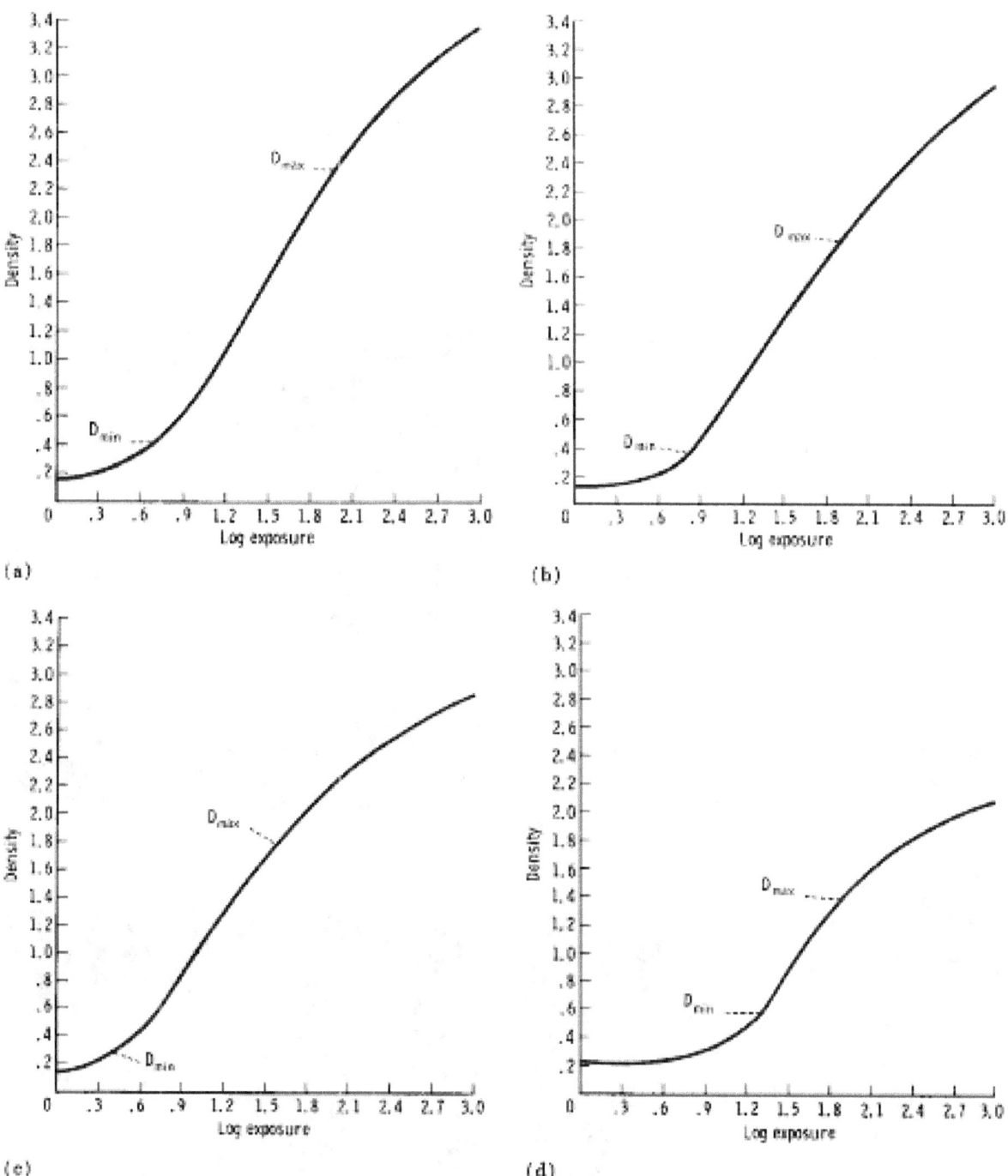

FIGURE 9-3.—Processing curves obtained from preexposed standard wedge. (a) Blue filter. (b) Green filter. (c) Red filter. (d) Infrared.

(1) Scanning by video film converter (VFC) of each set of pictures, including the preflight calibration step tablets (scanning spot sizes of 25, 40, and 50 μm have been used)

(2) Obtaining a VFC response curve and linearizing the $D \log E$ curve

(3) Converting all points (1 to 2×10^6 points per frame) to the log E domain, thus eliminating

any film sensitivity or processing differences between colors

(4) Registration of the frames of the two colors to be differenced

(5) Point-by-point subtraction and normalization of two frames

(6) Contrast stretching, up to a factor of 3, so that each step in log E becomes visible as a distinct shade of gray (the step size is approximately 2.5 percent in exposure)

(7) Replaying the picture onto film

The registration of the frames has been the major problem encountered. Each camera/filter combination exhibits different geometrical distortions and a slightly different magnification. These systematic distortions are coupled with near-random distortions that are due to lack of film flatness. Good registration can be facilitated only by brute-force rubber-sheet stretching programs.

Figure 9-4 shows frame 8438, a view of the Fra Mauro formation and the Apollo 13 landing site, taken with the blue-, green-, and red-filtered cameras. The dark corner in each picture is the result of obscuration by the edge of the hatch window. No noticeable density variations, other than an overall brightness difference, are visible. Two-color computer difference pictures, blue minus green and red minus blue, have been produced. The composites show no significant color differences over the entire frame, including the mare-highlands boundary. Some implications of this result are given in the following paragraphs.

A two-color difference picture of frame 8392

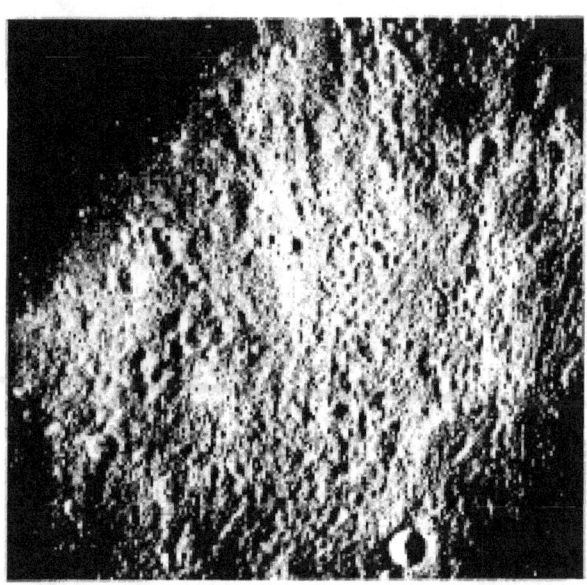

(b)

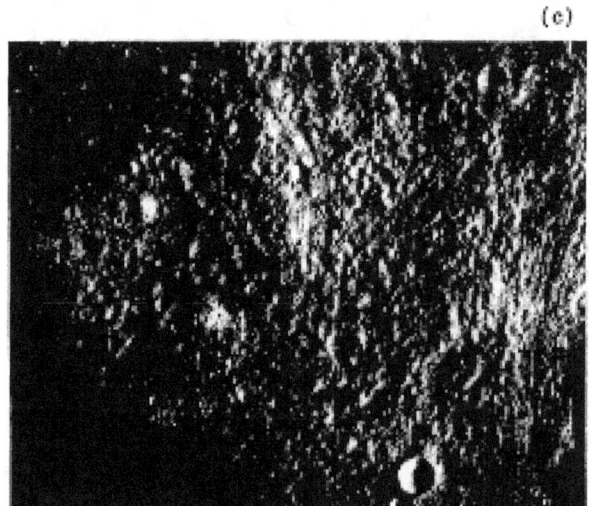

(c)

FIGURE 9-4. — Fra Mauro formation and Apollo 13 landing site. (North is at the top of the photos.) (a) Blue filter (AS12-56B-8438). (b) Green filter (AS12-56D-8438). (c) Red filter (AS12-56C-8438).

(fig. 9-5), which covers a portion of Mare Nubium and Lalande η, was constructed. Figure 9-6 is a contrast-stretched red-minus-blue print of frame 8392 in which dark areas are bluer and light areas are redder than an arbitrary point that was chosen to be gray.

Distinct color differences are evident, in particular between Lalande η and the surrounding mare and between what appears to be secondary-impact ejecta rays and the surrounding mare. Although the frames have not been calibrated for camera vignetting or adjusted to a lunar standard by Earth-based photometry, there is now, for the first time, clear evidence for local small-scale color variations on the lunar surface.

FIGURE 9-5. — Two-color difference picture of frame 8392 covering a portion of Mare Nubium and Lalande η.

Discussion

Differences in spectral reflectivity can be caused by several factors. Compositional variations give rise to color differences. However, subtle differences, such as recorded in frame 8392, may very well be caused by variations in the iron and titanium content of the glasses, as in the Apollo 11 sample described by Adams and Jones (ref. 9-5).

Particle size affects the slope of the reflectivity curve of rock materials. Typically, for basaltic materials in the laboratory, the sample becomes redder with decreasing particle size (ref. 9-7).

The effects of age on color appear to be insignificant, at least with respect to alteration of the optical properties of in situ materials by the lunar environment. No evidence for metal coatings or deposits of sufficient thickness to affect the optical properties was found on mineral grains in the Apollo 11 sample (ref. 9-5). On the basis of telescopic measurements, McCord (ref. 9-1) finds no evidence of an aging effect that causes color variations.

Frame 8392, which contains Lalande η, exhibits definite color differences, as shown in figure 9-6. Careful inspection reveals that color is not directly correlated with brightness, particularly as seen in the comparison of the boundary between the embayment and the mare on the east side of Lalande η and the boundary on the flat southeastern side of the same feature. Therefore, except in the deep shadows, the color variations are real and are not artifacts produced by photographic nonlinearities.

The white streaks in the mare appear to be impact ejecta plumes, probably produced by secondaries from the crater Lalande η. It is not apparent whether the light material is secondary projectile material or material that has been excavated from beneath the mare surface. The plumes are bluer than the surrounding mare, and their boundaries are sharp on a 100-m scale. The color difference runs contrary to the observation that in basaltic rocks, finer higher-albedo material is redder. Therefore, a compositional difference must be assumed to explain the bluer plumes.

If, indeed, the plume material has been excavated from below the mare surface, then the material must have been ejected from craters smaller than about 30 m in size, since no single plume-source crater is visible and since the smaller craters are below the resolution limit of the film. For such a case, the thickness of the dark surface material is limited to 7 to 10 m. On a small scale, the blue, bright plumes are similar to the feature Reiner γ, which exhibits a similar color difference.

The absence of color variations in frame 8438, particularly at the mare boundary, is somewhat

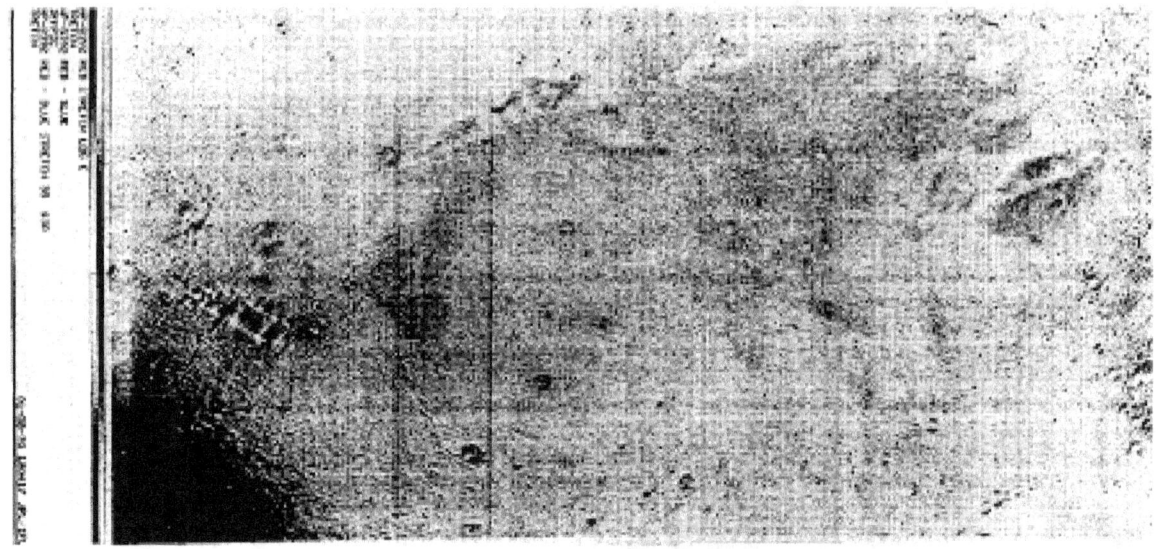

FIGURE 9-6.—Contrast-stretched red-minus-blue difference picture of the center portion of frame 8392. Dark areas are bluer and light areas are redder than an arbitrary point taken as neutral gray. The light vertical banding is an artifact introduced in the scanning process. Slight misregistration of the red and blue frames enhances crater boundaries. In addition, color information cannot be obtained from deep shadow areas; hence, those areas will appear anomalously colored.

surprising but not without precedent. McCord (ref. 9-1) has found a similar lack of contrast across some mare-upland boundaries between average areas 18 km in diameter. Since a definite albedo and morphological discontinuity are present, a color boundary would be expected. A mean particle-size difference has been suggested as the cause of the mare-upland albedo differences observed (ref. 9-8). However, such a size difference would be expected to result in a color difference. The analysis of the Apollo 13 samples should solve this dilemma. Furthermore, more detailed reduction and analysis of multispectral photographs will better define the power of this method in the interpretation of lunar surface geology.

Summary

The lunar multispectral photography experiment has yielded 142 black-and-white photographs, taken with blue-, green-, red-, and IR-filtered cameras, that are suitable for color-difference analysis. Two existing image data-reduction methods are being expanded to produce images that display greatly enhanced three-color contrast. Two-color difference pictures have been produced, and the method has been shown to be effective. The color enhancement of the Apollo 13 landing-site frame shows a somewhat surprising lack of color variation. The frame containing Lalande η exhibits color differences, the first such differences to be detected in high-resolution photography of the lunar surface, which probably can be attributed to compositional variations.

References

9-1. McCord, T. B.: Color Differences on the Lunar Surface. J. Geophys. Res., vol. 74, no. 12, June 15, 1969, pp. 3131-3142.

9-2. McCord, T. B.: Color Differences on the Lunar Surface. Ph. D. Dissertation, Calif. Inst. of Tech., 1968.

9-3. Heacock, R. L.; Kuiper, G. P.; Shoemaker, E. M.; Urey, H. C.; and Whitaker, E. A.: Ranger VII, Part II: Experimenters' Analyses and Interpretations. TR 32-700, JPL, Calif. Inst. of Tech., Feb. 1965.

9-4. McCord, T. B.: A Double-Beam Astronomical Photometer. Applied Optics, vol. 7, no. 3, Mar. 1968, pp. 475-478.

9-5. ADAMS, J. B.; and JONES, R. L.: Spectral Reflectivity of Lunar Samples. Science, vol. 167, no. 3918, Jan. 30, 1970, pp. 737-739.

9-6. BILLINGSLEY, F. C.; GOETZ, A. F. H.; and LINDSLEY, J. N.: Color Differentiation by Computer Image Processing. Photographic Science and Engineering, vol. 14, no. 1, Jan. 1970, pp. 28-35.

9-7. ADAMS, J. B.; and FILICE, A. L.: Spectral Reflectance 0.4 to 2.0 Microns of Silicate Rock Powders. J. Geophys. Res., vol. 72, no. 22, Nov. 15, 1967, pp. 5705-5715.

9-8. ADAMS, J. B.: Lunar Surface Composition and Particle Size: Implications from Laboratory and Lunar Spectral Reflectance Data. J. Geophys. Res., vol. 72, no. 22, Nov. 15, 1967, pp. 5717-5720.

ACKNOWLEDGMENTS

The authors wish to thank Jurrie van der Woude of the California Institute of Technology for his invaluable assistance in obtaining telescope test photography. Richard R. Baldwin of MSC made an outstanding effort in coordinating the hardware and flight plan integration. Noel T. Lamar of MSC was instrumental in expediting the stringent film processing requirements.

10. Preliminary Geologic Investigation of the Apollo 12 Landing Site

PART A

GEOLOGY OF THE APOLLO 12 LANDING SITE

E. M. Shoemaker,[a†] *R. M. Batson,*[b] *A. L. Bean,*[c] *C. Conrad, Jr.,*[c] *D. H. Dahlem,*[b]
E. N. Goddard,[d] *M. H. Hait,*[b] *K. B. Larson,*[b] *G. G. Schaber,*[b] *D. L. Schleicher,*[b]
R. L. Sutton,[b] *G. A. Swann,*[b] *and A. C. Waters*[e]

This report provides a preliminary description of the geologic setting of the lunar samples returned from the Apollo 12 mission. A more complete interpretation of the geology of the site will be prepared after thorough analysis of the data.

The Intrepid landed on the northwest rim of the 200-m-diameter Surveyor Crater (in which Surveyor 3 touched down on April 20, 1967) in the eastern part of Oceanus Procellarum. The landing site was at 23.4° W and 3.2° S, approximately 120 km southeast of the crater Lansberg and due north of the center of Mare Cognitum. The landing site is on a broad ray associated with the crater Copernicus, which is located approximately 370 km to the north.

The landing site is characterized by a distinctive cluster of craters ranging in diameter from 50 to 400 m (fig. 10-1). Informal names were given to these craters for use during the mission and have been adopted for this report. The traverses during the two extravehicular activity (EVA) periods were generally made on or near the rims of these named craters and on deposits of ejecta from the craters.

A total of 23 panoramas were taken during the Apollo 12 lunar stay to document the astronauts' traverses. These include partial panoramas taken from inside the lunar module (LM) through both LM windows, complete 360° panoramas taken from the surface at intervals throughout the traverse, and partial panoramas that were frequently taken in pairs for stereoscopic coverage of large features of particular interest. Panoramas taken from the LM windows are useful because of their high vantage point, even though their azimuthal field of view is less than 180°.

Complete panoramas were taken to record as much lunar surface detail as possible with a surface-based camera. When joined as mosaics, the panoramas provide accurate map control data in the form of horizontal angles. Control can be obtained analytically, with high precision, from measurements of glass-plate reproductions of the photographs or graphically, with moderate precision, by measuring the mosaics. Complete panoramas are more useful than broken or partial panoramas because complete panoramas provide an immediate check of error accumulation in measuring horizontal angles and because lunar directions can be determined accurately and independently of any other data from the location of the image of the Sun and of the image of the astronaut's shadow. This was one of the reasons that the crew was requested to take photographs looking into the Sun, even though poor photograph quality was anticipated.

Partial panoramas produce some of the same data as complete panoramas at a considerable saving of film. They are useful for photographic documentation of large features of geologic interest. When two partial panoramas are taken of the same feature from slightly different vantage points, pairs of photographs from the adjacent

[a] California Institute of Technology.
[b] U.S. Geological Survey.
[c] NASA Manned Spacecraft Center.
[d] University of Michigan.
[e] University of California.
[†] Principal Investigator.

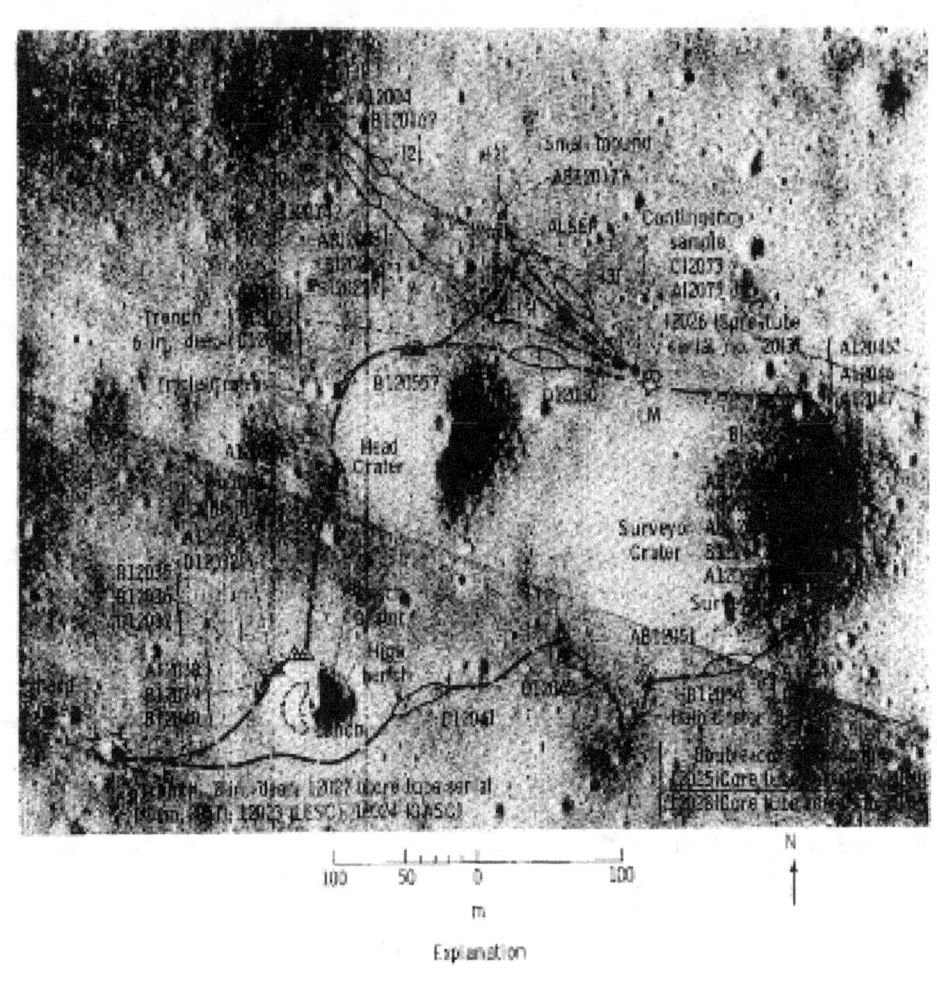

Explanation

First EVA Traverse
.... Both astronauts
—·—·— Conrad
———— Bean

Second EVA Traverse
———— Both astronauts
— — — Conrad
- - - - Bean

△ Photographic control station

*B12054 Sample locality. Number refers to sample number assigned in Lunar Receiving Laboratory. Letters
AB12056? refer to rock type. Queried where sample identification is uncertain. Rock types are as follows:

    A - fine-grained igneous rock
    B - medium-grained igneous rock
    AB - intermediate, fine-to-medium-grained igneous rock
    C - breccia
    D - fine-grained material

⬭ Sample locality. Circle indicates locality not accurately determined. Dot shows best estimate of location

(3) Number of rocks collected in sample locality

✴ diagrammatic sketch of fresh ray pattern around Sharp Crater

ALSEP Apollo lunar surface experiments package
LESC Lunar environment sample container
GASC Gas analysis sample container

FIGURE 10-1. — Traverse map for Apollo 12. (Compare with fig. 10-50.)

panoramas can be viewed stereoscopically, and precise photogrammetric measurements of the feature can be made.

Mosaics of the panoramas are presented in figures 10-2 to 10-11, because they provide a more comprehensive view of some parts of the landing site than do the individual photographs that are generally available. It is impossible to make undistorted panoramic mosaics of photographs taken with conventional cameras because the scene viewed by the camera is, in effect, a spherical surface. If the photographs constituting the panorama could be made to fit inside a sphere, images on adjacent photographs would match. Because this is an impractical format for a report, the prints have been mosaicked so that images on the horizon match and so that the horizon is level. No attempt was made to match foreground images.

Surface Features

Astronauts Conrad and Bean visited four craters larger than 50 m in diameter and many craters of smaller size. They described the characteristics of eight craters and collected a variety of material ejected from the craters. They also made numerous comments about smaller craters and about the surface features that lie between these smaller craters, including surface material that may be underlain by ray material from more distant craters (especially Copernicus). Therefore, the rock collections returned to Earth contain a variety of material ejected from local craters that were visited on the traverse and contain fine-grained material, some of which probably was derived locally and some of which may have come from far-distant sources.

Fragmental Material

The lunar surface at the Apollo 12 landing site is underlain by fragmental material, the lunar regolith, which ranges in size from particles too fine to be seen with the naked eye to blocks several meters across. Along several parts of the traverse made during the second EVA period, the astronauts found fine-grained material of relatively high albedo that, at some places, was in the shallow subsurface and, at other places, was at the surface. This light-gray material was specifically reported to be at the surface near Sharp Crater and a few centimeters below the surface near Head, Bench, and Block Craters. It is possible that some of this light-gray material may constitute a discontinuous deposit that is observed through telescopes as a ray of Copernicus.

Darker regolith material that generally overlies the light-gray material is only a few centimeters thick in some places, but probably thickens greatly on the rims of some craters. It varies from place to place in the size, shape, and abundance of its constituent particles and in the presence or absence of patterned ground. Most local differences are probably the result of local cratering events.

Many comments of the astronauts concern the large amount of glass that is contained in the regolith. Irregularly shaped, small fragments of glass and glass beads are abundant both on and within the regolith; glass is also splattered upon some of the blocks of rock at the surface and is found within many shallow craters.

Size-Frequency Distribution of Fragmental Material

The nominal resolution of the ALSCC photographs is approximately 0.1 mm, but there was some difficulty in distinguishing rounded, irregular grain aggregates from small topographic irregularities. Where it was possible to distinguish an individual aggregate, it was counted as a single particle.

The area of the photograph is 72 by 82 mm, but the results have been normalized to an area of 100 m^2 in order to compare the Apollo 12 particle count with the particle count made from Surveyor 3 (ref. 10-1) and Apollo 11 (ref. 10-2) photographs. The cumulative size-frequency distribution of particles between 1 and 2 mm in diameter near the LM is similar to that in the immediate vicinity of the Surveyor 3 spacecraft.

Neither the precise location nor the subject of the frame was transmitted, but the frame was probably taken generally north of the LM and probably 5 to 15 m from the engine bell. The regolith in this area appears to be relatively undisturbed, although the area may have accumulated dust sprayed from nearby footprints, or it may have been slightly swept by the descent engine. The size-frequency distribution of small particles on the surface of the regolith was studied from one photograph (frame 11) taken

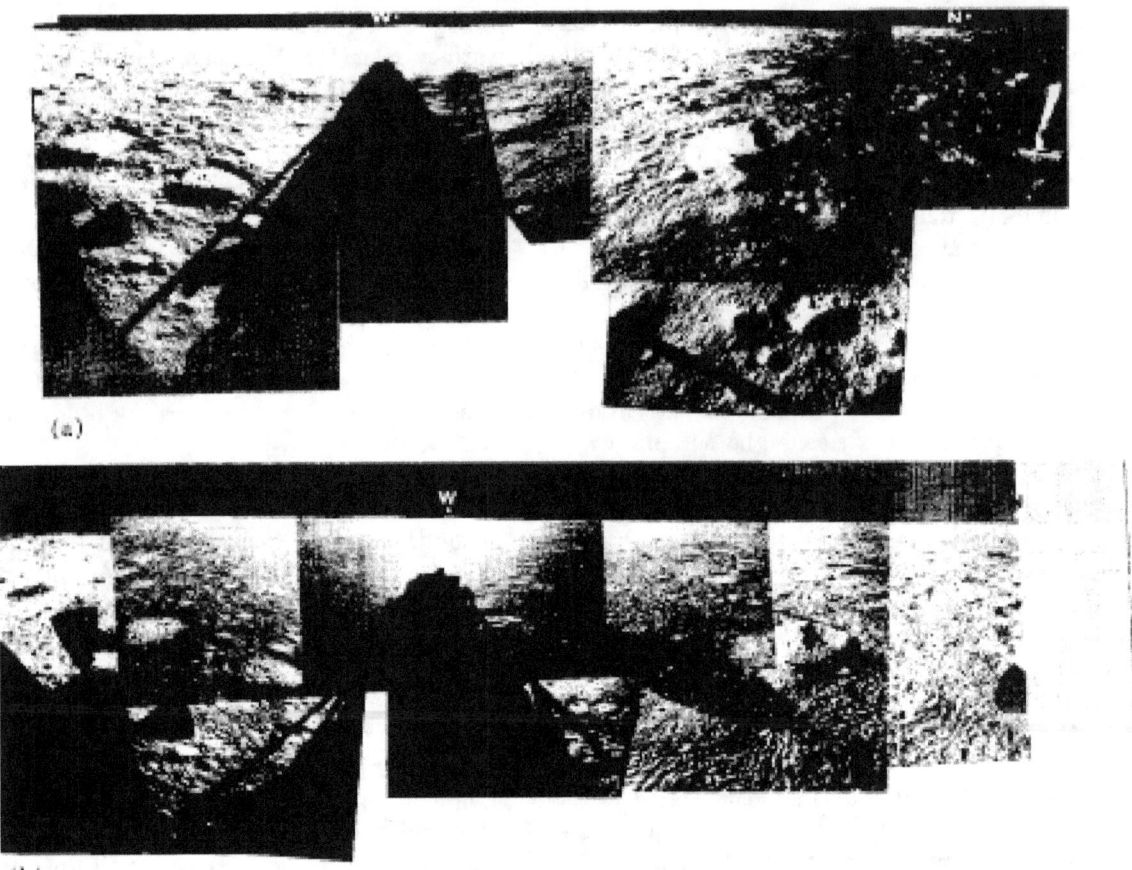

FIGURE 10-2.—Panoramas 1, 11, 12, and 23. (a) Panorama 1: View from both LM windows prior to EVA. (AS12-48-7022 to 7033) (b) Panoramas 11 and 12: View from both LM windows after the first EVA period. (AS12-47-7011 to 7021, AS12-46-6853 to 6868) (c) Panorama 23: View from both LM windows after the second EVA period. (AS12-48-7153 to 7171)

GEOLOGIC INVESTIGATION OF THE LANDING SITE

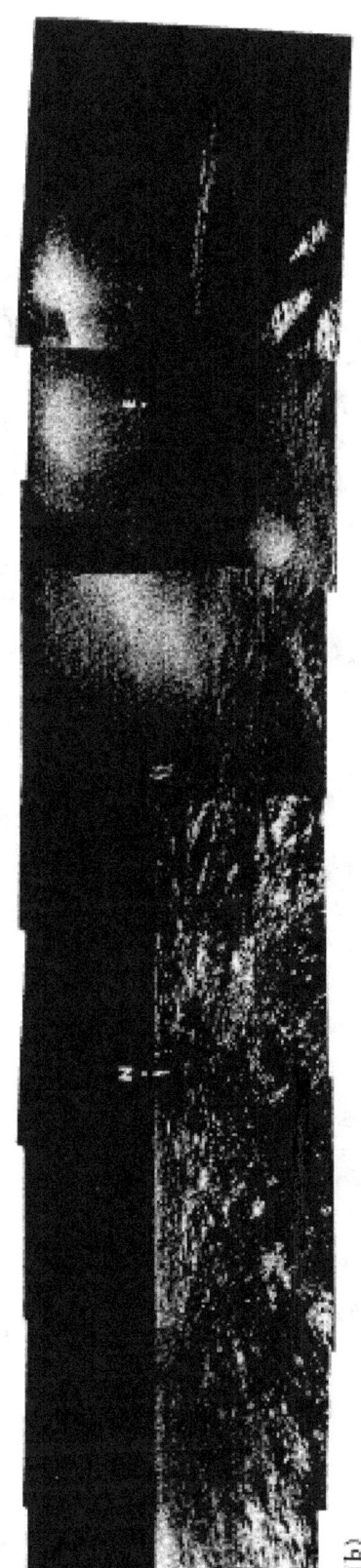

FIGURE 10-3.—Panorama 2: The landing site from 15 m west of the LM. (a) View looking south and west. (b) View looking north and east. (AS12-46-6730 to 6745)

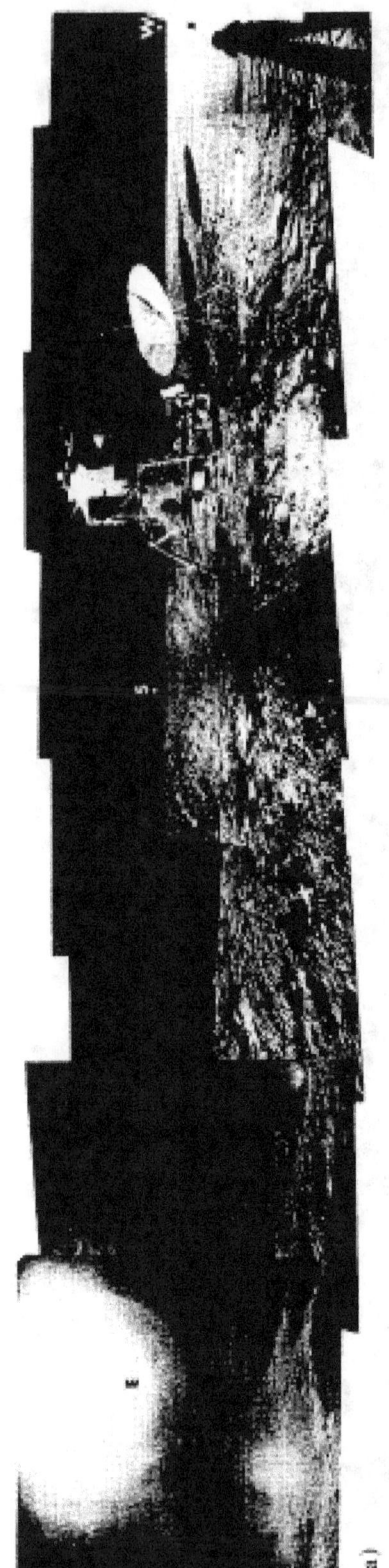
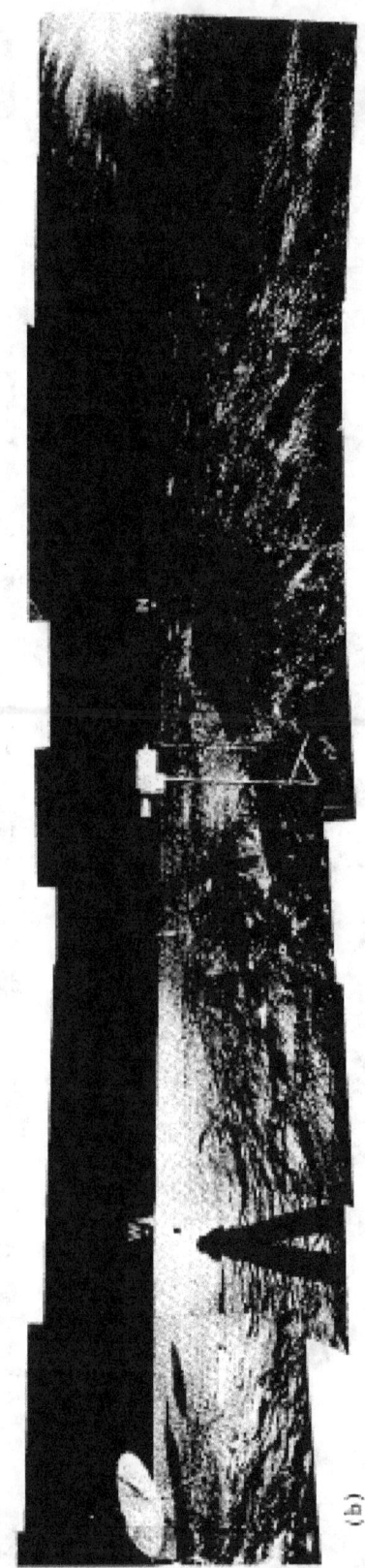

FIGURE 10-4.—Panorama 3: The landing site from 15 m northeast of the LM. (a) View looking east and south. (b) View looking west and north. (AS12-46-6746 to 6763)

GEOLOGIC INVESTIGATION OF THE LANDING SITE

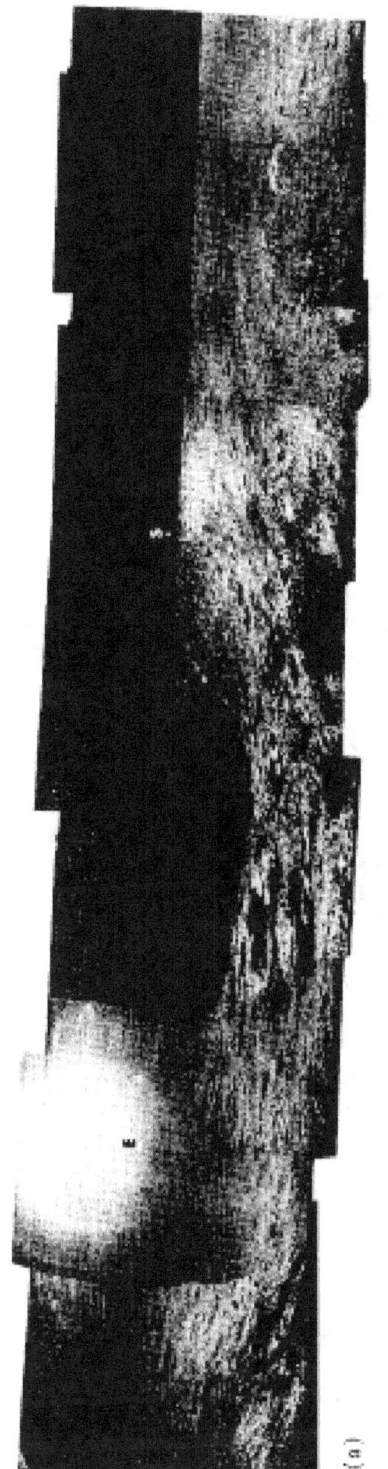

FIGURE 10-5.—Panorama 4: The landing site from 9 m southeast of the LM. (a) View looking east and south. (b) View looking south, west, and north. (AS12-46-6454 to 6482.)

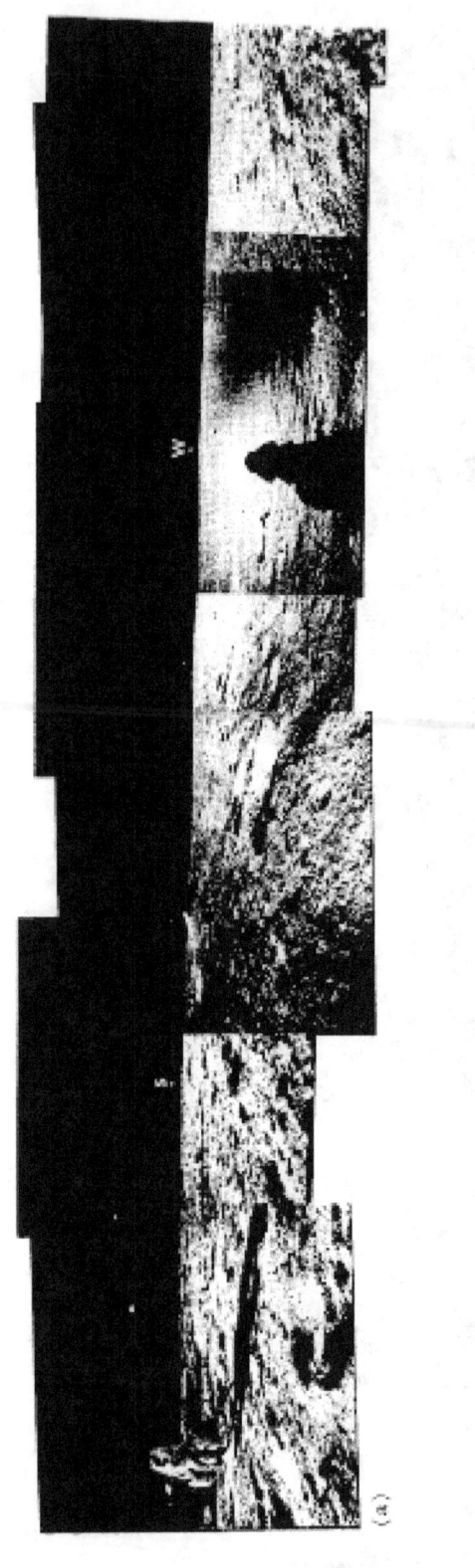
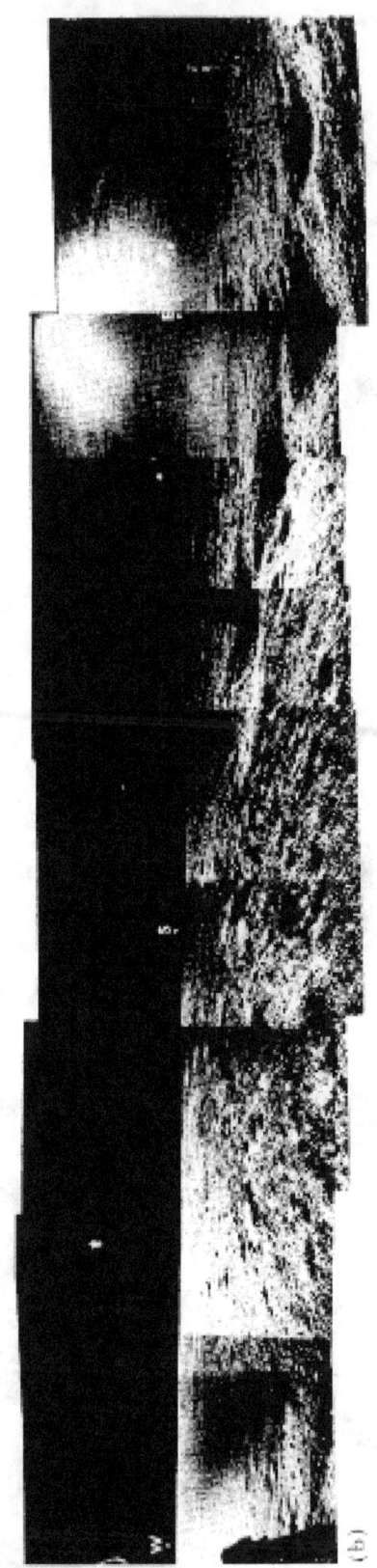

FIGURE 10-6.—Panorama 5: The ALSEP area prior to ALSEP deployment. (a) View looking south and west. (b) View looking south and east. (AS12-46-6796 to 6811)

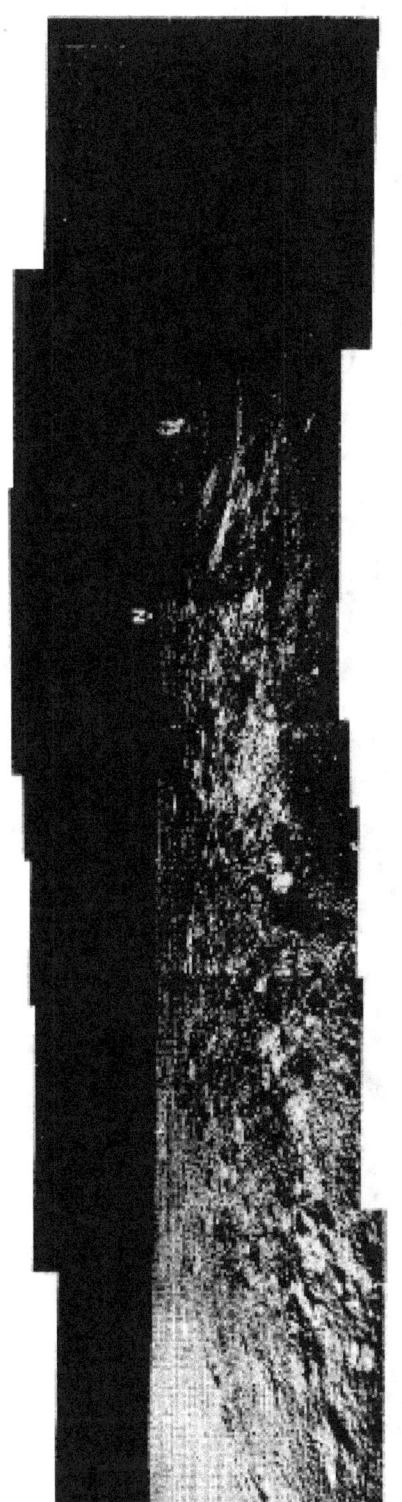
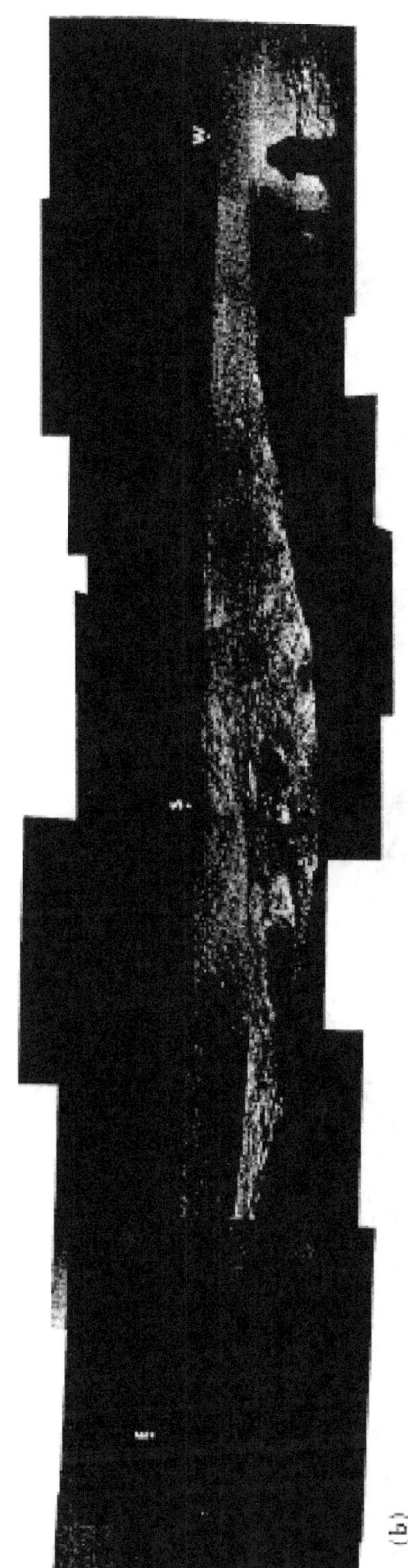

FIGURE 10.7.—Panorama 8: The landing site from 13 m west of the LM. (a) View looking north. (b) View looking east, south, and west. (AS12-47-6941 to 6980)

FIGURE 10-8.—Panorama 9: The landing site from 12 m east of the LM. (a) View looking north and east. (b) View looking south and west. (AS12-47-6961 to 6981)

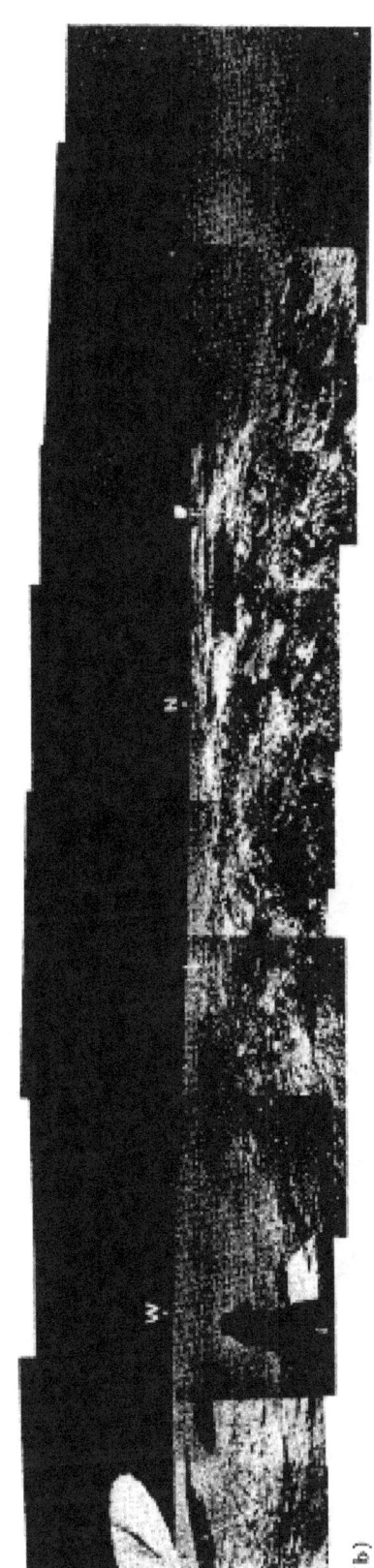

FIGURE 10-9.— Panorama 10: The landing site from 10 m northeast of the LM. (a) View looking south and west. (b) View looking west and north. (AS12-47-6952 to 7006)

FIGURE 10-10.—Panorama 16: The area to the northeast of Sharp Crater. (a) View looking south and west. (b) View looking west, north, and east. (AS12-49-7244 to 7262)

GEOLOGIC INVESTIGATION OF THE LANDING SITE

FIGURE 10-11. — Panorama 19: Area near Halo Crater and southwest of Surveyor Crater. (a) View looking south and west. (b) View looking west, north, and east. (AS12-49-7289 to 7311)

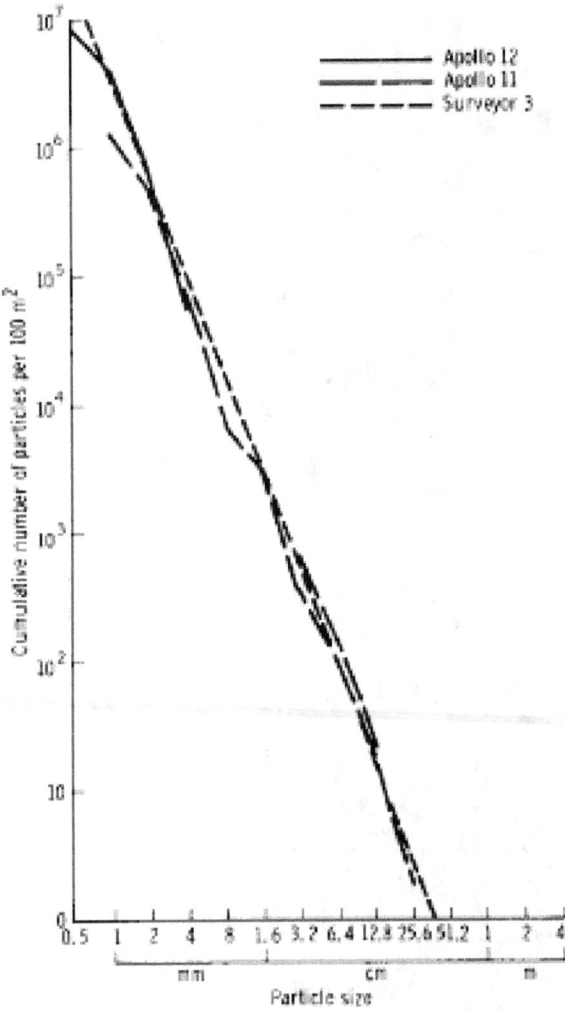

FIGURE 10-12.—Cumulative size-frequency distributions of surface particles at the Apollo 12 and 11 and Surveyor 3 sites. The dashed line is a power function fitted to the Surveyor 3 particle counts. The solid line and the dotted line are observed particle counts.

with the Apollo lunar surface closeup camera (ALSCC). Four hundred and ninety particles were counted and plotted on a cumulative size-frequency distribution diagram (fig. 10-12).

Linear Grooves

Much of the surface in the area of the geologic traverse made during the second EVA period is patterned by small, linear grooves. These grooves are visible on the returned photographs and were reported from several localities by the astronauts. They are similar in appearance to those visible on some of the Apollo 11 photographs (ref. 10-2).

The astronauts referred to the patterned ground as "trenches," "grooves," "lines," and "streaks." When referred to as "trenches," the grooves were estimated to be approximately 3 mm deep. The linear features were reported to trend generally north (north-northeast or northeast in Surveyor Crater) and were reported to occur in strips of patterned ground perhaps 30 m wide. During postmission debriefings of the crew, the strips of patterned ground were also reported to be north trending.

Examination of returned photographs shows an additional set of grooves that trend roughly west in the areas where the north-trending grooves are present (fig. 10-13). In addition to the grooves, north- and west-trending chains of small elongate depressions and small scarps are also present. At an azimuth of approximately 325° from the LM (fig. 10-13) is a nearly square crater, 4 or 5 m across, whose sides are parallel to the north- and west-trending grooves.

All of the linear features have a vertical relief generally less than 1 cm, are commonly approximately 2 cm wide, and are approximately 5 cm to 1 m long. A few of these features observed in the photographs are several meters long. The grooves, chains, and scarps cross small craters and other surface irregularities without apparent change in form or direction.

Similar linear features noted at the Apollo 11 site trend roughly northeast and northwest and have been interpreted as being caused by drainage of fine-grained material into fractures in the underlying bedrock (ref. 10-3). This would imply northeast- and northwest-trending joint sets in the bedrock of the Apollo 11 site and north- and east-trending joint sets in the Apollo 12 site bedrock. The lineated trips of ground reported by the crew probably reflect joint sets within larger fracture zones in the bedrock.

Craters

The Apollo 12 landing site contains a wide variety of craters; their characteristics can be seen in figure 10-1 and in the panoramas. The general pattern of small craters (from approximately 2 cm to several meters in diameter) is shown in the foreground of most of the pano-

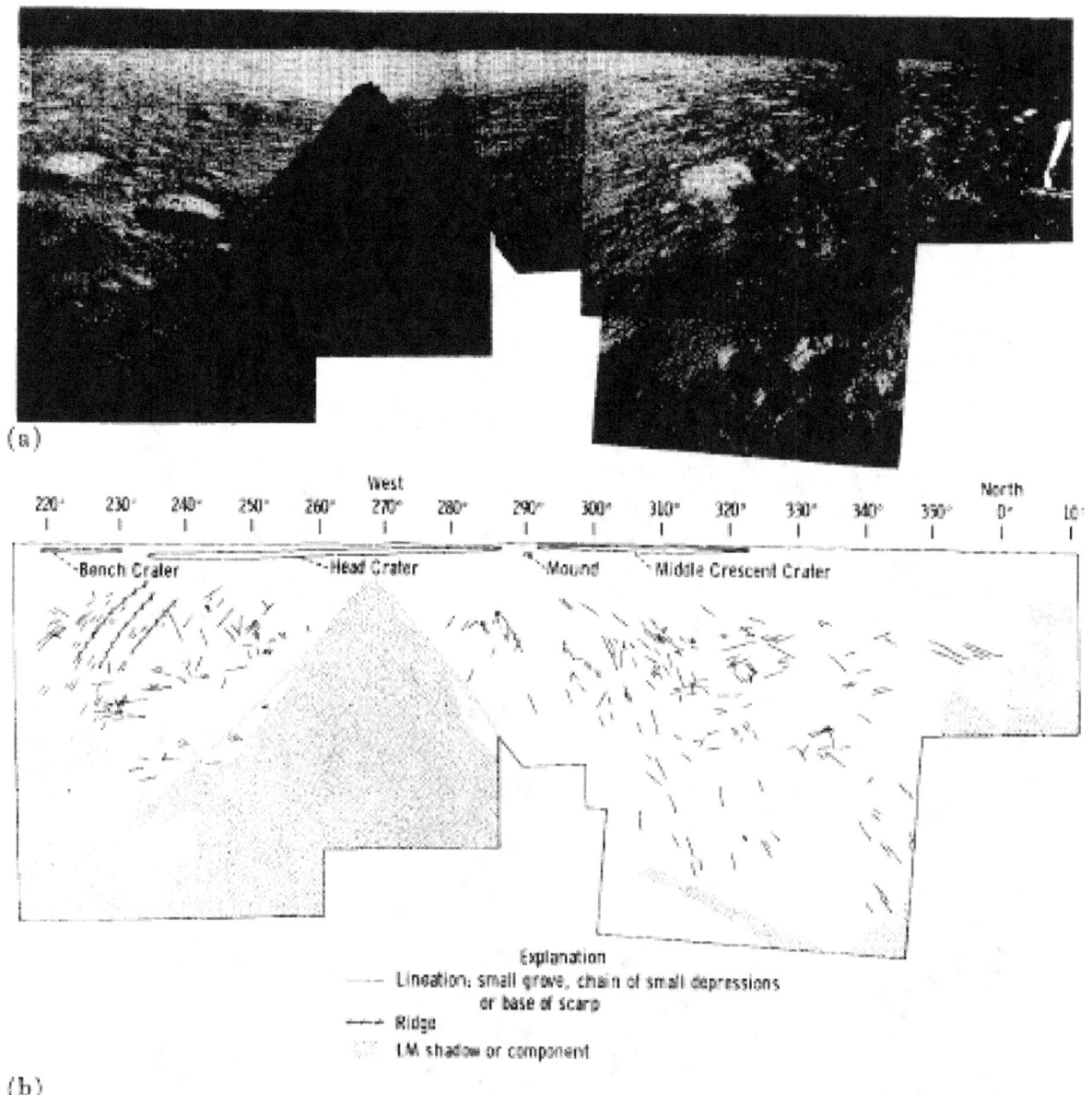

FIGURE 10-13. — LM window panorama. (a) View from the LM window. (b) Map showing distribution of linear features near LM.

ramas, but can be observed especially well in figure 10-2(a), which was taken at a low-Sun elevation angle.

The cross-sectional shapes of the craters range from very subdued, rimless depressions to very sharp, well-defined craters. Middle Crescent Crater, which can be seen in figure 10-14, is a large subdued depression with its blocky areas concentrated inside the rim. Smaller subdued depressions are shown in figure 10-2(a). These smaller depressions range in size from less than a meter across to the one approximately 25 m across that is centered near the tip of the LM shadow (fig. 10-2(a)).

Sharp, well-defined craters range from fresh craters less than a meter across and a few centimeters deep to craters approximately 13 m across and 3 m deep, such as Sharp and Block Craters.

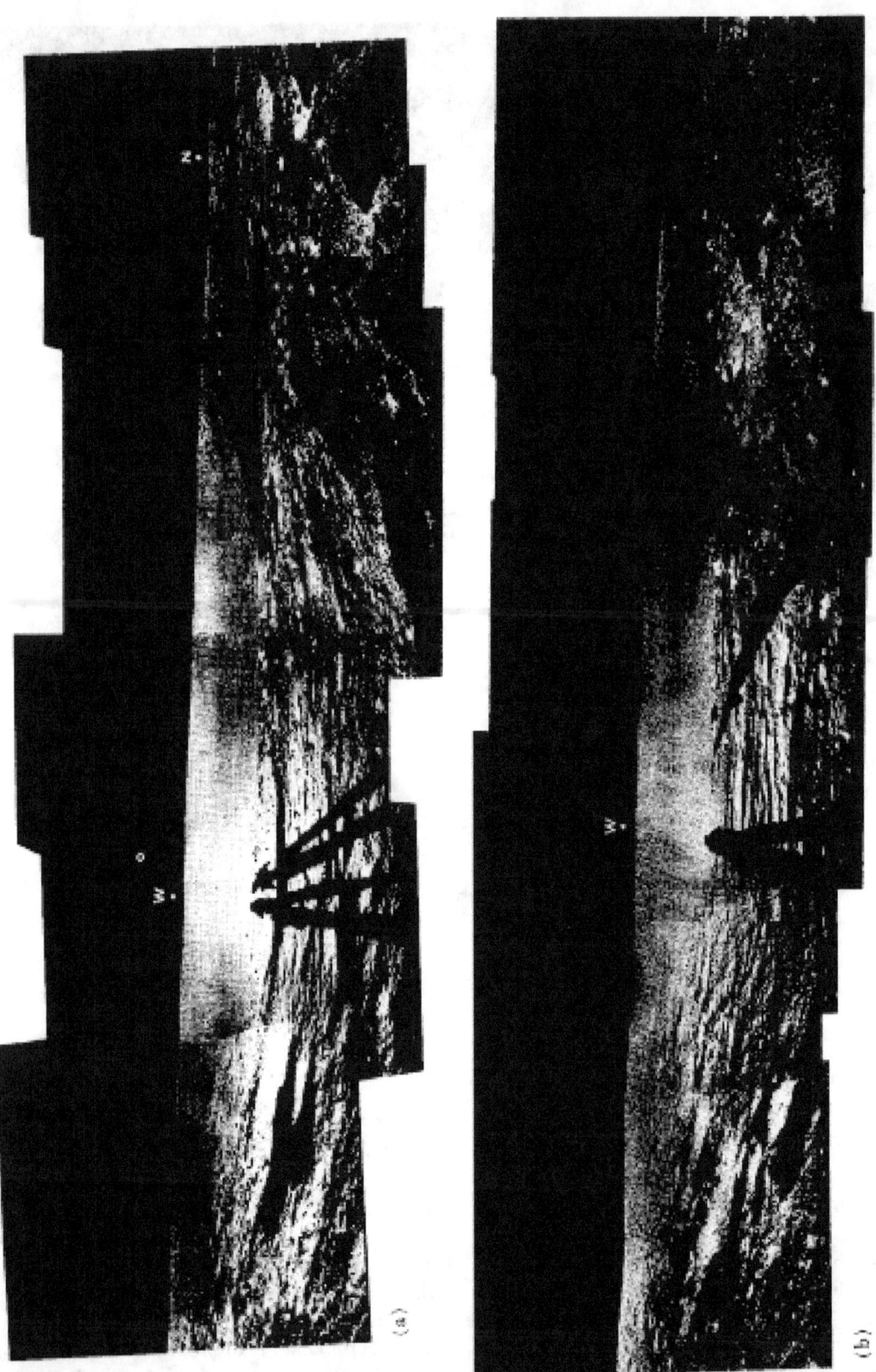

FIGURE 10-14.—Panoramas 6 and 7 (partial), showing Middle Crescent Crater. A stereoscopic pair of panoramas. (a) Right-hand member of the stereoscopic pair. (AS12-48-6835 to 6864) (b) Left-hand member of the stereoscopic pair. (AS12-46-6845 to 6852)

The small, fresh craters have rubbly rims, apparently comprising aggregates of fine-grained material and small rock fragments derived from the upper few centimeters of the regolith. An example of this type of crater is shown in figure 10–2(a) next to the LM shadow. Both Sharp and Block Craters are fresh, but differ in the distribution of their associated ejecta. Sharp Crater (fig. 10–15) has a rubbly bottom and inner walls, but its rim surface is similar to the general regolith surface in that the rim consists of fine-grained material with some scattered blocks. The freshness of Sharp Crater is suggested by its radiating pattern of high-albedo material described by the astronauts. In contrast to Sharp Crater, Block Crater (fig. 10–16) has a very blocky rim and ejecta blanket (fig. 10–17, background); the freshness of this crater is suggested by the large abundance of angular blocks.

Many craters are intermediate in shape—between the subdued depressions and the fresh, sharp craters. These intermediate-type craters range in size from several centimeters across (foreground of most panoramas) up to larger craters like Head (fig. 10–18), Surveyor (figs. 10–16 and 10–17), and Bench Crater (fig. 10–19). These craters are characterized by fairly smooth rims and bottoms. Bench Crater is characterized by a distinct bench high on its northeastern side. The bench, which may be a resistant layer within or under the regolith, is shown on figure 10–1 by the shadow pattern. A bench near the bottom of the crater, not clearly visible on figure 10–1, is shown in figure 10–19. This lower bench may be another resistant layer, or it may be the result of mass wasting of the crater walls.

The larger craters at the Apollo 12 landing site are probably widely different in age. The

(a)

(b)

FIGURE 10-15. — Panoramas 17 and 18, showing Sharp Crater. A stereoscopic pair of panoramas. (a) Right-hand member of the stereoscopic pair. (b) Left-hand member of the stereoscopic pair.

FIGURE 10-16. — Panoramas 21 and 22 (partial), showing Block Crater. A stereoscopic pair of panoramas. (a) Right-hand member of the stereoscopic pair. (AS12-48-7140 to 7143) (b) Left-hand member of the stereoscopic pair. (AS12-48-7144 to 7147)

age sequence from oldest to youngest is interpreted as follows:[1]

(1) Middle Crescent Crater
(2) Surveyor and Head Craters
(3) Bench Crater
(4) Sharp, Halo, and Block Craters

Rock fragments collected from the rims of these craters may be expected to have a wide range of exposure ages, which are related, in part, to the ages of the craters.

Northwest of the LM is the largest crater visited, the 400-m-diameter Middle Crescent Crater. On looking down into the crater, the astronauts noticed huge blocks on the crater wall, which were probably derived from the local bedrock. Large rock fragments in this crater probably have been exposed since the crater was formed and probably represent the deepest layers excavated at the Apollo 12 landing site.

Both rounded and angular blocks litter the

[1] From unpublished maps prepared for use on board the Apollo 12 spacecraft by P. J. Cannon and T. N. V. Karlstrom of the U.S. Geological Survey.

FIGURE 10-17.— Panorama 20 (partial), showing the area near the Surveyor 3 spacecraft. (AS12-48-7101 to 7106)

(a)

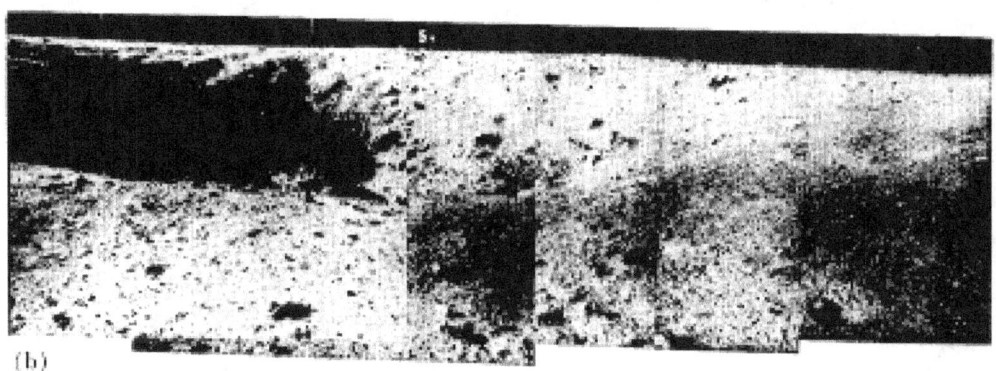

(b)

FIGURE 10-18.— Panoramas 14 and 15 (partial), showing Bench Crater. A stereoscopic pair of panoramas. (a) Left-hand member of the stereoscopic pair. (AS12-49-7223 to 7228) (b) Right-hand member of the stereoscopic pair. (AS12-49-7229 to 7233)

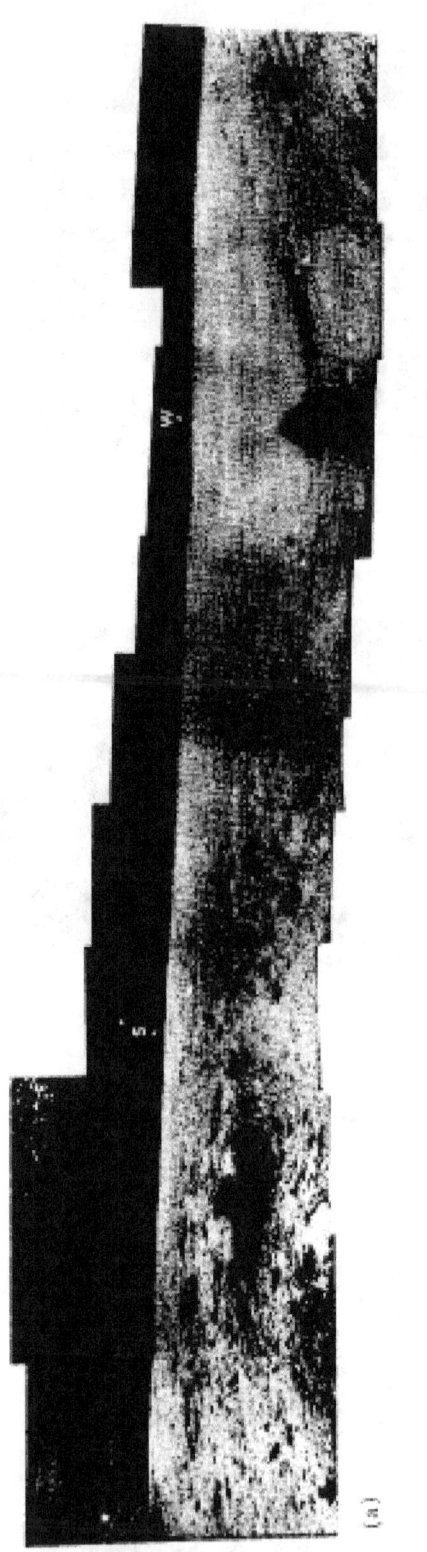 

FIGURE 10-19.—Panorama I3. Head Crater viewed from the west rim. (a) View looking south and west. (b) View looking north and east. (AS12-49-7201 to 7216)

surface of the rims of Head and Bench Craters. Some rocks appeared to be coarse grained; to the astronauts, the coarse-grained rock crystals were clearly visible. Many rocks on the rim of Bench Crater were reported to be splattered with glass.

Samples were collected from three small, very fresh, blocky-rimmed craters that apparently penetrate through the regolith into underlying materials. These craters are Sharp Crater, approximately 14 m across and 3 m deep; Block Crater, approximately 13 m across and 3 m deep; and an unnamed crater, 4 m across and approximately 1 m deep, that lies on the south rim of Surveyor Crater just north of Halo Crater.

Sharp Crater (fig. 10-15) has a rim 0.66 m high that is composed of material with high albedo. This material has been splashed out radially around the crater and is softer than the normal regolith. A core tube driven into the rim of the crater penetrated the ejecta without difficulty. Samples collected near the center may show the youngest exposure ages. Sharp Crater appears to have just barely penetrated the regolith. A terrace near the crater floor is probably controlled by the subregolith bedrock at a depth of approximately 3 m.

At Block Crater (fig. 10-16), high on the north wall of Surveyor Crater, nearly all the ejected blocks are sharply angular, which suggests that the crater is very young. Many of the blocks clearly show lines of vesicles similar in appearance to vesicular lavas on Earth. The blocks are probably derived from the older, coarse blocky ejecta deposit underlying the rim that resulted from the Surveyor Crater event. The regolith at Block Crater may be a meter or less thick.

The 2-m-diameter blocky crater on the southern rim crest of Surveyor Crater may have been excavated in the old rim deposit of Surveyor Crater at a depth of less than 0.5 m. In this blocky crater, the regolith may be very thin. It is also possible that some of the blocks in this small crater were derived from a low-velocity (secondary) impacting projectile.

Size-Frequency Distribution of Craters

The relative age of the Apollo 11 and 12 landing sites can be derived from the size-frequency distribution of craters in the landing areas. The underlying assumption for determining the age difference is that older surfaces have a greater cumulative number of larger craters. Figure 10-20 suggests that the Apollo 11 landing site is older than the Apollo 12 landing site because more craters greater than a few hundred meters in diameter were observed at the Apollo 11 site.

The size-frequency distribution of craters from several kilometers to a few centimeters in diameter was determined for the Apollo 12 landing site from Lunar Orbiter 3 and 4 photographs and from one Apollo 12 EVA photograph (NASA AS12-48-7075). The cumulative number of

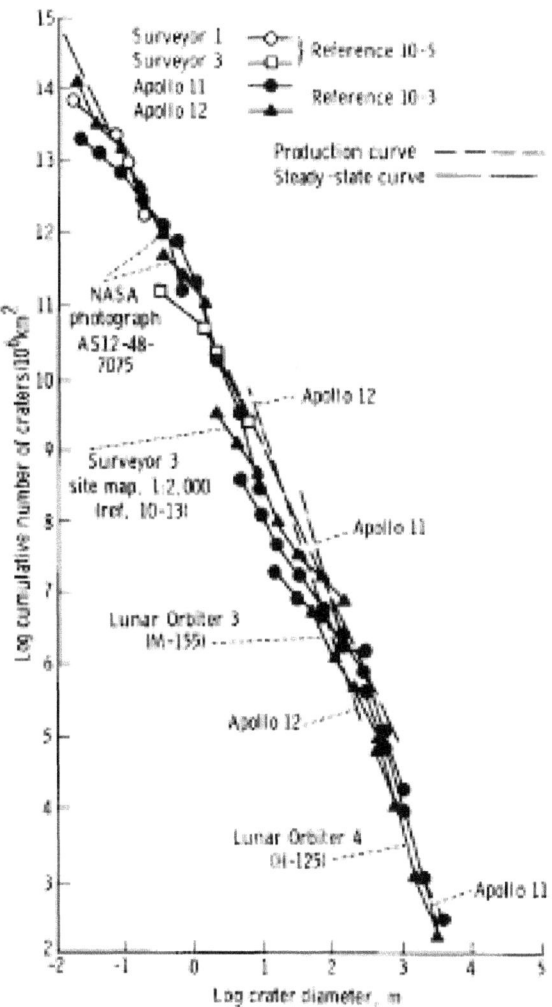

FIGURE 10-20. — Cumulative size-frequency distribution of craters on the lunar surface in the vicinity of Apollo 12 compared to crater distributions for Surveyor 1 and 3, and Apollo 11 sites.

craters in each counting area was normalized to an area of 10^6 km^2. The size-frequency distribution of craters at the Apollo 12 site is compared with the distribution at the Apollo 11 and Surveyor 1 and 3 sites in figure 10–20 and is compared with the distribution of small craters derived from Ranger 7, 8, and 9 data (ref. 10–4) in figure 10–21.

The distribution of craters less than 100 m in diameter roughly follows the function, $F = \Phi c^\mu$, where F is the cumulative number of craters with diameter equal to or higher than c per 10^6 km^2, c is any given crater diameter, Φ is a constant $(10^{10.9}/m^{-2})10^6$ km^2, and μ is a constant (-2.00). This function, determined initially from Ranger photographs (ref. 10–4), has been found to fit very closely the distribution of small craters at all Surveyor landing sites (ref. 10–5); it also closely fits the distribution of small craters at the Apollo 11 and 12 landing sites (fig. 10–21). The function $F = \Phi c^\mu$ is interpreted as the steady-state distribution of craters produced by repetitive bombardment of level surfaces on the Moon.

The distributions of craters larger than a few hundred meters in diameter in the vicinity of the Apollo 11 and 12 landing sites do not follow the steady-state distribution, but can be fitted closely by the power function $F = \chi c^\lambda$. For Apollo 11, χ is $(10^{12.9}/m^{-2.93}) 10^6$ km^2, and λ is -2.93. For Apollo 12, χ is approximately $(10^{12.3}/m^{-2.86}) 10^6$ km^2, and λ is approximately -2.86. The intersection of $F = \Phi c^\mu$ with $F = \chi c^\lambda$ (fig. 10–21) is the upper-limit crater diameter of the steady-state distribution in each area, designated c_s on the figure. On the basis of the observed size-frequency distributions, the Apollo 11 site is older than the Apollo 12 site. This relative age difference is shown by the ratio of the cumulative number of craters with diameters of 1 km. The ratio of the number of these size craters between the Apollo 11 and 12 landing sites averages 2.37:1 and ranges between 1.5:1 and 3.16:1.

Crystalline Rocks and Microbreccias

One of the notable differences between the collection of rocks obtained at the Apollo 12 landing site and the collection obtained at the Apollo 11 landing site (Tranquility Base) is the ratio of crystalline rocks to microbreccia. At the Apollo 12 site, the rocks collected are predominantly crystalline, whereas at Tranquility Base, approximately half the rocks collected were crystalline and half were microbreccia. This difference is probably attributable to the fact that the rocks collected at the Apollo 12 landing site were primarily on or near crater rims. On the crater rims, the regolith is thin or only weakly developed, and many of the rocks observed are probably derived from craters that have been excavated in bedrock that is well below the reg-

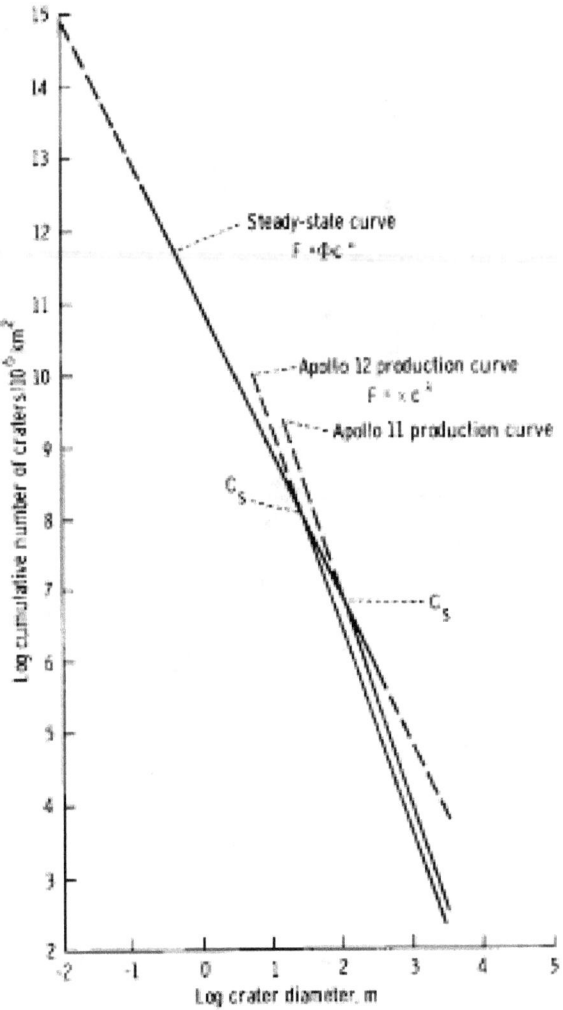

FIGURE 10-21.—Size-frequency distribution of large craters at the Apollo 11 and 12 landing sites compared with the steady-state size-frequency distribution of craters determined from Ranger 7, 8, and 9 photographs.

olith. Tranquility Base was on a thick, mature regolith, where many of the observed rock fragments were produced by shock lithification of regolith material and were ejected from craters too shallow to excavate bedrock.

Mounds

Two mounds (figs. 10-22 and 10-23) in the area north of Head Crater were noted and photographed by the crew. Both mounds are visible on the high-resolution Lunar Orbiter photographs and are located on figure 10-1. These mounds are probably clumps of regolith material that were slightly indurated by impact and ejected by the impact from one of the nearby craters, possibly from Head Crater. Bombardment by meteoritic material and by secondary impacts and, possibly, the effects of diurnal temperature changes have probably caused sloughing of the sides of the mounds, resulting in their present, rather smooth form.

Samples

Three types of sampling activities were conducted during the mission. The first was the collection of the contingency sample (early in the first EVA period) in the vicinity of the LM. The second was the collection of the selected sample (late in the first EVA period) after deployment of the Apollo lunar surface experiments package (ALSEP) in the vicinity of the mounds and Middle Crescent Crater. Also during this time, a core tube was driven near the LM. The third sampling activity was the

FIGURE 10-22. — Southeast view of large mound. Size and frequency of small particles on the mound seem to decrease toward the bottom of the mound, suggesting that the particles are moving slowly downslope. (AS12-46-6795)

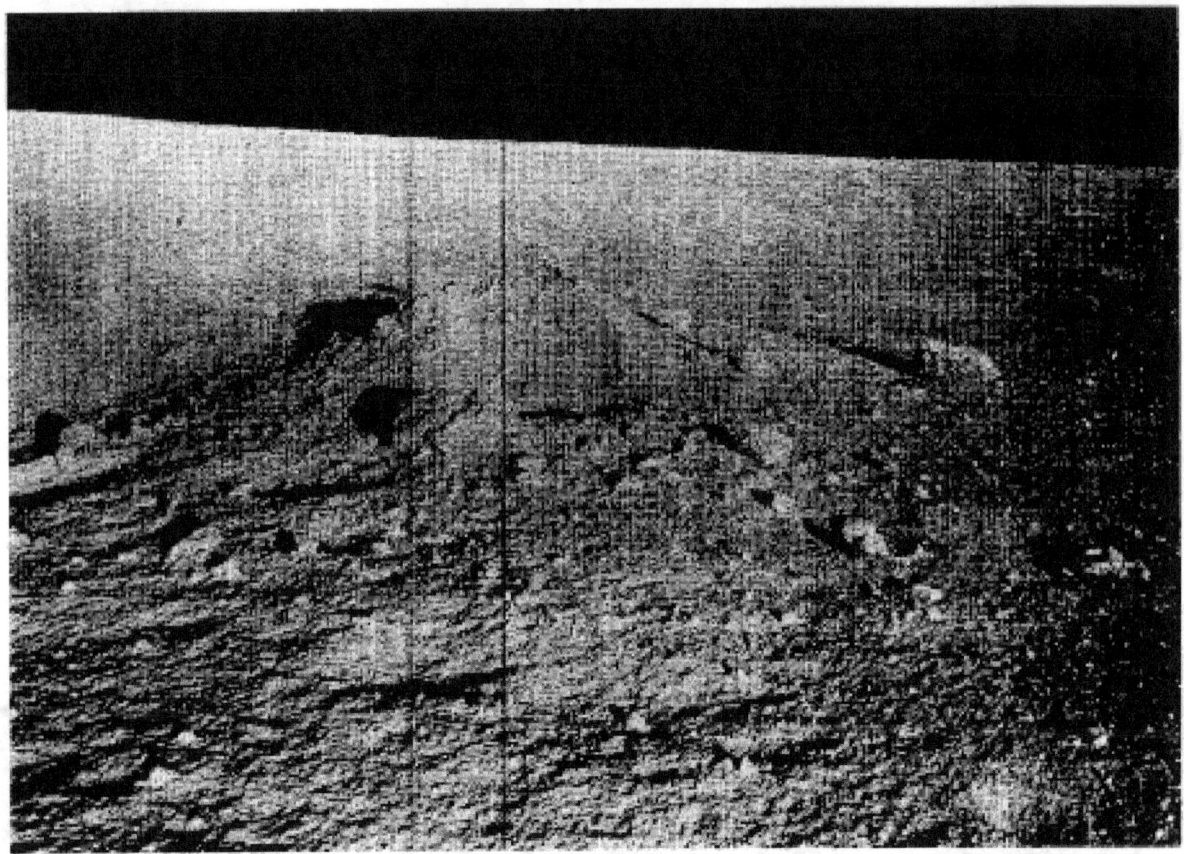

FIGURE 10-23. — Northwest view of small mound. The depression in the near side of the mound may have been caused by a secondary impact. (AS12-46-6823)

collection of the documented sample, which was collected along the traverse made during the second EVA period. In addition to a variety of rock and soil samples, the documented sample included one single-core tube, one double-core tube, the special environment sample, the gas analysis sample, and the totebag samples.

By studying the voice transcript and the returned surface photographs, a search has been made to determine the location of each specimen collected. To date, 19 samples have been located, and 13 have been tentatively located. In addition, the orientations have been determined for five samples and have been tentatively determined for seven samples.

Table 10-I is a cross-reference of all samples taken in the first and second EVA periods; tentative locations and lunar surface photographs that show the samples in the sequence in which they were collected are keyed to ground elapsed time in the table. Tentative identification of rocks is based on a combination of the astronauts' descriptions and correlation of rock characteristics as seen in photographs taken on the lunar surface and in the Lunar Receiving Laboratory (LRL). Samples indicated in the left column of the table as "?(rock)" have not been identified. Samples with an LRL sample number followed by a question mark indicate an uncertainty of the sample number, but the reference to the actual sample is strongly suggested. Further study of the surface photographs may permit additional identification and orientation of samples.

Contingency Sample

The contingency sample was taken in full view of the sequence camera on and near the southeast rim of a 6-m-diameter crater approximately 15 m northwest of the LM (figs. 10-13 and 10-24). The sample was collected in six dis-

GEOLOGIC INVESTIGATION OF THE LANDING SITE 137

TABLE 10-I. *Cross-reference table for lunar samples, photographs, and transcript*

(a) Contingency sample

LRL sample no.[a]	Comments	Location	Lunar surface photographs	Ground elapsed time, day hr:min	Confidence in identification and orientation
12073	Largest rock of contingency sample, labeled "C" in figures 10-47 and 10-48.	Rim of small crater 15 m northwest of LM.	AS12-46-6719 (before/during collection) to AS12-46-6723 (after collection), sequence camera photographs. (See figs. 10-24, 10-26, 10-27, 10-29, and 10-31.)	4:19:26 to 4:19:28	Identification: high Orientation: moderate
12075	Olivine basalt, labeled "A" in figure 10-47.	Rim of small crater 15 m northwest of LM.	AS12-46-6719 (before/during collection) to AS12-46-6723 (after collection), sequence camera photographs. (See figs. 10-24, 10-25, 10-28, and 10-30.)	4:19:38 (in the bag)	Identification: high Orientation: moderate
12070 to 12077	Total contingency sample.	Rim of small crater 15 m northwest of LM.	AS12-46-6719 (before/during collection) to AS12-46-6723 (after collection), sequence camera photographs.	4:19:46	—

(b) Selected sample

LRL sample no.[a]	Comments	Location	Lunar surface photographs	Ground elapsed time, day hr:min	Confidence in identification and orientation
(? rock)	"Here's a rock they'll be glad to see in Houston.... it's rather soft." (LMP)	Near small mound, after ALSEP deployment.	Photographs of small mound and surrounding area, AS12-46-6822 to AS12-46-6825, sample not recognized.	4:22:02	—
12017? (rock)	"... all that glass spatter on it." (CDR)	Near small mound.	Photographs of small mound and surrounding area, AS12-46-6822 to AS12-46-6825, definitely recognized.	4:22:06	Identification: moderate (from description only) Orientation: —
(? rock)	"Here's another big rock." (CDR)	On the way from the small mound to the large mound.	No photograph.	4:22:06	—
(? rock)	"Okay. Grab her up, Pete." (LMP)	Near large mound.	Photographs of large mound, AS12-46-6827 and AS12-46-6828; sample (?) in AS12-46-6829 and AS12-46-6830; sample not recognized.	4:22:08	—
12063? (rock)	"Here's a black rock." (CDR) Sample 12063 is notably dark gray or black and may be mostly ilmenite.	From south side of large mound.	Very tentatively identified in AS12-46-6831 and AS12-46-6832.	4:22:09	Identification: moderate (from description) Orientation: —
12021? (rock)	"Let's get another from in there." (LMP)	From near base (?) of east side of large mound.	Tentatively identified in AS12-47-6832. (See figs. 10-34 and 10-35.)	4:22:09	Identification: moderately high Orientation: moderately high

Footnotes at end of table.

138 APOLLO 12 PRELIMINARY SCIENCE REPORT

TABLE 10-I. *Cross-reference table for lunar samples, photographs, and transcript* – Continued

(b) Selected sample—Continued

LRL sample no.[a]	Comments[b]	Location	Lunar surface photographs	Ground elapsed time, day/hr/min	Confidence in identification and orientation
12022 (rock)	"I'm trying to knock a piece of that off ... I wouldn't be surprised if this was microbreccia." (LMP)	From northeast side of large mound.	Tentatively identified in AS12-47-6933. (See figs. 10-36 and 10-37.)	4:22:10	Identification: moderately high Orientation: moderately high
(? rock)	No description.	Near top center of large mound?	Photographs of large mound; view looking south, AS12-47-6934 (showing top) and AS12-47-6935; sample not recognized.	—	—
12014? (rock)	"There's a rock for you." (CDR)	From rim of "fresh-looking" crater approximately 61 m from rim of Middle Crescent Crater.	AS12-46-6833, view looking south; crater shadow in photograph; sample very tentatively identified.	4:22:17	Identification: low (from description) Orientation: —
12015? (rock)	"I made a dent on this rock." (CDR)	Near southeast rim of Middle Crescent Crater.	AS12-46-6834, view looking north; sample not identified in photographs.	4:22:17	Identification: low (from description) Orientation: —
12006 or 12012? (rock)	"Wow, look at that rock. I'd like to —." (CDR)	On southeast rim of Middle Crescent Crater, just before taking photographic panorama.	AS12-46-6835, view looking south; sample very tentatively identified.	4:22:17	Identification: low Orientation: low
12004 (rock)	"I was just looking over this rock down here; it looks like it came —." (LMP)	Near panorama site at Middle Crescent Crater.	Tentatively identified in AS12-47-6836 and AS12-47-6837, view looking northeast. (See figs. 10-32 and 10-33.)	4:22:19	Identification: moderately high Orientation: moderately high
12016? (rock)	"... can't pick it up with the tongs." (CDR) Sample 12016 is the largest rock in the selected sample.	Near panorama site at Middle Crescent Crater.	Sample not identified, but may be in AS12-47-6838.	4:22:20	Identification: low (from description only) Orientation: —
(? rock)	"Try that one." (CDR) "That's a good one." (LMP)	Southeast of rim of Middle Crescent Crater.	Sample not identified, but may be in AS12-47-6939.	4:22:21	—

GEOLOGIC INVESTIGATION OF THE LANDING SITE 139

Sample	Quote	Location	Photograph	Time
(? rock)	"Let's get this real good one." (LMP)	Southeast of rim of Middle Crescent Crater.	Sample not identified, but may be in AS12-47-6939.	4:22:21
(? rock)	"I'm getting some up here." (LMP)	On the way back to ALSEP from Middle Crescent Crater.	No photograph.	4:22:23
(? rock)	"All right. Here's one right here." (CDR) "Okay. Let me get a photograph of it." (LMP)	Near ALSEP on return from Middle Crescent Crater.	AS12-47-6940, view looking south; sample not identified.	4:22:23
(? rock)	"There's a good one... step in and take the picture." (LMP)	Approximately 91 m northwest of LM on return from Middle Crescent Crater.	Apparently not photographed; LMP said, "Forgot the picture."	4:22:24
(? rock)	"There's a good rock.... never saw one like that before." (CDR) Green color mentioned.	Approximately 91 m northwest of LM on return from Middle Crescent Crater.	No photograph.	4:22:25
(? rock)	"They're a little different. They're more the gabbro type."	Within 91 m northwest of LM on return from Middle Crescent Crater.	No photograph.	4:22:25
Glass beads (with sample 12001 or 12003)	"Look at that, a pure bead of glass." (CDR) "They appeared to be black or green and approximately ⅜ in. in diameter to the astronauts.	Nearing LM from northwest on return from Middle Crescent Crater.	No photograph.	4:22:26 - 4:22:27
12026 (core tube serial no. 2013)	"... now it's full length, and let me take a picture of it and that will be it." (LMP)	Near LM, but precise location not determined from photographs.	AS12-47-7098 and AS12-47-7099, view looking north.	4:22:36
12001 and 12003 (soil)	"Dump some dirt in that bag..." (LMP)	Near LM at end of first EVA period.	No photograph.	4:22:44

140   APOLLO 12 PRELIMINARY SCIENCE REPORT

TABLE 10-I. *Cross-reference table for lunar samples, photographs, and transcript* — Continued

(e) Documented sample

LRL sample no.[a]	Comments[b]	Location	Lunar surface photographs	Ground elapsed time, day:hr:min	Confidence in identification and orientation
12065? (rock in tote bag)	"Man, have I got the grapefruit rock of all grapefruit rocks. It's got to come home in the spacecraft; it'll never fit in the rock box." (CDR)	Between ALSEP and north rim of Head Crater according to CDR; possibly picked up near Surveyor 3.	No photograph.	5:12:04	—
12030 (bag 1-D, fines)	Glass-covered fragments from 3-ft-diameter crater. "I'm picking them up with the tongs, but ... they don't seem to hold together too well ...". (LMP)	Between LM and north rim of Head Crater according to LMP; location not precise.	AS12-48-7043 and AS12-48-7044 (before collection) and AS12-48-7045 (after collection).	5:12:13	—
(No bag 2-D) 12031 (bag 3-D, rock)	"... This rock is very typical of all the fragments around here." (LMP)	Trench site approximately 15 m inside north rim of Head Crater.	AS12-48-7048, AS12-49-7189, and AS12-49-7190 (before collection), and AS12-48-7050 (after collection); sample 12031 identified and oriented in surface photographs. (See figs. 10-38 and 10-39.)	5:12:20	Identification: high Orientation: high
12032? (bag 4-D, fines with rock fragments)	(Apparently taken after bag 6-D; see entry at 5:12:40.)	Note: Sample bag number is apparently out of sequence.	—	5:12:40	—
12033 (bag 5-D, fines)	"... dig as deep as you can, then give me a sample right out of the bottom .... I'll put it in sample bag number 5-D." (LMP) Lighter gray color.	Trench, 15 cm deep; approximately 15 m inside north rim of Head Crater.	Trench site photographs: AS12-49-7191 to AS12-49-7196, AS12-48-7049, AS12-48-7051, and AS12-48-7052.	5:12:25	—
12034 (bag 6-D, rock)	"Let's sample that rock that I dug up from down deep. Let me get a picture of it first." (CDR)	Trench at Head Crater; rock came from depth of approximately 15 cm.	AS12-49-7195 and AS12-49-7196; sample identified in surface photographs.	5:12:26	Identification: high Orientation: — (buried)

GEOLOGIC INVESTIGATION OF THE LANDING SITE 141

Sample	Description	Location	Notes	Time	Identification/Orientation
12053? (rock)	"This rock is about 6 in. in diameter. The bottom part is gray, about half of it . . ." (LMP) "All right, let me have that." (CDR) Two different rocks were photographed. Description by crew suggests that disturbed rock was picked up.	A short distance west of trench site at Head Crater; on the way to triple craters.	AS12-49-7197, AS12-49-7198 (rock A, standing on end beneath gnomon in figs. 10-47 and 10-48 is sample 12053), AS12-49-7199, AS12-49-7200, and AS12-48-7053 to AS12-48-7055 show a disturbed rock with gray bottom is not the same rock as the one beneath gnomon. (See figs. 10-47 to 10-49.)	5:12:30	Identification: moderate Orientation: low
12052 (rock)	LMP described the rock as partly rounded with dirt adhering to it and typical of others nearby.	At panorama site west on rim of Head Crater.	AS12-48-7059, AS12-49-7217, and AS12-49-7218. Sample was identified in photographs; however, it was apparently rolled into position before photography. (See figs. 10-43 and 10-44.)	5:12:34	Identification: high Orientation: high (original orientation not known)
12053? (rock)	"It won't fit in there . . . the rock's too big . . ." (LMP) ". . . put it in here and we've got a nice picture of it . . ." (CDR)	Northwest rim of Bench Crater.	AS12-48-7063, AS12-48-7064, AS12-49-7234, and AS12-49-7235. Sample was identified in photographs; however, it was tilted slightly before photography.	5:12:40	Identification: high Orientation: high (original orientation not known)
12032? (bag 4-D, fines with rock fragments)	". . . a couple of small rocks . . . I don't think they appeared in the photo, but . . . they're] typical of other rocks around here." (CDR)	Northwest rim of Bench Crater.	No photograph.	5:12:40	—
12035? (bag 7-D, rock fragment)	"Pete's picking up a small piece of this [fractured] rock (LMP). I'm trying to get a piece that's fractured right off the middle." (CDR)	Northwest rim of Bench Crater.	AS12-48-7064 and AS12-49-7236 to AS12-49-7230; sample not identified in photographs.	5:12:43	— (from description only)
12036 (bag 8-D, rock) and 12037 (bag 8-D, fines)	A rock fragment from the fractured rock is included in the bag containing soil collected from beside the fractured rock.	Northwest rim of Bench Crater	AS12-49-7236 to AS12-49-7239; sample not identified in photographs.	5:12:44	Rock: — (from description only) Fines: —

TABLE 10-I. *Cross-reference table for lunar samples, photographs, and transcript*—Continued

(c) *Documented sample*—Continued

LRL sample no.	Comments	Location	Lunar surface photographs	Ground elapsed time, day:hr:min	Confidence in identification and orientation
12038 (bag 9-D, rock)	"...that big piece right there... It's got spattered glass or something all over it." (CDR) This comment could refer to sample 12063 from the totebag.	West rim of Bench Crater.	AS12-49-7240 and AS12-49-7241; sample not identified in photographs.	5:12:48	—
12039 and 12040 (bag 10-D, rocks)	"Okay. That's a good rock, and that one fills that one up." (LMP) This comment could refer to sample 12038, which alone filled bag 9-D.	West rim of Bench Crater.	AS12-49-7240 and AS12-49-7241; samples not identified in photographs. Also, AS12-49-7242 and AS12-49-7243; samples beside tool carrier(?).	5:12:49	— (from description only)
12023 (LESC*)	"Fill the big container with dirt... this dirt came from about 8 in. down." (CDR)	Trench on east rim of Sharp Crater.	Trench site at Sharp Crater; AS12-49-7276, AS12-49-7277, AS12-48-7057, and AS12-49-7278. (Photograph shows LMP holding container.)	5:13:02	—
12027 (core tube serial no. 2011)	"We're driving it all the way in pretty easy." (LMP)	Bottom of 8-in. trench on east rim of Sharp Crater.	AS12-48-7068, AS12-48-7069, AS12-49-7279, and AS12-49-7280 (core tube in trench); and AS12-48-7070 (trench after core tube collection).	5:13:04	—
12024 (GASC*)	"We need some little rock fragments for the gas analysis sample." (LMP) Included fines with rock fragments, glass, and "shiny" fragments.	Near east rim of Sharp Crater.	No photograph.	5:13:08	—
12041 (bag 11-D, fines)	"There's a beautiful round ⅜-in. glass ball we must have..." (CDR)	East of Bench Crater en route to Halo Crater. Apparently near a small sharp crater: "Watch that crater behind you. Don't step back." (CDR) Precise location not known.	No photograph.	5:13:14	—

GEOLOGIC INVESTIGATION OF THE LANDING SITE 143

12042 (bag 12-D, fines)	"Ok, we'll take a couple of dirt samples . . ." (LMP)	In area of "wrinkled texture," approximately 20 m northwest of Halo Crater; precise location not known.	AS12-48-7074 to AS12-48-7076 and AS12-49-7282 to AS12-49-7284.	5:13:19	—
12025 (core tube serial no. 2010) and 12028 (core tube serial no. 2012) (double-core tube)	"We've got a double. Now, the question is can we pull it out?" (LMP)	North rim of a 10-m-diameter crater approximately 25 m south of Halo Crater.	AS12-49-7285 to AS12-49-7289 and AS12-48-7077.	5:13:28 (capped at 5:13:36)	—
Sample not returned (bag 13-D)	A rock sample with dirt around it.	Near site of panorama 19, south of Halo Crater.	AS12-49-7312.	5:13:39	—
12054 (rock)	"It has a big glass splotch on it." (CDR)	South rim of Surveyor Crater.	AS12-49-7313 and AS12-49-7314; sample identified and oriented in photographs. (See figs. 10-45 and 10-46.)	5:13:41	Identification: high Orientation: high
12051 (rock)	"Look at the shear face on that rock. Something whistled by it or something." (CDR)	Rim of small, sharp crater on south rim of Surveyor Crater.	AS12-49-7318 and AS12-49-7319 (before collection); AS12-49-7320 (after collection); sample identified and oriented in photographs. (See figs. 10-40 to 10-42.)	5:13:45	Identification: high Orientation: high
12043 (bag 14-D, rock); 12044 (bag 14-D, fines and glass dumbbell)	"Al, grab a shot of that bead of glass there and we'll bag it . . . along with that good-looking rock." (CDR)	South rim of Surveyor Crater; before starting onto crater to Surveyor 3.	AS12-48-7082 and AS12-48-7083; glass identified and rock tentatively identified in photographs.	5:13:51	Rock — Identification: low Orientation: low Class — Identification: high Orientation: high
12060 and 12061 (chips in tote-bag fines)	"We have an extra sample for you. The scoop has dirt in it." (LMP)	Surveyor 3 scoop.	Photographs of Surveyor 3.	5:14:29	—

144   APOLLO 12 PRELIMINARY SCIENCE REPORT

TABLE 10-I. *Cross-reference table for lunar samples, photographs, and transcript* — Continued

(c) Documented sample—Continued

LRL sample no.[a]	Comments[b]	Location	Lunar surface photographs	Ground elapsed time, day:hr:min	Confidence in identification and orientation
120565? (rock)	"Get us a platy one? Where's the one with lines in it?" (CDR)	Surveyor 3 site.	Not identified in photograph of Surveyor 3 area.	5:14:31	— (from description only)
12063? (rock in totebag)	"Let me get this in the bag, too." (CDR) Rock sample 12063 may have been picked up at Bench Crater, along with samples 12038, 12039, and 12040.	Surveyor 3 site.	Not identified in photograph of Surveyor 3 area.	5:13:32	— (from description only)
12062? (rock from totebag)	"...these rocks...all have fillets of soil around them...shall we grab this one right here?" (CDR) Sample taken.	Surveyor 3 site.	Very tentatively identified in AS12-48-7139. (Photograph shows rock with fillet.)	5:13:32	Identification: low Orientation: low
12064? (rock from totebag)	"Right here's the one — the square one, Pete." (LMP)	Surveyor 3 site.	Not identified in photographs.	5:13:33	Identification: moderate (from description only) Orientation: —
12045, 12046, and 12047 (bag 15-D, rocks)	"...put two or three rocks in here...I'll photograph them, and we can see what you took." (LMP)	North rim of Block Crater.	AS12-48-7148 to AS12-48-7150; samples not identified in photographs.	5:13:43	— (from description only)

[a] Question marks indicate uncertainty of sample number.
[b] Quotation marks signify astronaut comments from the transcript; otherwise, these are author comments. Commander is abbreviated CDR; lunar module pilot is LMP.
[c] Lunar environment sample container.
[d] Gas analysis sample container.

FIGURE 10-24.—Position of the six contingency scoop motions. Scoops 1 and 6 included LRL sample no. 12075 and sample no. 12073, respectively. Larger crater at top of photograph is approximately 6 m in diameter. Enlargement of photograph taken before first EVA period from right LM window. (AS12-48-7031)

tinct scoop motions and consisted of 1.9 kg of selected rock fragments (three greater than 5 cm in longest dimension), fine-grained material, and at least one glass bead. Locations of areas scooped are identified in sequence-camera photographs taken from the LM windows prior to the first EVA period, and all six sample scoop marks are documented on surface photographs (NASA AS12-46-6719 to AS12-46-6723) taken by Astronaut Conrad. Scoops 1 and 6 included rock fragments that were visible on the lunar surface from the LM windows prior to sampling; these fragments are suggested to be samples 12075 and 12073, respectively.

Samples 12075 and 12073 were identified from the 16-mm sequence-camera film in the series of before-and-after photographs shown in figures 10-25 to 10-27. Details of the scoop marks are shown in figures 10-28 and 10-29. Samples 12075 and 12073, which appear in Hasselblad photograph AS12-48-7031 (fig. 10-24), were collected from a group of small rocks alined roughly northeast between two small craters located 15 m northwest of the LM. Most of the rocks seem to have fillets banked against them. Many small craters that dot the surface near the line of rocks may have contributed to the fillets; it is not certain whether any of the rocks were the projectiles that caused the craters. The tentatively suggested orientations of samples 12075 and 12073 are shown in figures 10-30 and 10-31, respectively.

Selected Sample

The selected sample was collected in 1¾ hr (late in the first EVA period) in the area northwest of the LM (fig. 10-1). The sample consisted of 14.8 kg of selected rock fragments and fine-grained material, including glass beads. The sample was collected primarily in the vicinity of Middle Crescent Crater, on and near the two mounds located approximately 120 m and 160 m northwest of the LM, and in the vicinity of the ALSEP. One core tube, included in the previously mentioned weight, was driven near the LM.

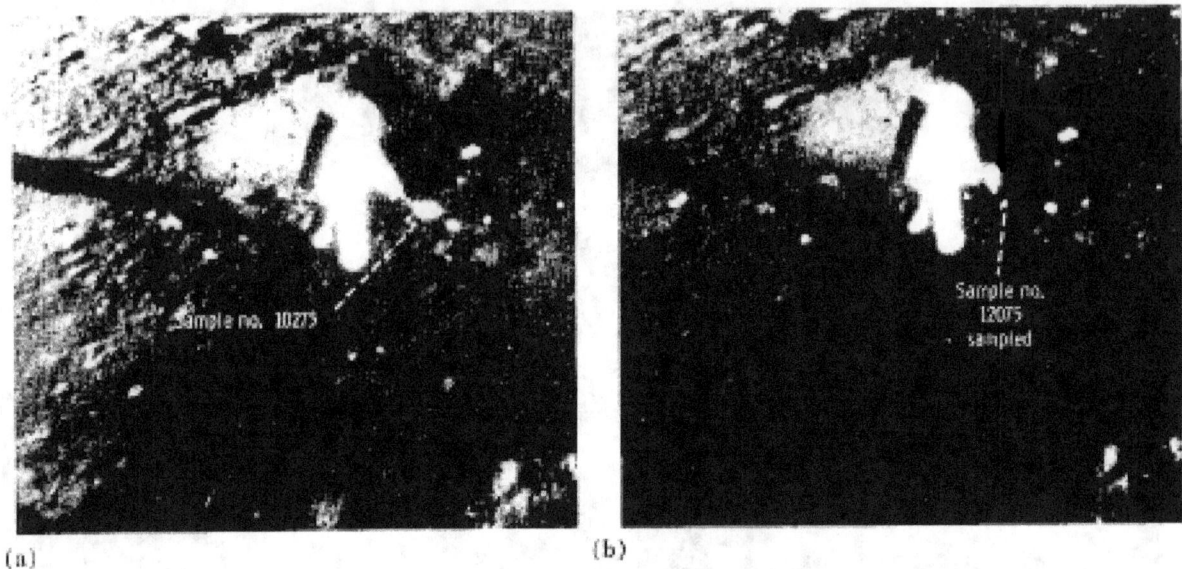

FIGURE 10-25. — Scoop 1 of the contingency sample collection. (a) Astronaut Conrad immediately prior to scoop 1 of the contingency sample collection, which included sample no. 12075. (Enlargement of 16-mm sequence-camera film.) (b) Astronaut Conrad immediately after collecting sample no. 12075 during scoop 1 of the contingency sample collection. (Enlargement of 16-mm sequence-camera film.)

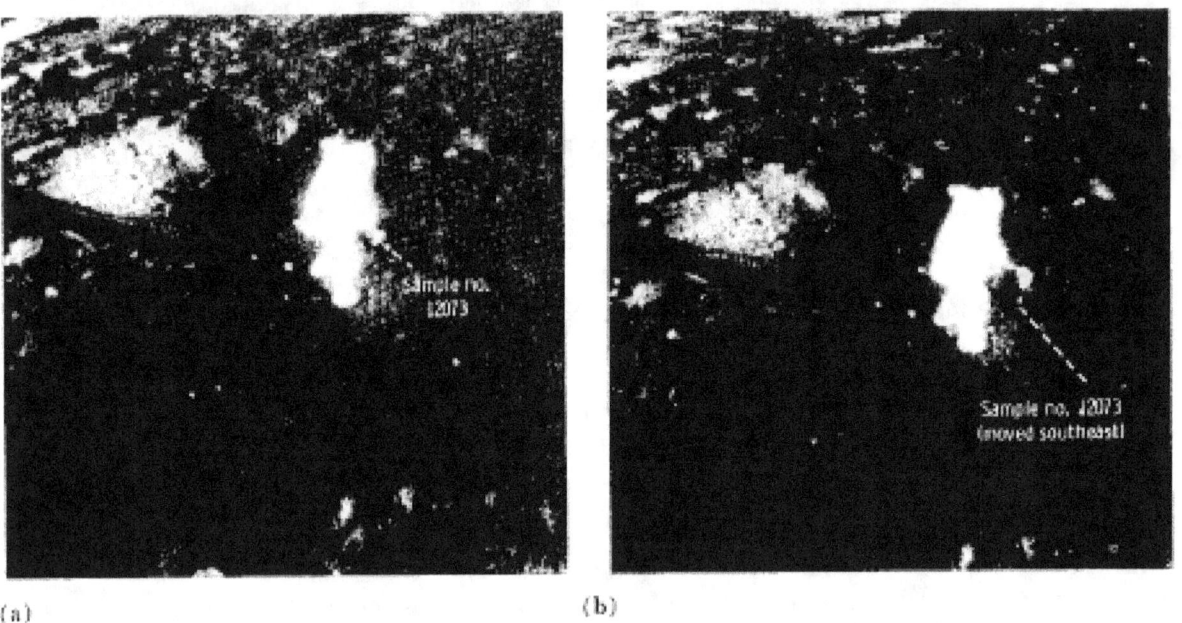

FIGURE 10-26. — Scoop 5 of the contingency sample collection. (a) Astronaut Conrad immediately prior to scoop 5 of the contingency sample collection, in which he was attempting to collect sample no. 12073. (Enlargement of 16-mm sequence-camera film.) (b) Astronaut Conrad immediately after scoop 5 of the contingency sample collection; sample no. 12073 has been moved to the southeast. (Enlargement of 16-mm sequence-camera film.)

(a)

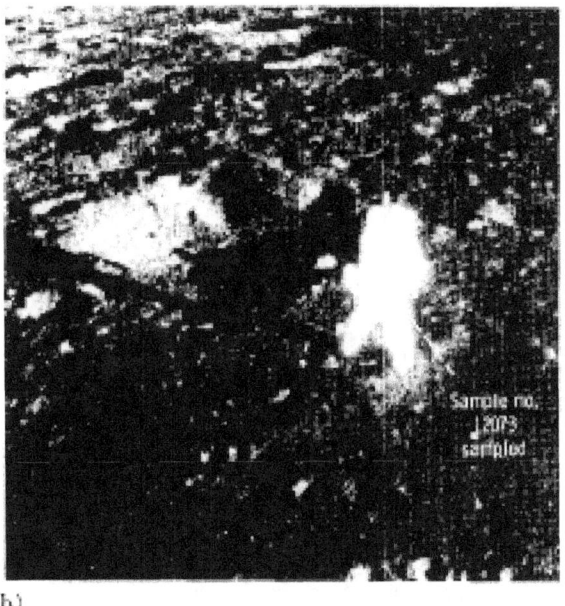

(b)

FIGURE 10-27. — Scoop 6 of the contingency sample collection. (a) Astronaut Conrad immediately prior to scoop 6 of the contingency sample collection. Note sample no. 12073 directly below sample scoop bag. (Enlargement of 16-mm sequence-camera film.) (b) Astronaut Conrad immediately after scoop 6 of the contingency sample collection, which included sample no. 12073. (Enlargement of 16-mm sequence-camera film.)

FIGURE 10-28. — Area of scoops 1 to 3 of the contingency sample collection. Arrows indicate direction of scoop motions. Sample no. 12075 was included in scoop 1 as indicated by the X-mark. (AS12-47-6719)

FIGURE 10-29. — Area of scoops 4 to 6 of the contingency sample collection. Scoop 5 attempted to collect sample no. 12073, but the scoop motion moved the sample to the southeast (indicated by X-marks and arrow). Scoop 6 was successful in collecting sample no. 12073. Other arrows indicate direction of scoop motions. (AS12-47-6722)

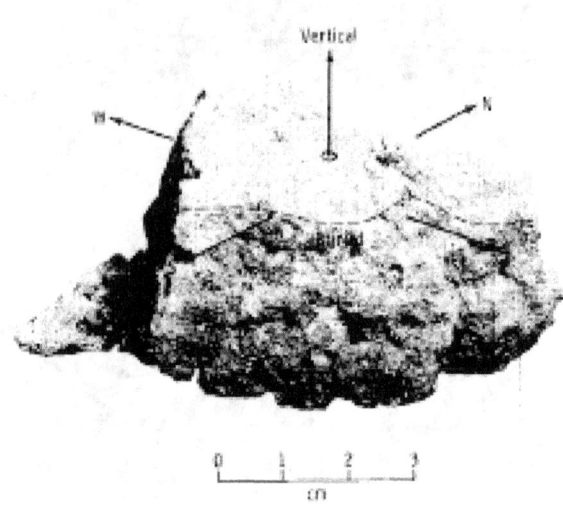

FIGURE 10-30.—Sample no. 12075, showing suggested selenographic orientation. Note excellent burial mark on the southeast side. Well-developed fillet, mentioned by Astronaut Conrad upon approaching area for scoop 1 of the contingency sample collection, was present on the southeast face. (LRL photograph S-69-61490)

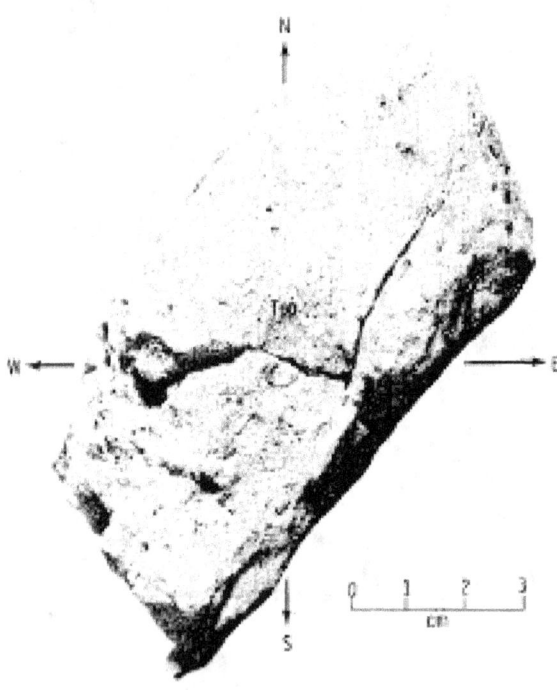

FIGURE 10-31.—Sample no. 12073, showing suggested selenographic orientation. Specimen was broken before LRL photography, thus making a more precise orientation impossible. (LRL photograph S-69-61080)

Three selected-sample rocks (samples 12004, 12021, and 12022) have been tentatively identified and have been oriented in their presampling position on the lunar surface. Three other rocks (samples 12006, 12008, and 12014) have been tentatively identified, but not oriented in their presampling positions, from surface photographs. Table 10-II gives areal distribution of the selected sample, as inferred from the study of surface photographs and the verbal transcript. Specific LRL sample numbers are given for several rocks where identification is aided by description.

TABLE 10-II. *Areal distribution of selected sample*

Sample location	No. and type of samples	LRL sample no.[a]
Small mound north of ALSEP site	2 rocks	12017? (1 unidentified)
Large mound south of ALSEP site	5 rocks	12006? 12021 12022 (2 unidentified)
On or near rim of Middle Crescent Crater	7 rocks	12004 12008? 12014? 12015? 12016? (2 unidentified)
Intercrater area near ALSEP site and northwest of LM	6 rocks 1 sample of fines with glass	([b])

[a] Question marks indicate uncertainty of sample number.
[b] No samples identified to date.

Sample 12004. Rock sample 12004 (weight, 585 g; dimensions, 9 by 8 by 4 cm) has been identified on NASA photograph AS12-47-6936 as the rock that is resting on end in the regolith at the panorama site near the rim of Middle Crescent Crater (fig. 10-32). The sample was also correlated with LRL photograph S-69-62023 by irregularities on the vertical exposed face and by its very nearly flat, east-facing face. The suggested orientation of sample 12004 is shown in figure 10-33.

Sample 12021. Sample 12021 (weight, 1876 g; dimensions, 14 by 12 by 8 cm) has been tentatively identified on NASA photograph AS12-47-6932 as being collected from the area near the

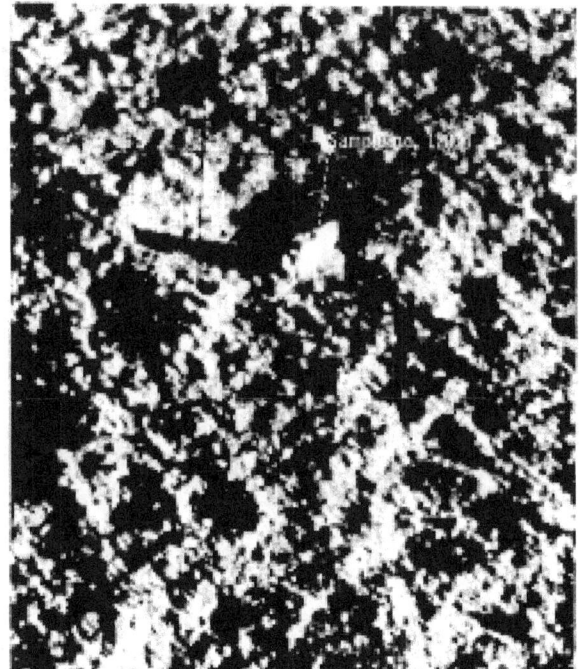

FIGURE 10-32.— Sample no. 12004 resting on end on the southeast rim of a small crater. Area is near the photographic panorama site at Middle Crescent Crater. (Enlargement of AS12-47-6936)

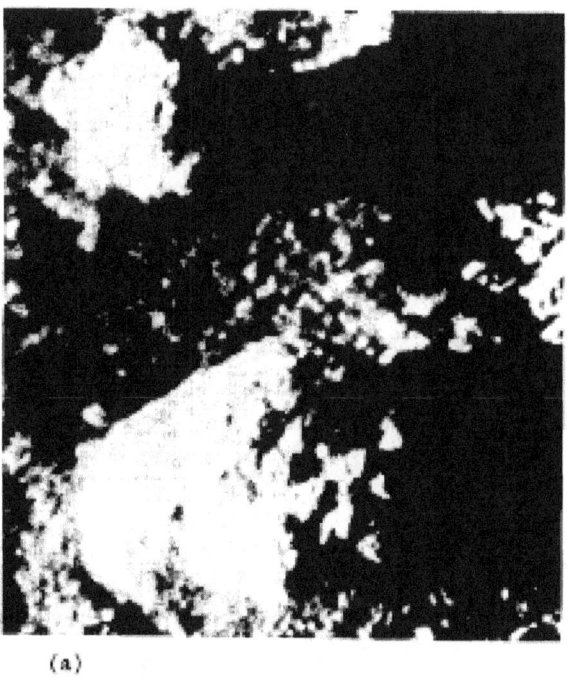

(a)

(b)

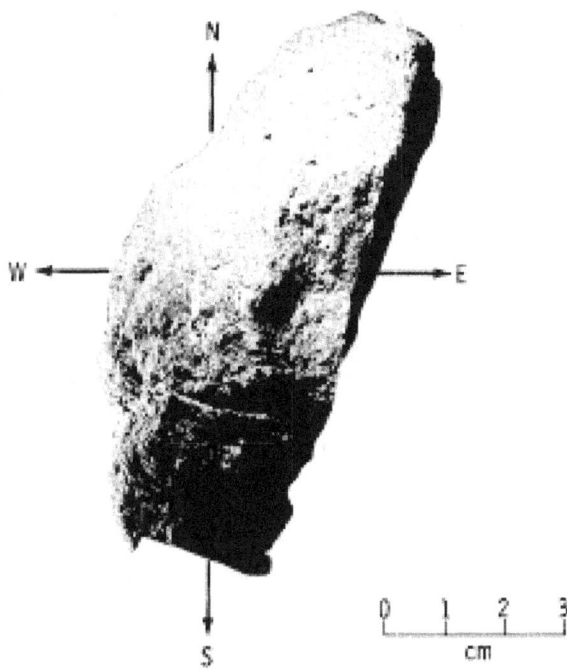

FIGURE 10-33.— Sample no. 12004, showing suggested orientation. (LRL photograph S-69-62023)

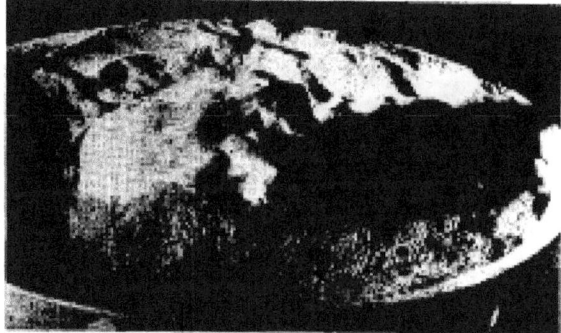

FIGURE 10-34.— Sample no. 12021. (a) Sample before collection. Rock is located near the large mound approximately 120 m northwest of the LM. It is still uncertain whether the other rock fragment in the photograph was sampled. (Enlargement of AS12-47-6932) (b) LRL model of sample 12021 showing approximate reconstruction of lunar lighting.

eastern base of the large mound shown in figure 10-22, located 120 m northwest of the LM and north-northeast of Head Crater. The photograph shows the sample immediately prior to its being collected with the sample tongs (fig. 10-34). The other two smaller rock fragments in the photograph were not identified as collected samples. The orientation of sample 12021 is shown in figure 10-35.

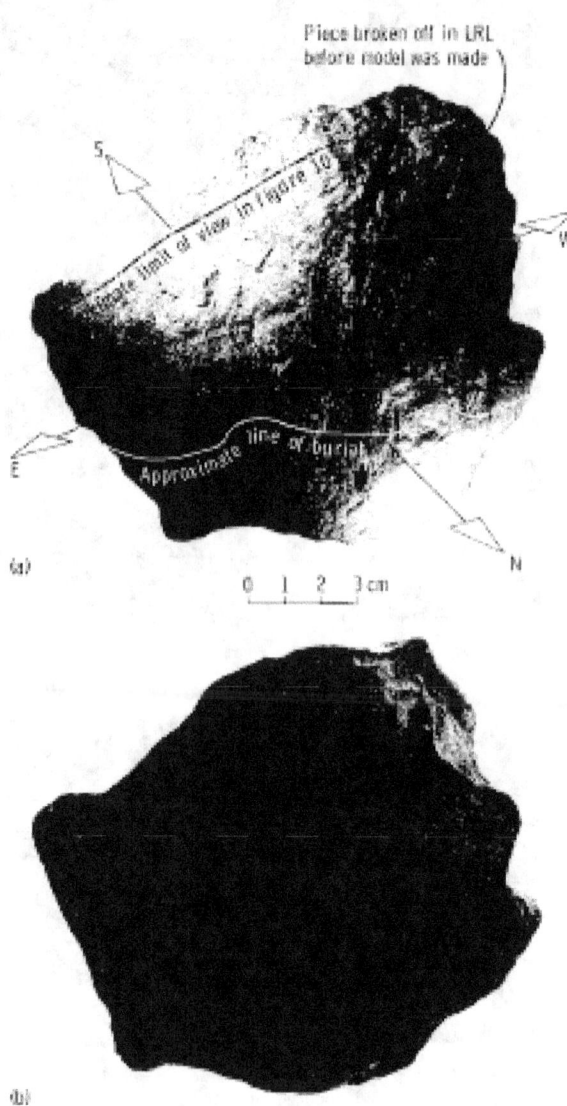

FIGURE 10-36. — Sample no. 12022 before collection. The sample was collected on the southeast side of the large mound that was located approximately 120 m northwest of LM. Note the burial of the rock. (Enlargement of AS12-47-6933)

FIGURE 10-35. — Sample no. 12021, (a) suggested orientation. (LRL photograph S-69-61986) (b) LRL model of sample 12021. Holes are bubbles formed when model was made.

Sample 12022. Sample 12022 is another large specimen (weight, 1864 g; dimensions, 14 by 9.5 by 7 cm) that has been located within the northeast side of the large mound. The sample is shown beneath the tongs, approximately 0.66 m below the top of the mound on NASA photograph AS12-47-6933 (fig. 10-36). The diagnostic triangular shape and characteristic depression in the exposed portion of the sample greatly facilitated identification and orientation. The sample may have been broken by Astronaut Bean during sampling. He mentioned at the time of sampling that he tried to break off a piece from the mound or from a rock. Sample 12022 was correlated with LRL photograph S-69-61999. The suggested orientation of the sample is shown in figure 10-37.

Documented Sample (Including Totebag Sample)

The second EVA period was devoted to sample collection and geologic observations in a traverse that followed a preplanned course of approximately 1450 m. The documented sample consisted of approximately 17.6 kg of rocks, scooped soil samples, and core samples collected primarily from the rims of craters, as indicated on the sample location map (fig. 10-1). The documented sample comprised 21 rocks, seven scooped soil samples, three core-tube samples (two of which were joined end-to-end and driven their full lengths, i.e., the double-core tube), and two vacuum-sealed samples for environmental study and gas analysis. Four of the large individual rocks and a small amount of

GEOLOGIC INVESTIGATION OF THE LANDING SITE

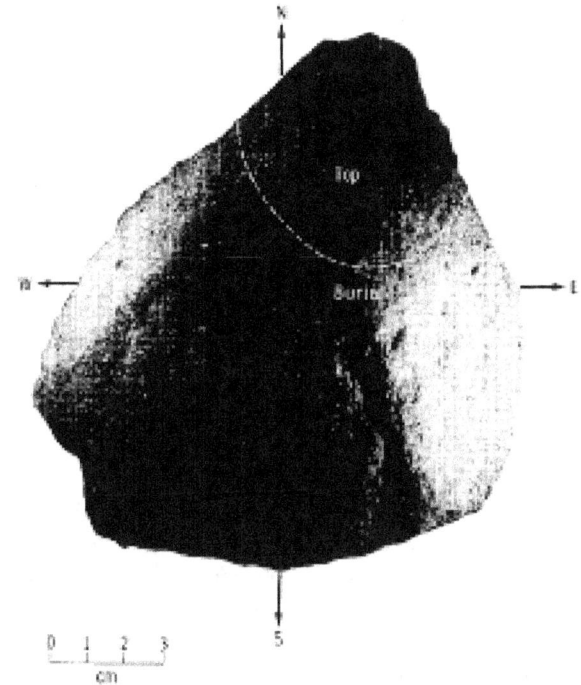

FIGURE 10-37.— Sample no. 12022, showing suggested selenographic orientation and burial. Diagnostic hole in northern side visible in figure 10-36 (AS12-47-6933). Note characteristic triangular shape. (LRL photograph S-69-61999)

TABLE 10-III. *Areal distribution of documented sample*

Sample location	No. and type of samples	LRL sample no.[a]
On or near rim of Head Crater	5 rocks	12031 12034 12052 12055
	2 fines	12030 12033
On or near rim of Bench Crater	6 rocks	12035 12036 12038 12039 12040 12053
	3 fines	12032 12037 12041
On or near rim of Sharp Crater	Core LESC[b] GASC[c]	12027 12023 12024
On rim of 10-m-diameter crater south of Halo Crater	Core (from double-core tube)	12025[d] 12028[e]
On or near rim of Surveyor Crater	10 rocks	12043 12045 12046 12047 12051 12054 12056? 12062? 12063? 12064? 12065?
	2 fines	12042 12044

[a] Question marks indicate uncertainty of sample number.
[b] Lunar environment sample container.
[c] Gas analysis sample container.
[d] Core tube serial no. 2010.
[e] Core tube serial no. 2012.

soil from the returned scoop of Surveyor 3 (not included in the previously mentioned soil count) were brought back in the totebag. All other documented samples were returned in a sealed metal sample return container. Table 10-III lists the areal distribution of the documented samples.

Sample 12031 (field sample 3D). Sample 12031 has been identified with assurance in Hasselblad photographs AS12-49-7189, AS12-49-7190, and AS12-48-7048 (fig. 10-38). The documentation of this sample is precise because photographs were taken both before (NASA AS12-48-7048) and after (NASA AS12-48-7050) collection and because the sample was put into a prenumbered bag (3D) and identified when it was picked up. The sample was collected from a small cluster of rocks approximately 2 m southeast of the trench that was dug in the northwest rim of Head Crater. Nearly half of the sample was buried in the fine-grained regolith. There is the suggestion of a small fillet banked on the north and northeast sides of the rock. Other rocks nearby have fillets with predominantly similar orientations, indicating a possible source of material to the north or northeast. The orientation of sample 12031 is shown in figure 10-39.

Figure 10-38 illustrates the orientation of rock sample 12031 at the time of sampling. The top was one of the two flattest and least-pitted surfaces of the sample (the other surface being the face shown in LRL photograph S-69-61811). The conclusion is made that the present orientation was attained relatively recently, as compared to the past exposure history. However, the amount of burial (approximately half) and the development of minor fillets indicate an accumulation of fine-grained material around the rock since the time of its last tumble.

FIGURE 10-38. — Sample no. 12031 before collection. (Enlargement of AS12-48-7048)

Figure 10-39 is an orientation diagram of rock sample 12031 as it was exposed on the lunar surface. An unfolded box shows orthogonal views of the top and sides of the rock, using photographs of the rock taken in the LRL. Lines indicate the degree of burial on each side. The oblique view at the lower left shows the orientation of the sample before it was collected.

Sample 12051 (loose specimen in documented sample box). Sample 12051 is one of the distinctive large fragments collected on the second EVA period traverse. The sample is part of the blocky ejecta from a fresh 4-m-diameter crater on the south rim of Surveyor crater, and its location is well shown in Hasselblad photographs AS12-49-7318 (fig. 10-40) and AS12-49-7319 taken by Astronaut Bean. The sample was picked up by hand by Astronaut Conrad while Astronaut Bean steadied him with a strap from the totebag. The site from which the sample was removed is shown in Hasselblad photograph AS12-49-7320 (fig. 10-41). This sample is

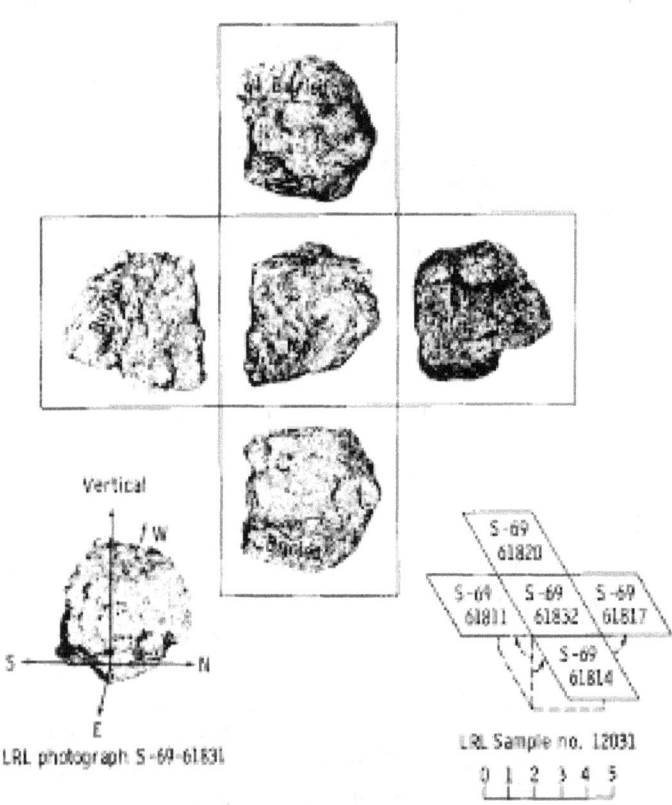

FIGURE 10-39. — Sample no. 12031, orthogonal views and orientation diagram. LRL photograph numbers are shown in the schematic diagram.

Figure 10-40. — Sample no. 12051 before collection. (Enlargement of AS12-49-7318)

Figure 10-41. — After collection of sample no. 12051. Note that the shape of the remaining hole is very similar to the shape of the top view of rock sample no. 12051, as shown in figure 10-42. (AS12-49-7320)

characterized by a sheared surface on one side that was described by the astronauts. Part of both the sheared surface and the opposite, convex surface of the sample were buried in the regolith. The suggested orientation of sample 12051 is shown in figure 10-42.

The surface of sample 12051 exhibits three faces with distinctly different shapes and weathering characteristics, indicating different exposure ages. Surface A (fig. 10-42) is rounded and covered by abundant, small glass-lined pits; surface B is apparently a fracture surface that displays only a few small pits; surface C is flat and appears to be a freshly broken surface without pits. At the time of sampling, approximately half of this rock was buried, standing on its small end, in the regolith.

Surface B on rock sample 12051 appears to be an older fracture face that probably resulted from blow by impact at an earlier time. The minor pitting and weathering of surface B indicates postfracture exposure at the lunar surface prior to the event that formed surface C and buried surface B.

The rounded, heavily pitted surface A indicates a long exposure of rock sample 12051 at the

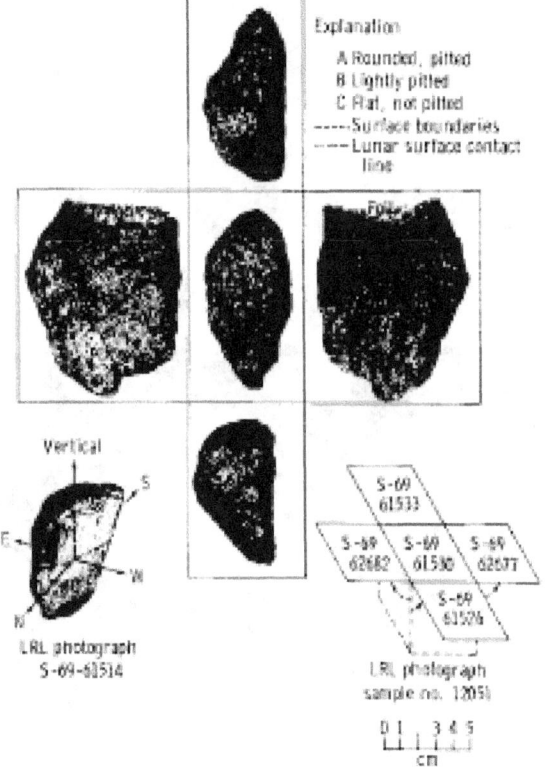

Figure 10-42. — Sample no. 12051, orthogonal views and orientation diagram. LRL photograph numbers are shown in the schematic diagram.

lunar surface. The two broken surfaces, B and C, of different ages attest to multiple bombardment and tumbling of the rock at or near the surface of the regolith.

The occurrence of this rock, along with other rounded and angular rocks, in the rim of the rocky fresh crater indicates high probability that the sample represents ejecta from the small crater—with the ejecta being caused by either a fragmented impacting body or displaced regolith or subregolith at the impact site. In either case, the unpitted shear face probably resulted from the impact that formed the small crater. The attitude of the sample when found suggests that the rock fell into place with enough surrounding fine material to hold it in an otherwise unstable position. Determining the exposure age of the portion of the flat side that was above the lunar surface might date the small crater.

Sample 12052 (loose specimen in documented sample box). Sample 12052 is a large rock, rounded on one side, that was collected from the west rim of Head Crater. The rock is well shown in Hasselblad photographs AS12–49–7217 (fig. 10–43) and AS12–49–7218 taken by Astronaut Conrad and Hasselblad photograph AS12–48–7059 taken by Astronaut Bean. Apparently, the rock was moved from its original site and set down again on the surface before the photographs were taken. Therefore, although the sample can be oriented with precision with respect to its position in the photographs, this orientation probably does not represent the orientation of the rock before it was disturbed. It appears that the lower angular part of the rock was partially buried in the regolith, but the rock is resting on the surface in the photographs. A drag mark nearby indicates the place from which the rock has been removed. At least two pieces on one side of the sample were broken off after the rock had been photographed on the lunar surface and before it was photographed in the LRL. Figures 10–43 and 10–44 show sample 12052 as it was photographed on the lunar surface and in the LRL, respectively, with nearly identical orientations and shadow characteristics.

Sample 12053 (loose specimen in documented sample box). This specimen is an angular fragment that was collected from the northwest rim of Bench Crater. It appears to be part of a field of coarse fragments on the crater rim which probably have been ejected from the crater. The specimen is readily identified in Hasselblad photographs AS12–49–7234 and AS12–49–7235 taken by Astronaut Conrad, and AS12–48–7063, taken by Astronaut Bean. The rock appears to have been rotated out of its original position, however, before the first photographs were taken so that the original orientation is uncertain.

Sample 12054 (loose specimen in documented sample box). The unusual glass-coated rock sample 12054 was collected a short distance south of the fresh 4-m-diameter crater with which sample 12051 is associated, and the astronauts believed that sample 12054 may have been ejected from this small fresh blocky crater. The sample is well shown in Hasselblad photographs AS12–49–7313 to AS12–49–7315 (fig. 10–45) taken by Astronaut Bean. In these photographs, the sample can be seen to be resting essentially on the surface. Three sides of this rock are coated with glass, and two sides are free of glass. When photographed, sample 12054 was resting on one glass-free side, and the other glass-free side was oriented toward the southeast. The

FIGURE 10-43.— Sample no. 12052 before collection. (Enlargement of AS12-49-7217)

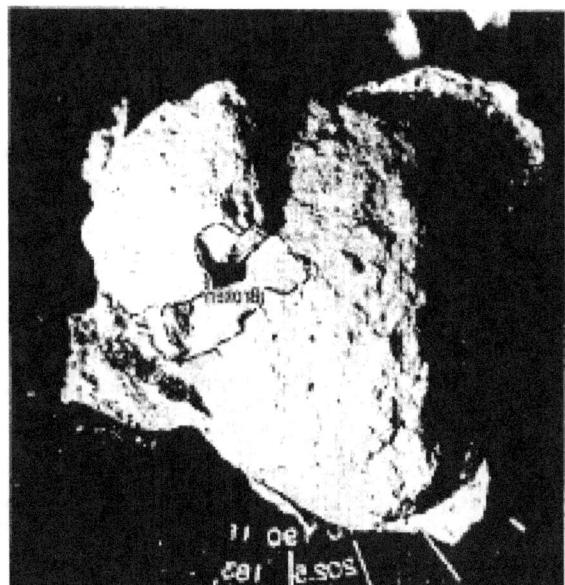

FIGURE 10-44. — Sample no. 12052, showing lunar orientation reconstructed in the LRL.

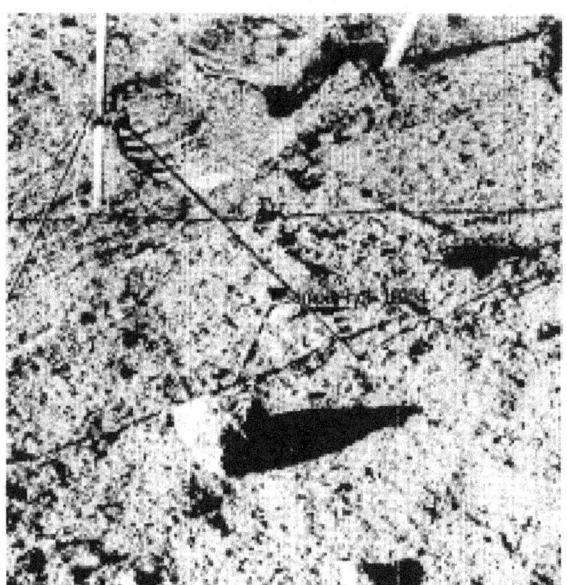

FIGURE 10-45. — Sample no. 12054 before collection. (Enlargement of AS-49-7315)

glass may have been sprayed onto the rock as it was ejected from an impact crater, and, by coincidence, the rock landed on a side not coated with glass. Figure 10-46 shows a reconstruction in the LRL of the orientation and lighting as seen in the lunar surface photograph of the sample.

FIGURE 10-46. — Sample no. 12054, showing lunar orientation reconstructed in the LRL.

FIGURE 10-47. — Sample no. 12055, showing sample undisturbed before collection. Rock B is the same rock as the disturbed rock B in figure 10-48. Letters A to C are for reference in this figure and figure 10-48. (AS12-49-7197)

field relations suggest that the rock was sprayed with glass in the position in which it was found on the surface and that the spray came from the west. Alternatively, as the rock was found to exhibit shock damage in the laboratory, the

FIGURE 10-48. — Disturbed rock B, showing a distinct two-toned gray on its near end. (AS12-48-7055)

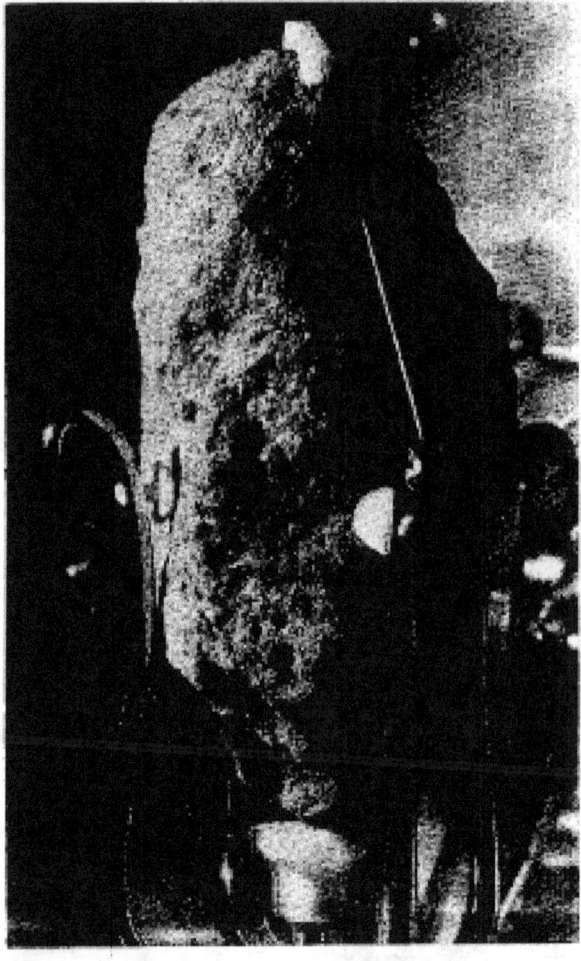

FIGURE 10-49. — Sample no. 12055, showing approximate lunar orientation reconstructed in the LRL.

Sample 12055 (loose specimen in the documented sample box). Sample 12055 is a tabular rock that is shown apparently undisturbed before sampling in Hasselblad photographs AS12-49-7197 (fig. 10-47) and AS12-49-7198. Another rock, different from the one shown beneath the gnomon, appears in Hasselblad photographs AS12-49-7199 and AS12-49-7200 (fig. 10-48) taken by Astronaut Conrad and also in Hasselblad photographs AS12-48-7053 to AS12-48-7055 taken by Astronaut Bean. There has been some confusion about which rock was actually picked up. Verbal descriptions by the astronauts suggest that the disturbed rock is the one that was sampled. However, experimental lighting of sample 12055 in the LRL (fig. 10-49) indicates a distinct similarity between sample 12053 and the rock standing upright beneath the gnomon (fig. 10-47). The laboratory specimen of sample 12055 was broken between the time of sampling and the time of photography in the LRL. This breakage occurred in the shadowed side of the rock as photographed (fig. 10-47) on the lunar surface beneath the gnomon. Fragments of sample 12055 may comprise some of the chips from the sample Return Container.

Sample 12065 (specimen from the totebag). Sample 12065 has not been identified in either the Hasselblad photographs or from descriptions given by the astronauts. The sample location is shown tentatively on the traverse map (fig. 10-1) as being in the area of Surveyor 3.

PART B

PHOTOMETRIC AND POLARIMETRIC PROPERTIES OF THE LUNAR REGOLITH

H. E. Holt[a] and J. J. Rennilson[b]

Photometric Properties

Several special, as well as general, photometric and polarimetric studies of the lunar regolith in the vicinity of the Apollo 12 landing site have been undertaken by means of the black-and-white photographs obtained on the geologic traverse during the second extravehicular activity (EVA) period. Areas of special interest for photometric study were the Surveyor 3 footpad imprints and the lunar material disturbed by the Surveyor surface sampler. A search was made for changes that occurred during the 30 months since the Surveyor 3 landing. Other studies were undertaken to verify the nature of the lunar photometric function at the centimeter scale and to compare macroscopic textures of the surface with the observable photometric function.

The photometric characteristics of lunar materials were measured around each Surveyor spacecraft and were used to discriminate different materials. At all Surveyor sites, darker material was revealed wherever the fine-grained surface layer was disturbed. The report, by the Apollo 12 astronauts, of lighter-colored material at some depth below the surface was the first observation of this type during seven lunar landings (five Surveyor missions and Apollos 11 and 12).

Photometric data were obtained from SO-267 black-and-white film by microdensitometry of a first-generation film positive and by scanning with a flying-spot scanner and digitization of selected photographs. The luminances were calculated from the transfer characteristic curve of the processed film furnished by the Photographic Technology Laboratory at the NASA Manned Spacecraft Center, Houston, Tex., and from the reported camera settings. Photographs that include the gnomon with its 10-step gray scale provided a verification of the transfer characteristic curve. The luminances observed by using the gray-scale steps were compared with the luminances computed from the preflight goniophotometric calibration of the gnomon, the lunar photometric angles for each photograph, and an assumed solar illuminance of 13 000 lumens. Corrections for iris, filter, and frame shading were applied to the luminance values computed for selected areas.

The albedo (normal luminance factor) of selected areas along the second EVA traverse was measured to establish the range of albedo variation. Preliminary measurements show that the albedo of undisturbed fine-grained surface material in intercrater areas ranges from 8.8 to 10.1 percent. The brightest material, with an albedo of 11.2 percent, occurs around and inside Sharp Crater. Most measurements were made around the shadow of the astronauts' heads and shoulders, where the phase angles were 2.5° to 4°. The photometric data were then extrapolated to the zero phase angle to determine the normal albedo.

The reflectance of various rocks within a 6° to 30° phase angle was measured. Rock reflectance varied from 10 to 210 percent more than the reflectance of fine-grained material adjacent to the rocks. The estimated normal albedo of selected rocks ranges from 12 to 16 percent. The greatest difference between the luminance of the rocks and the luminance of the fine-grained material adjacent to the rocks generally occurs at phase angles of 40° to 90°, and the least contrast occurs near the zero phase angle. This suggests that the rocks have a more nearly Lambertian photometric function than does the fine-grained material.

Preliminary photometric measurements were made of lunar materials near the Surveyor 3 spacecraft by using NASA photographs AS12-48-7110 to AS12-48-7113 (indicated as panorama 20 on fig. 10-50). In table 10-IV, the

---
[a] U.S. Geological Survey.
[b] California Institute of Technology.

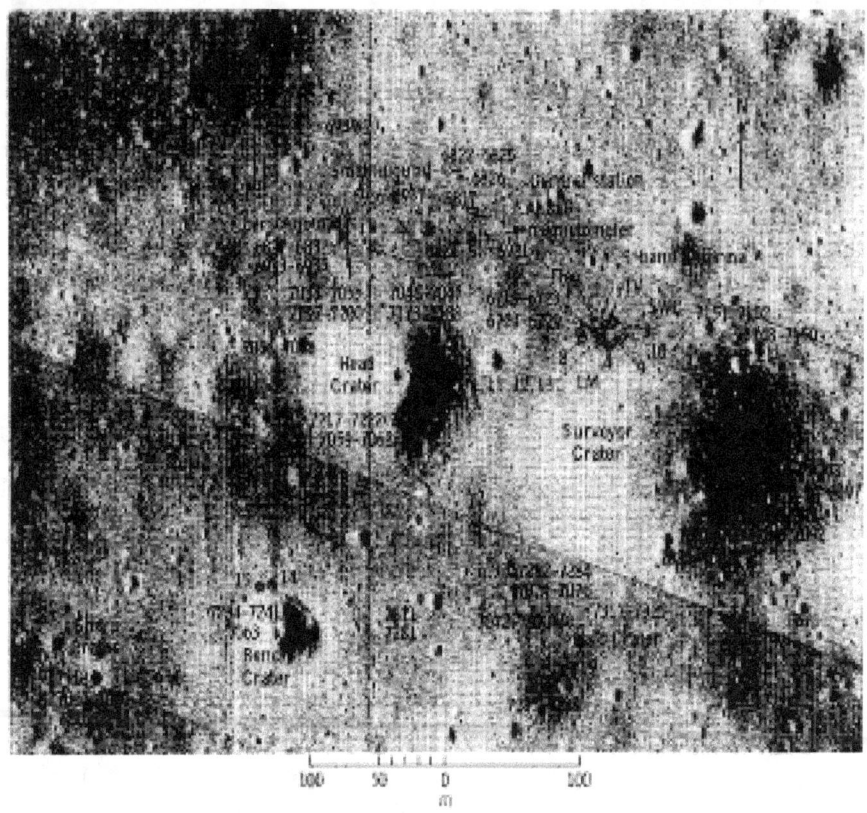

FIGURE 10-50. — Map showing location of photographs taken on first and second EVA periods. Compare with figure 10-1.

TABLE 10-IV. *Albedo of materials measured from Surveyor 3 and Apollo 12 photographs*

Material	Albedo, percent	
	Apollo 12, Nov. 1969	Surveyor 3, Apr. 1967
Undisturbed fine-grained lunar material	9	8.5
Surveyor-disturbed material	7.4	7.6
Astronaut-disturbed material	7	—
Rock	12 to 16	—

photometric properties of undisturbed, disturbed, and compressed lunar fine-grained material, as measured from the Apollo 12 photographs, are compared to those measured by the Surveyor 3 camera. The only significant difference between the Surveyor measurements and those obtained from the Apollo 12 photographs was for the compressed material.

Compressed fine-grained material was observed in the imprint of Surveyor 3 footpad 2, and the disturbed material adjacent to the compressed material was ejected by the footpad impact. The estimated normal albedo of the undisturbed material near the footpad imprint is 9 percent, and the albedo of the disturbed material is 7.4 percent (or approximately 15 percent darker), which is in close agreement with comparable measurements of 8.5 and 7.6 percent, respectively, made through the use of the Surveyor 3 television camera photographs (ref. 10-1). As measured from Apollo 12 photographs, at phase angles of 88° to 92°, the compressed material in the footpad imprint area reflects approximately 25 percent less light than nearby undisturbed material. At phase angles of 40° to 60°, the Surveyor 3 television camera photographs recorded 30 percent more light reflected from the footpad imprint area than from adjacent undisturbed surfaces.

Compressed fine-grained material was observed in the imprint of the Surveyor 3 footpad 2, and the disturbed material adjacent to the compressed material was ejected by the footpad impact. The estimated normal albedo of the undisturbed material near the footpad imprint is 9 percent, and the albedo of the disturbed material is 7.4 percent (or approximately 15 percent darker), which is in close agreement with comparable measurements of 8.5 and 7.6 percent, respectively, made through the use of the Surveyor 3 television camera photographs (ref. 10-1). As measured from Apollo 12 photographs, at phase angles of 88° to 92°, the compressed material in the footpad imprint area reflects approximately 25 percent less light than nearby undisturbed material. At phase angles of 40° to 60°, the Surveyor 3 television camera photographs recorded 30 percent more light reflected from the footpad imprint area than from adjacent undisturbed surfaces.

The astronauts walked over the area near Surveyor 3 footpad 2 and produced footprints of freshly compressed and disturbed material. Luminance measurements from NASA photographs AS12-48-7112 and AS12-48-7113 indicate that the material kicked outward from the astronaut footprint has an estimated albedo of 7 percent, approximately 20 percent darker than the adjacent undisturbed lunar surface. The astronaut boot imprint of compressed material appears to have approximately 40 percent greater reflectance at an 80° phase angle than does the undisturbed material. This measurement for the astronauts' bootprints is similar to the Surveyor 3 measurements of the albedo of its own footpad imprint.

The similar albedo measurements for undisturbed and disturbed fine-grained lunar materials near Surveyor 3 obtained from the Surveyor 3 television camera photographs and from the Apollo 12 photographs indicate that little, if any, measurable change has occurred in the photometric properties of these materials during the 30 months that Surveyor 3 was on the lunar surface before Apollo 12 landed. The optical properties indicate that the lunar surface in this area has not received a new covering of dust or been mechanically altered by the lunar environment during the 30 months. The significant change in the reflectance of the Surveyor 3 footpad imprint over the 30-month span may have been caused by microscopic mechanical alteration of the compressed surface.

Astronaut observations and comparisons of the Apollo 12 photographs of the Surveyor 3 spacecraft to known preflight appearance and to Surveyor 3 television camera photographs of spacecraft components did not reveal any major changes in the expected appearance of the spacecraft components, except for color. The hop-and-skip landing of Surveyor 3 was known to have caused some dusting of the spacecraft. The television camera mirror was known to be more highly dusted near its top than its bottom, and the more specular-reflecting spacecraft components had significant reflectance. The Apollo 12 photographs show similar relationships; the specular areas are still quite specular, and the white-painted areas still have relatively high reflectance. Even mirrors on top of the electronic compartments show rather high specular reflectance in NASA photograph AS12-48-7117. The closeup photograph of the television camera mirror reveals an image on the mirror of the camera housing and the filter wheel assembly. An overall qualitative assessment of the Apollo 12 photographs of the Surveyor 3 spacecraft suggests that only minor changes have occurred in the appearance of the spacecraft components during its 30 months on the lunar surface. The fine material adhering to the spacecraft is probably lunar dust blown up during the Surveyor 3 landing.

Near Head, Bench, Sharp, and Block Craters, lighter colored material was exposed in areas scuffed by the astronauts' boots and in a shallow trench. The lighter colored material shown in NASA photograph AS12-48-7052 has an estimated normal albedo of 11 to 13 percent, and the undisturbed surface has an albedo of 8.8 percent. Other areas, when disturbed, show darkening of 10 to 15 percent of the normal albedo. The lighter colored material may represent fresh-crushed rock or ray material ejected during some cratering event and subsequently covered by darker, normal regolith material.

Photographs taken down-Sun, which include the shadow of the astronaut's head, permit the measurement of lunar surface reflectance from approximately a 20° phase angle to the zero phase angle (near the head shadow) and then out toward a 20° phase angle again. Several microdensitometer scans were made across NASA photographs AS12–49–7255 and AS12–49–7257. The film density luminances were calculated and plotted relative to the phase angle. The resultant curves (fig. 10–51) are similar to the telescopically observed lunar photometric function (ref. 10–6). Slight differences appear near the zero phase angle, where the curves obtained from measurements of the Apollo 12 photographs are not as sharply peaked as are the curves calculated from telescopic measurements.

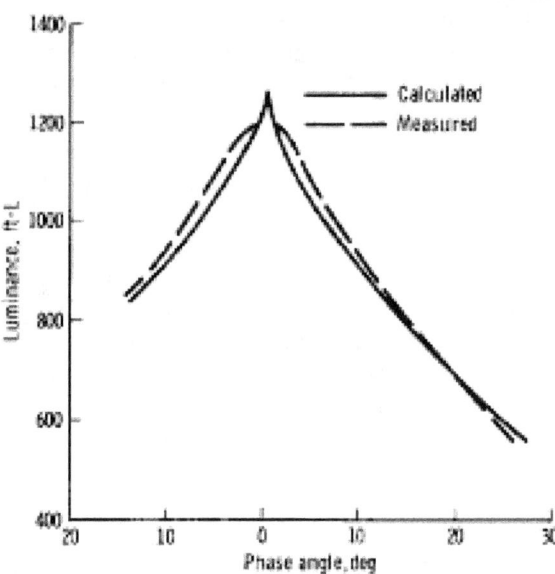

FIGURE 10-51.—Comparison of calculated luminance and measured luminance values (based on AS12-49-7255, scan 1).

Polarimetric Properties

A polarizing filter was attached to the front of Astronaut Conrad's camera at the start of the second EVA period in order to record polarization measurements. Photographs of the lunar surface were taken with the polarizing filter rotated sequentially in front of the camera lens, while the aperture and other camera conditions were held constant. The procedure called for rotating the polarizing filter to three detent positions corresponding to angles of 0°, 45°, and 90° clockwise with respect to a horizontal reference plane. If the camera was not moved, the variation observed in the three photographs in the apparent radiance of the same image element is due to a polarized component in the light incident on the filter. A greater degree of polarization, or percentage of linearly polarized light in the light scattered from the lunar surface, produces a greater variation between photographs in the apparent radiance of image elements.

Light scattered from the lunar surface is partially plane polarized at most phase angles. The degree of polarization of sunlight scattered from fine-grained areas of the lunar surface has been determined by telescopic and Surveyor 6 and 7 measurements to depend principally upon the phase angle. The degree of polarization of light scattered from the fine-grained lunar surface on the maria (Surveyor 6) varies from 0 to 19 percent. Rock surfaces observed on the maria and in the highlands exhibit a greater variation of polarizing properties than does the fine-grained material. A few rock surfaces are similar in their polarizing properties to the fine-grained material adjacent to the rocks, whereas other rock surfaces produce a maximum polarization of the scattered light that is up to several times the polarization produced by fine-grained material. The degree of polarization of light scattered from rock surfaces varies as a function of phase angle, and the peak polarization observed from rock surfaces that produce strong polarization effects occurs between a 120° to 124° phase angle. The polarization properties of lunar rocks probably depend on rock type. Crystalline rocks may be expected to produce higher peak polarization than microbreccias.

Preliminary measurements for analysis of polarization have been made on NASA photographs AS12–48–7077 to AS12–48–7079, taken near Head Crater. These photographs, together with a gray scale, were scanned by the film reader of the Image Processing Laboratory at the Jet Propulsion Laboratory. The scanner had a 50-μm spot size, and each photograph was digitized to 64 gray levels. The gray scale was read with the same spot size, and an averaged matrix of values was used to plot the light-transfer characteristics. By using this characteristic, the digital photographs were then converted to a luminance domain. Cross-correlation programs were then applied to register the frames.

The phase angle was approximately 90° for the areas processed for preliminary calculations of the polarization. At this phase angle, the degree of polarization of the fine-grained material varied from 9 to 15 percent, whereas the light reflected from rock surfaces varied from 2 to 24 percent in polarization. The plane of partial polarization is approximately perpendicular to the phase plane. These preliminary polarization measurements show that the observed rocks differ from one another in their polarizing properties and that some of the rocks differ markedly from the fine-grained material.

PART C

MECHANICAL PROPERTIES OF THE LUNAR REGOLITH

R. F. Scott,[a] W. D. Carrier,[b] N. C. Costes,[c] and J. K. Mitchell[d]

This report presents, from a soil mechanics point of view, the results of a preliminary examination of the returned data, photographs, and soil samples of the Apollo 12 mission. The various features of the descent, landing, and the extravehicular activity (EVA) periods of the astronauts are compared with those observed at the Apollo 11 landing site. Some comments are also appended regarding the appearance of and conditions around the Surveyor 3 spacecraft visited by the Apollo 12 astronauts. The report on soil properties at the Apollo 11 landing site (ref. 10–7) gives a brief summary of prior observations of the lunar surface and may be referred to for detailed comparison with the results of this study. The events of the Apollo 12 landing are described in chronological sequence.

Descent and Touchdown

Visibility Problems From Surface Erosion

The descent film made during the Apollo 12 lunar module (LM) approach shows a considerable amount of movement of the lunar surface material. The movement reached such a level that, in the final stages of the descent, no surface features were visible. The astronauts reported a loss of visibility at this time. This occurrence poses a potential hazard to future lunar landings, and it is highly desirable to evaluate the causes of this loss of visibility. The Apollo 11 and 12 spacecraft followed different descent profiles to land in different regions of the Moon, and in addition, the thrust of the Apollo 12 LM was higher than that of the Apollo 11 LM by approximately 5 percent. The amount of erosion at the two sites may be different because the descent profiles, the surface soil, the thrusts, or any combination of these factors was different for the two missions. To explain this possible erosion difference, a detailed analysis of all features related to erosion of the lunar surface by the descent engine is required. To date, only a few preliminary considerations have been examined. The following paragraphs describe some of the observations that will establish a framework in which more detailed studies of the relationship between descent profile and soil characteristics may be made.

Descent

The descent profiles of the Apollo 11 LM and the Apollo 12 LM differed considerably in the last 60 m. The Apollo 11 LM descended at a rate of approximately 0 to 6 m/sec to a height (as

[a] California Institute of Technology.
[b] Manned Spacecraft Center.
[c] Manned Space Flight Center.
[d] University of California.

measured from the surface to a level plane through the footpads) of approximately 1.5 to 2.4 m; then, the LM paused at this elevation for 13 sec before descending to the surface in 3 sec. By comparison, the Apollo 12 LM made the last portion of the descent at a rate of approximately 0.5 m/sec, with no pauses. On Apollo 11, the descent propulsion system (DPS) engine was not turned off until approximately 1 sec after footpad contact. On Apollo 12, the engine was shut down, according to Astronaut Conrad, as soon as the contact probes touched the lunar surface. The probe contact occurs at a footpad height of approximately 1.5 m above the surface. The last few feet of descent of the Apollo 12 LM, therefore, took place as a hindered free fall as the thrust of the DPS engine decayed after shutdown.

Although final information on the spatial profile of the Apollo 12 descent is not yet available, the data at hand indicate a considerable difference between the Apollo 12 and 11 descents. The lateral velocity of the Apollo 11 LM was relatively high, approximately 0.9 m/sec, for most of the final 20 or 30 sec of flight, while the Apollo 12 spacecraft approached at a lateral rate of approximately 0.5 m/sec and slowed down to slightly more than 0.3 m/sec as it approached the landing site. Thus, the Apollo 12 spacecraft traversed a much shorter lateral distance over the surface during the final 20 or 30 sec of descent than did the Apollo 11 LM. It can be inferred that the lunar surface landing area for Apollo 12 was exposed for a longer time to the blast of the descent engine than was the corresponding area of the Apollo 11 landing.

To determine the difference between the observed behavior of the lunar surface during the two descents, a detailed examination of individual frames of the Apollo 11 and 12 descent films was made. In this study, the spacecraft heights at earlier stages in the descent were determined first by internal evidence in each frame (camera geometry, spacecraft dimensions, and known crater dimensions) and then compared with heights deduced from the framing rates of the cameras and from the known descent profiles. Partly as a consequence of this analysis, it was found that the framing rate of the Apollo 11 descent film was 1.8 m/sec, rather than 3.7 m/sec, as previously reported. As planned, the framing rate of the Apollo 12 camera was 3.7 m/sec. Because close agreement was found between the heights determined by the two methods at higher altitudes, the framing-rate/descent-profile technique was employed with some confidence for the later stages of descent, when the surface was partially or totally obscured. The results of the evaluation are presented in table 10–V. Loss of visibility was never as complete on the Apollo 11 descent as it was on the Apollo 12 descent. From table 10–V, it can be seen that the altitudes at which various events occurred on Apollo 12 descent are considerably greater than those at which similar events occurred in the Apollo 11 mission, as deduced from the descent films.

Erosion of the surface by the engine exhaust depends on a number of mechanical properties such as cohesion, grain size and bulk density of the soil, and the angle of friction of the granular material. (Erosion is more sensitive to cohesion and grain size and less sensitive to friction angle and bulk density.) If two soils both exhibit the same response (penetration depth) to an astronaut's boot, and if one soil has less cohesion and a higher friction angle than the other, the soil with less cohesion will be much more sensitive to erosion by a rocket engine. The following paragraphs of this section show that the gross mechanical properties of the lunar surface material at the two landing sites are not greatly different, in terms of the depths of astronaut bootprints, penetration of the spacecraft into the surface, and operation of various tools. The data available on the variation of lunar surface material properties are not sufficient at present to allow a decisive conclusion as to the relationship of mechanical properties of the lunar material to rocket erosion.

As seen in a later portion of this section, laboratory examination of the soil returned from the lunar surface by the two missions indicates that the soil in the Apollo 12 core tubes possesses a substantially larger proportion of particles in the fine size range. The sieving technique, however, was changed for the Apollo 12 analysis, and this new technique should result in a greater breakdown of soil clumps and aggregates. Therefore, it is not clear, at present, if there is any fundamental differences between the material. The distribution and proportion of particles larger

TABLE 10–V. *Comparison of Altitudes and Times at Which Similar Events Occurred During Apollo 11 and 12 Descents*

Event	Altitude, m		Time to touchdown, sec	
	Apollo 11	Apollo 12	Apollo 11	Apollo 12
First signs of blowing dust	24	33.5	65	52
Streaking fully developed	4.5	9	21	21
Loss of visibility	2.7	7.3	15	17

than 0.1 mm in diameter from the different Apollo 12 core tubes were similar. The Apollo 12 soil appears to become coarser with depth, but it is not known if this affects the erosion problem.

At present, therefore, the primary differences between the two landing sequences are the descent profiles and the effects on the rocket gas/surface interaction caused by the different descent profiles. The rocket gas/surface interaction problem will be examined elsewhere in detail to determine the extent of the contributions of two processes, which, at present, are analyzed separately: (1) particle entrainment by the gas flowing over the surface of the soil and (2) pressure changes in the soil caused by the flow of gas into and through the voids in the soil. The first of these processes, entrainment, is analyzed essentially as a time-independent phenomenon; whereas the second process is analyzed as a transient effect. If only the first process is operating on the surface, the rate of erosion depends almost entirely on the height of the engine nozzle above the surface. If both processes are at work, which is most probably the case, the erosion rate depends upon the height of the engine nozzle above the surface and the time during which the engine nozzle remained at this point above the surface. With gas flow through the soil, the erosion rate will increase with time for any given nozzle height.

Landing

Following engine shutdown, when the footpads were approximately 1.5 m above the lunar surface, the spacecraft fell (as the engine thrust decayed) until the footpads made contact. The impact was relatively gentle, with stroking of the main shock absorbers limited to a few centimeters at most. The spacecraft came to rest oriented with the $+Z$ axis approximately 20° clockwise from due west and tilted approximately 4° toward the $-Y$ axis (or to the south). As shown in figure 10–52, all the footpads except the $-Y$ footpad penetrated the surface only a small distance, on the order of 1 or 2 cm. The $-Y$ footpad penetrated deeper (approximately 10 cm) and disturbed the surface material to a greater extent than did the other footpads. The appearance of the surface around the $-Y$ footpad can be seen in figure 10–53. Typical penetration of the other footpads is shown in the photograph of the $+Y$ footpad (fig. 10–54). The depression adjacent to the footpad and appearing below the bent contact probe in figure 10–54 was apparently not caused by the impact and bounce of the $+Y$ footpad, but is a natural surface crater. This fact is indicated by the track of the contact probe seen in the foreground of figure 10–54. It is evident in other photographs, which are not included in this report, that this groove extends 1 or 2 m to the left of the area shown in the photograph, indicating the motion of the spacecraft in the last 1 to 2 sec of descent. The position of the groove is consistent with the location of the contact probe on the footpad, and no other groove appears in a similar position with respect to the small crater mentioned previously. It was observed on the Apollo 11 landing that the usual effect of the descent-engine exhaust gas is to accentuate the surface disturbance caused by a probe rather than to cover up the disturbance; therefore, it is unlikely that a track was formed and subsequently obscured. The broken probes indicate that the spacecraft was traveling in a direction slightly north of west in the final stages of descent.

FIGURE 10-52.—Area under LM and around footpads. (AS12-46-6777)

FIGURE 10-53.—Penetration of −Y footpad. (AS12-47-6901)

FIGURE 10-54. — Crater, +Y footpad, and area under LM descent engine. (AS12-47-6906)

As in the Apollo 11 photographs, Apollo 12 photographs show that the surface under the descent engine and adjacent to the footpads (fig. 10-54) was apparently swept by the descent-engine exhaust gas, although more particles seem to have been left on the surface in the vicinity of the Apollo 12 LM than under the Apollo 11 LM. This may have been a result of the different shutdown conditions. In a number of photographs that show an area of lunar surface 12 to 15 m to the east of the center of the LM along its approach track, such as figure 10-55, a path appears that is clearly different from the surrounding surface and apparently occurs along the approach path. This path seems to be a result of the surface disturbance caused by the exhaust gas during descent. According to the descent trajectory, the spacecraft engine nozzle was at a height of 9 to 12 m from the surface at a position corresponding to the right-hand portion of the area shown in figure 10-55.

Observations During the EVA Periods

During the EVA periods, information relating to the physical characteristics and the mechanical behavior of the lunar surface material at the Apollo 12 landing site was extracted from various astronaut activities, including the following:

(1) Initial familiarization and adjustment to the lunar environment

(2) Trenching and collection of rock, soil, and core-tube samples

(3) Deployment of the solar-wind composition (SWC) experiment, the U.S. flag, and the Apollo lunar surface experiments package (ALSEP)

(4) Observations and photography relating to the LM landing interaction with the lunar surface

The general topography of the Apollo 12 landing site is characterized by a gently rolling surface that includes several large, subdued craters and many smaller craters with raised rims.

FIGURE 10-55. — Lunar surface, 40 to 50 ft east of the LM, showing surface erosion track. (AS12-46-6781)

In general, the surface material at the Apollo 12 site can be described as a medium-gray, slightly cohesive, granular soil that is composed largely of bulky grains in the silt to fine-sand size range, with scattered glassy rocks and coarse rock fragments. Large rock fragments ranging in size from several centimeters to several meters across and varying in shape from angular to subrounded are sparsely strewn throughout this matrix material. Coarse blocks occur abundantly on and around a few craters. Most of the rocks are partially buried with fillets of fine-grained material built up around them. The soil is generally similar in appearance and behavior to that encountered at the Apollo 11 site and at the Surveyor equatorial landing sites. There is, however, some variability in soil conditions at different points along the geologic traverse shown in figure 10-1, as well as some features and aspects of behavior that are unlike those found at the Apollo 11 site.

The astronauts concluded that in terms of soil texture and behavior, there were three different and distinct areas along the geologic traverse made during the second EVA period. These three areas were:

(1) *In the vicinity of the LM and out to Head Crater.* In this region, the soil was of moderate compactness and provided good support for the astronauts and the experiment packages.

(2) *In the vicinity of Sharp Crater.* The soil was softest near this crater. This soil could be trenched easily, as shown in figure 10-56, and footprints in this area were the deepest.

(3) *In the region around Halo and Surveyor Craters.* According to the astronauts, the soil was firmest in the area around Halo and Surveyor Craters. In this area, the soil had the appearance of a dusty surface that had been lightly rained upon, as may be seen in figure 10-57. The material in this area was described as being more cohesive and coarser than in other areas. The bootprint in the lower portion of the photograph indicates that this material compacts under load in the same manner as the other types of surface material encountered.

Figure 10-58 is a view of the area west-north-

GEOLOGIC INVESTIGATION OF THE LANDING SITE 167

FIGURE 10-56.— Trench approximately 8 in. deep in the soft material on the east rim of Sharp Crater. (AS12-48-7067)

FIGURE 10-57.— Lunar surface in the vicinity of Halo Crater showing "lightly rained on" texture of undisturbed material and material compacted by an astronaut's footstep. (AS12-49-7284)

west of the LM taken after the EVA periods; the figure shows the deployed ALSEP in the background and the darkened trails left by the astronauts where they passed along the surface. This characteristic difference between the undisturbed surface and the soil directly beneath was also observed at the Apollo 11 and Surveyor 3 sites. The shallow depth of footprints in the foreground of figure 10-58 indicates that the soil near the LM was reasonably firm and incompressible.

Color variations were evident within small zones, and the material did not appear to be homogeneous with depth at all locations. Figure 10-59 shows a zone of lighter colored material exposed in a small trench on the northwest rim of Head Crater. Where this lighter colored material occurs, it appears to have clumped up more as a result of disturbance; however, the overall mechanical behavior of the lighter colored material seems to be quite similar to that of the darker material.

Texture and General Appearance of Undisturbed Lunar Surface

The lunar surface texture is characterized locally by linear grooves approximately 0.3 cm deep, such as those northwest of the LM (fig. 10-13) near the Middle Crescent Crater, the outer slopes of Sharp Crater, near Halo Crater, and at and near the Surveyor 3 spacecraft. These lineations are described in part A of this section.

Figure 10-60 shows the texture and appearance of the disturbed lunar surface material at the ALSEP deployment site, looking cross-Sun. The material at that site appeared to be loose and fluffy and, according to Astronaut Bean, was difficult to compact by merely stepping and tamping on it. The fine-grained surface material had a powdery appearance and was easily kicked free as the astronauts moved on the surface. During the Apollo 11 EVA, Astronauts Armstrong and Aldrin noted the ease with which fine-grained material was set in motion while they were walking on the lunar surface. They also noted that the particles moved along ballistic trajectories according to a pattern that depended on the angle of impact of the boot with the lunar surface.

Adhesion

The tendency of the loose, powdery surface material to move easily in the lunar vacuum and 1/6g environment imposed operational problems that were augmented by the fact that the same

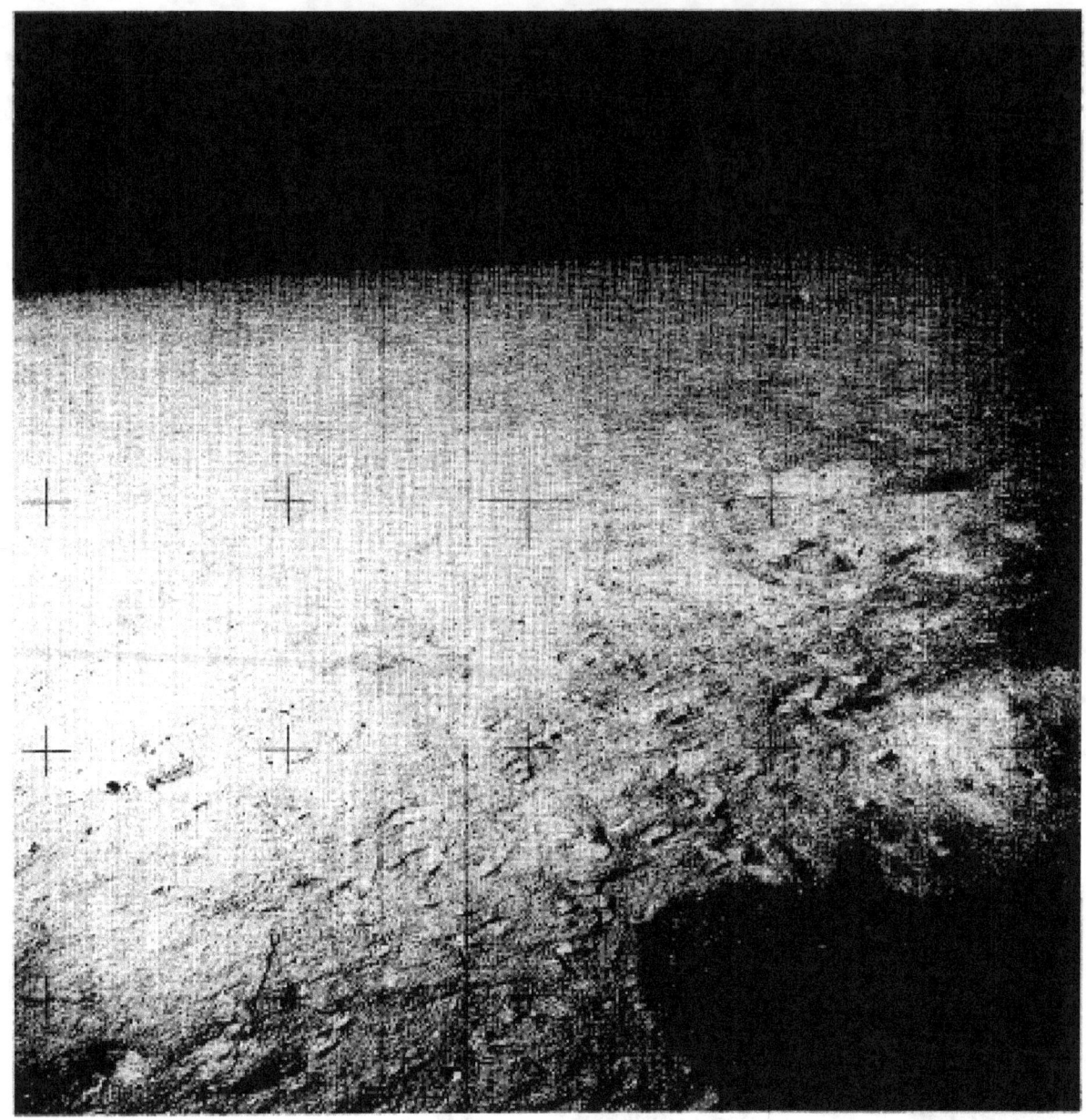

FIGURE 10-58. — View of the area west-northwest of the LM after the EVA periods, showing deployed ALSEP in the background and darkened trails left by the astronauts. (AS12-48-7169)

material also exhibited adhesive characteristics that resulted in a tendency for the material to stick to any object with which it came into contact. As a consequence, equipment and spacesuits became coated, and housekeeping problems arose from the dust brought aboard the LM at the conclusion of EVA periods.

Fine-grained material adhered to the astronauts' boots and spacesuits, the television cable, the lunar equipment conveyor, ALSEP components, astronaut tools, sample return containers, the color chart, and the cameras and camera magazines. Figure 10–61 shows fine-grained material accumulated on the radio thermoelectric

FIGURE 10-59. — Small trench near northwest rim of Head Crater. Note zone of light-colored material just beneath the surface. Firmness of the ground in this area is indicated by the shallow bootprints. (AS12-48-7052)

FIGURE 10-61. — Astronaut unloading RTG fuel capsule from LM. Note the general texture of surficial fine-grained material disturbed by astronaut activities and the fine-grained material accumulated on the RTG pallet and adhering to the tapes lying on the surface and to the astronaut's boots, suit, and gloves. (AS12-46-6789)

FIGURE 10-60. — General appearance and texture of lunar soil disturbed by astronaut activities at the ALSEP deployment site, looking cross-Sun. (AS12-46-6817)

generator (RTG) pallet, adhered to the astronauts' suits, boots, and gloves, and adhered to the tapes extending from the scientific equipment bay of the LM and to the tapes lying on the lunar surface. The fine, powdery material also adhered to lunar rock samples that were brought back to Earth and left a trace of fine dust that coated the core tubes that were returned to the Lunar Receiving Laboratory (LRL).[2] This adhesion, however, did not offer any resistance to pulling of objects inserted into the lunar surface, such as the SWC experiment staff, the flagpole, or the core tubes.

It appears that under the shirt-sleeve atmosphere (5 lb/in.^{2}) of the command module (CM), the fine, dusty material lost its adhesive characteristics. In connection with this phenomenon, Astronaut Conrad commented that camera magazines coated with lunar soil when stowed were clean when removed from their bags in the CM a few days later.

[2] See page 178.

Cohesion

The soil possesses a small, but finite, amount of cohesion. The fine-grained material tends to form clods in both its natural and disturbed state. For example:

(1) Figure 10-53 shows clumps of fine-grained material "bulldozed" away from the +Y pad as a result of the interaction between the LM and the lunar surface during landing.

(2) Figures 10-62 and 10-63 show weakly cemented clumps and blocks of fine-grained material from two distinct mounds on the lunar surface (figs. 10-22 and 10-23). According to Astronaut Conrad, the mounds appeared to contain very compacted, fine clumps similar to dried cement or sandstone. These clumps disintegrated when squashed in the palm.

Initially loose, fluffy material readily compacts under load and retains the detail of a deformed shape, as may be seen in figures 10-30 and 10-31 obtained from the contingency sample collection area. Under the existing illumination, compacted surfaces gave a glossy appearance. The rubber-like texture of deformed and compacted surface material can also be seen in the bootprints shown in the foreground of figure 10-64. The grainy

FIGURE 10-63. — Closeup view of small mound. Weakly cemented block of fine-grained material in foreground is approximately 0.5 m across. (AS12-46-6824)

structure of the compacted fine-grained material is clearly shown, however, in high-resolution stereoscopic photographs of bootprints (fig. 11-3) obtained by use of the ALSCC.

The material could stand unsupported at a height of at least a few centimeters on vertical slopes developed either by trenching, as shown by the grooves dug with the contingency samples (figs. 10-30 and 10-31), by the holes made by the core tubes that remained intact upon the removal of the tubes, or by lateral scuffing and compression, as can be seen from the ridges shown in figure 10-64.

Frictional Characteristics

As at the Apollo 11 site, the lunar soil encountered during this mission derives a major portion of its strength from interparticle friction. This is evidenced by the fact that material resistance to deformation increases considerably with confinement. The relevant material properties can be assessed from the following observations:

(1) The penetrations of the footpads were small, 5 to 7.5 cm, except for the −Y footpad, which penetrated 10 to 12 cm. These penetra-

FIGURE 10-62. — Weakly cemented fine-grained material and rock fragments on top of large mound. (AS12-46-6832)

FIGURE 10-64.—Fillet material around partially exposed rock. Note the ridges developed by scuffing and lateral compression of the fine-grained material and the deeper bootprints at the toe as a result of uneven distribution of astronaut weight on boot. (AS12-49-7181)

tion values correspond to static bearing pressures of 55.2×10^3 dyn/cm^2 to 75.8×10^3 dyn/cm^2.

(2) The depth of penetration was small, approximately 1 cm of the astronauts' boots on a level surface (figs. 10-61 and 10-64). Softer spots were found, however, on the rims and slopes of relatively fresh, small craters (fig. 10-65).

Strength and Deformation Characteristics

As in the Apollo 11 mission, no special soil mechanics testing or sampling devices were included in the Apollo 12 EVA hardware, and no other force- or deformation-measuring devices were used during the surface activities. Accordingly, the strength and deformation characteristics of the lunar soil could be determined only by indirect means, such as from observations made on the appearance of the material and its interaction with objects of known weight and geometry that came in contact with the lunar surface. From analyses based on such indirect means, it appears that, although during the second EVA period the astronauts noted three distinct areas in terms of soil consistency, compaction, and firmness, the mechanical behavior of the soil is, in general, consistent with the behavior that would be expected for a soil having properties characteristic of the soils studied at the Apollo 11 site (ref. 10-7) and the Surveyor equatorial landing sites (ref. 10-8). Such a material has a density on the order of 1.5 g/cc, an angle of internal friction of 35° to 39°, and a cohesion of 3.5×10^3 to 7×10^3 dyn/cm^2. The unit bearing capacity is certainly considerably in excess of the pressure of 7×10^4 to 14×10^4 dyn/cm^2 exerted by the astronauts and the LM footpads. Although, in the general area visited during the first EVA period, many of the footprints appear to have resulted from soil compression, in several instances footprints 5 to 8 cm deep are accompanied by bulging of the surrounding surface. An example of this type of soil deformation can be

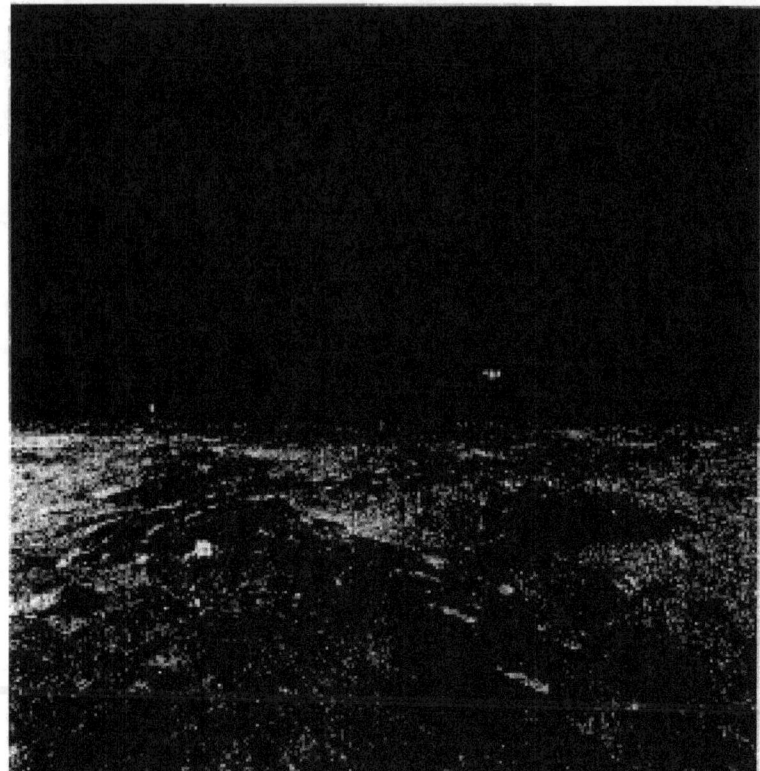

FIGURE 10-65.—Astronaut bootprints near small crater with soft rim. Note the large amount of sinkage at the toe of the bootprints in the foreground, resulting from uneven distribution of astronaut weight on boot, and the deep-seated bootprints at the rim of the crater. (AS12-47-6978)

FIGURE 10-66.—Astronaut bootprints developed as a result of uneven distribution of astronaut weight on boot. Note striation pattern on lunar surface, following an east-west direction, possibly caused by LM engine exhaust gas impingement. (AS12-46-6780)

seen in the footprints shown in the foreground of figure 10-66 and near the rim of the crater shown in figure 10-65. Lateral soil bulging that accompanies imprinting or trenching action can also be seen in the photographs showing the interaction of the LM footpads with the lunar surface (fig. 10-54) and in the trenches dug by the contingency sampler (figs. 10-30 and 10-31). Such behavior reflects deformation dominated by shear effects, rather than by compression, and is consistent with the behavior observed in some of the bearing tests conducted by the surface sampler during the Surveyor 3 and 7 missions.

Subsurface Conditions

The granular surface material extends at least to the maximum depth probed by the astronauts. The astronauts reported little change in texture or consistency with depth, although observations in the LRL suggest that some differences in grain size and color with depth did exist at the core-tube sites. (See part A.) No difficulty was encountered by Astronaut Conrad in scooping "a whole bagful of dirt" as a contingency sample from a small crater in front of the LM (figs. 10-30 and 10-31), nor were there any problems reported in collecting selected soil samples in the area located between the ALSEP deployment site, the east rim of the Middle Crescent Crater, and the LM.

Both the flagstaff and the SWC experiment staff were pushed to depths comparable to the penetrations of the same staffs at the Apollo 11 site. From the distance above the surface of the knurled markings on the flagstaff (ref. 10-7, fig. 4-8), the penetration of the flagstaff is estimated to be approximately 17 cm. From the distance above the surface of the knurled markings on the SWC experiment staff, the SWC experiment staff was estimated to penetrate to a depth of approximately 15 cm. The final deployment position was approximately 65 m from the LM. Less difficulty was encountered in core-tube driving in the immediate vicinity of the LM than at the Apollo 11 site. Inasmuch as the bits of the core tubes used in the Apollo 12 mission had straight inside walls, whereas those used in the Apollo 11 mission were flared inward 15°, the ease with which the tubes were driven at the Apollo 12 site may, to a great extent, result from the design of the core bit, which does not force the soil sample to deform and compress as it fills the tube.

A trench, approximately 20 cm deep, was dug near the rim of Sharp Crater by using the scoop. No difficulty was encountered in digging, and the trench depth was limited only by the length of the extension handle. The trench remained open and stable, although the top edges could be crumbled easily. Analysis shows that if the soil possesses a density, friction angle, and cohesion of the magnitudes postulated, then vertical walls several times the height of the trench walls should remain stable. A core tube was driven without difficulty to its full length (35 cm) beneath the bottom of the trench. It was reported that this tube could almost have been pushed in without a hammer.

A double-core tube was driven near Halo Crater to a depth of approximately 70 cm. Figure 10-67 is a view taken at the completion of

FIGURE 10-67. — Double-core tube at the completion of driving. No disturbance of the surrounding surface is evident as a result of driving the core tube. (AS12-49-7288)

FIGURE 10-68. — Surveyor 3 with the LM in the background. Footpad 1 is to the left, footpad 2 is in the foreground, and the surface sampler extends to the right. (AS12-48-7100)

driving. It may be seen that there is essentially no disturbance of the lunar surface adjacent to the core tube. No hard material was encountered during the driving of any of the core tubes, and the core tubes were easily withdrawn from the ground in each case. The astronauts reported that the core tubes were augered between blows, although both astronauts also indicated augering was probably not necessary. The influence of augering on the recovered samples is not known. Core-tube holes remained open after withdrawal of the tubes.

Soil Conditions at the Surveyor 3 Site

Surveyor 3 with the LM in the background is shown in figure 10–68. Examination of the photographs taken at this site suggests that the surface has undergone little change in the past 2½ yr. For example, figure 10–69 shows the waffle-textured print of footpad 2 even more clearly than the original Surveyor photographs. The detail shown in figure 10–69 indicates clearly that little change other than darkening of the imprint could have taken place during the past 2½ yr and that any depositional or erosional processes must be slow relative to this time span.

Surveyor 3 footpads 1 and 3 are shown in figures 10–70 and 10–71, respectively. It should be noted that footpad 1 was not visible and footpad 3 was only partially visible to the Surveyor 3 television camera. Footpad 1 left a waffle imprint similar to that of footpad 2. Material deposited on the top of footpad 3 appears to be of two different types: light and dark colored.

FIGURE 10-69.—Surveyor 3 footpad area, showing footpad prints formed during touchdown hop. Note fresh appearance of the footpad prints and the waffle pattern caused by the footpad honeycomb. (AS12-48-7110)

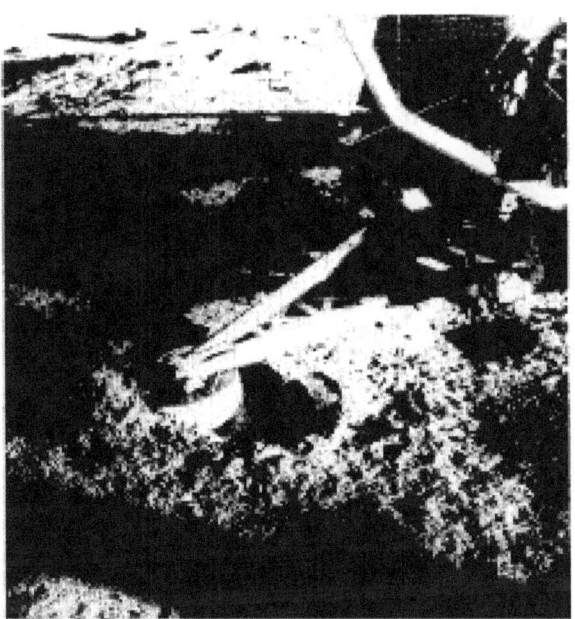

FIGURE 10-70.—Imprint of Surveyor 3 footpad 1, showing waffle pattern. (AS12-48-7119)

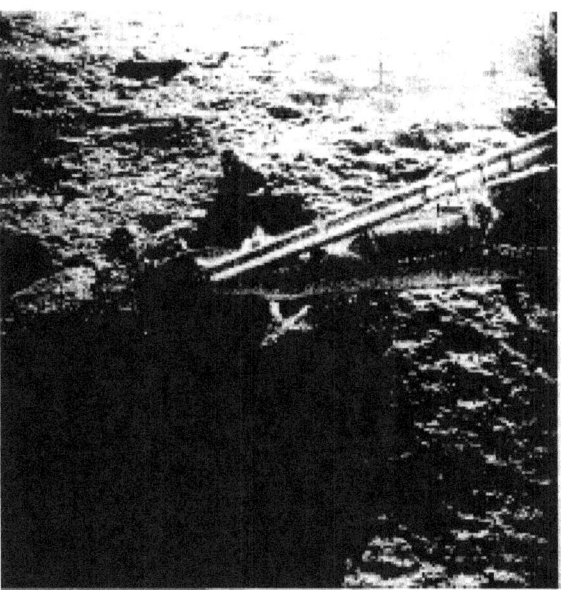

FIGURE 10-71.—Surveyor 3 footpad 3. (AS12-48-7124)

The significance of this is not yet understood. The astronauts reported that the white portions of the Surveyor 3 spacecraft were a light-tan color, and it appeared that some of this coating was dust. The possibility that at least some of the coating was caused by blowing dust produced by the LM landing cannot be eliminated at present. Further studies of the photographs and possibly of the returned components should clarify this question.

Views of the area of Surveyor 3 surface sampler operations are shown in figures 10-72 and 10-73. For comparison with the photographs, a plane view of the area is available in figure IV-4 of reference 10-1 that gives the locations of all tests performed during the first lunar day of the Surveyor 3 mission. Unfortunately, the slope of the surface on which Surveyor 3 rests and the low-Sun angle at the time of the astronauts' visit have the result that much of the surface-sampler test area is in shadow. The major features (trenches and some impact and bearing-test points) are visible, however, and a comparison of the lunar surface details around impact test 1 (as seen in the left foreground of fig. 10-72 and near the left of fig. 10-73), for example, is possible. In the preliminary examinations to date, no change has been detected.

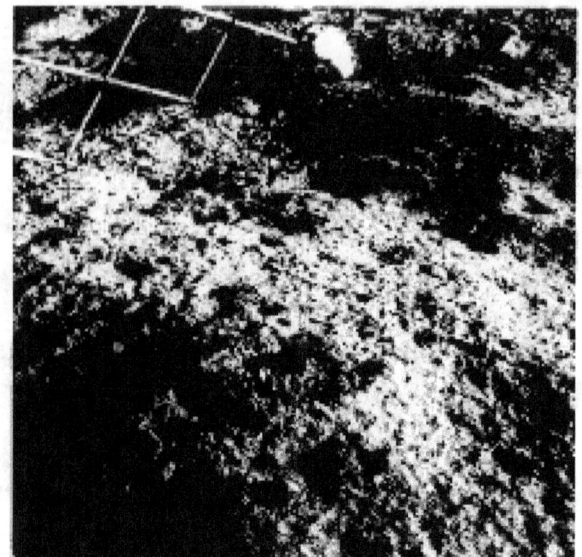

FIGURE 10-72. — Surface under Surveyor 3 soil mechanics surface sampler, showing trenching and bearing-test activities. (AS12-48-7107)

FIGURE 10-73. — Soil mechanics surface sampler operations at the Surveyor 3 site. (AS12-48-7129)

Astronaut Mobility

The soil conditions at the Apollo 12 landing site did not pose any problems on astronaut mobility. As at the Apollo 11 site, the lunar surface at the Apollo 12 site provided adequate bearing strength for standing, walking, loping, and jumping; and sufficient traction for starting, turning, and stopping. Neither astronaut experienced any slippery surfaces such as those reported by the Apollo 11 crew. Walking on the surface tended to compact the loose virgin material. Accordingly, astronaut mobility was enhanced in previously walked-on regions.

Walking through craters only a few feet in diameter that had relatively soft slopes with angles ranging up to 15° did not appear to be a problem. Such a crater is shown in figure 10-74. This crater is approximately 8 m in diameter and 1 to 2 m deep and is located between the magnetometer and the other deployed ALSEP components. Bootprints left by Astronaut Bean as he walked through this crater can be seen below the ALSEP central station shown by the standing astronaut and to the left of the magnetometer (fig. 10-74). Astronaut Bean remarked:

> As I walked through it, it was quite a bit softer in the bottom. I was prepared to find this softer soil, and I did . . . I did not see how far I was sinking in the soil. I only noticed that the bottoms of [this type] of craters were softer than the edges.

Walking on crater slopes on which the soil was firm and compact, such as the one at the Surveyor landing site, also proved to be no problem. Astronaut Bean remarked that he would not hesitate to attempt a steeper slope. However, coming up the slope involved more work because of the restriction imposed by the sloping ground on the astronauts' movements. The astronauts decided not to go to the bottom of Bench Crater because of the steepness of the walls, and there is evidence that the surface material on some crater walls may be of marginal stability. Figure 10-75 is a view looking west into Sharp Crater indicating material that appears to have slid downslope. Traversing such zones could be hazardous. Distinct evidence of sliding was observed by the astronauts in other craters as well.

Conclusions

The following conclusions may be made from the evaluation of available data.

(1) In spite of local variations in soil texture, color, grain size, compactness, and consistency, the soil at the Apollo 12 site is similar in appear-

GEOLOGIC INVESTIGATION OF THE LANDING SITE

FIGURE 10-74. — Small crater (8 m in diameter and 1 to 2 m deep) with soft fine-grained material at the slopes and bottom, located at the ALSEP deployment site. Note bootprints left by Astronaut Bean as he walked through this crater. (AS12-47-6921)

FIGURE 10-75. — Westward view into Sharp Crater, showing evidence of instability of unconsolidated surface material. (AS12-49-7274)

ance and behavior to the soils encountered at the Apollo 11 and the Surveyor equatorial landing sites.

(2) Although the deformation behavior of the surface material appears, in general, to involve both compression and shear effects, the conclusion drawn from the Surveyor 3 mission results—that the soil at the Surveyor 3 landing site is essentially incompressible—is consistent with the consistency, compactness, and average grain size of the soil at the Surveyor 3 site, as assessed during the Apollo 12 EVA.

(3) The adhering dust appears to have been the most aggravating operational problem during the Apollo 12 EVA.

(4) Compaction of loose, very fine-grained surface material for the purpose of supporting scientific instruments with low settlement tolerances as well as other foundation purposes may present problems in the future.

(5) There appears to be no problem associated with shallow excavation or small embankment construction on lunar surface sites with material formations similar to those studied to date.

(6) There appears to be no direct correlation between crater slope angle and consistency of soil cover. The consistency of the soil cover depends mainly upon the geologic history of lunar terrain features and local environmental conditions.

Preliminary Examination of Samples in the LRL

General Description

Considerably less soil material was returned from the Apollo 12 mission than from the Apollo 11 mission. The total weight of the lunar material with a grain size smaller than 2 mm is approximately 4 kg, as compared to 11 kg from Apollo 11. The small quantity returned from the Apollo 12 mission complicates comparison of the samples from the two landing sites. Nonetheless, the majority of the Apollo 12 soil samples are visually identical to the Apollo 11 soil samples. That is, the soil is a charcoal gray with a slight brown tinge. A significant portion of the soil is finer than the unaided eye can distinguish. The soil adheres in a fine layer to everything that comes in contact with it, including stainless steel tools, Teflon bags, and rubber gloves.

The first core-tube sample to be opened and examined in the LRL nitrogen cabinets was the core sample taken during the first EVA period in the vicinity of the LM. (See fig. 10-1.) This core-tube sample (serial no. 2013) was similar in appearance to the two core-tube samples taken during Apollo 11 (serial nos. 2007 and 2008); it was a uniform medium-gray to medium-dark-gray color, and no individual particles were visible. Fine-reflecting surfaces were present over approximately 10 percent of the sample area; these reflecting surfaces gave the sample a slight sparkly appearance. The sample retained its cylindrical shape while resting in a horizontal trough, thereby indicating that the soil retained some cohesion (fig. 10-76). Probing the sample with a spatula revealed that the fine particles tended to form clumps up to 0.5 cm in size; a vertical face 1 cm high could be cut across the diameter of the sample. There were transverse cracks across the sample, which might indicate different zones within the lunar soil sample depth. (Three sections were found to have slightly different grain size distributions.) On the other hand, the cracks may be the result of the rotation of the core tube while the astronaut was in the process of taking the sample on the lunar surface.

Some Apollo 12 samples were different from any obtained during the Apollo 11 EVA. Docu-

FIGURE 10-76. — Core-tube sample (serial no. 2013) taken during first EVA period. (LRL photograph S-69-60359)

mented sample 5D (LRL sample 12033), taken from a trench dug in the northwest quadrant of Head Crater (fig. 10-1), has a distinctly different color from the other soil samples in that it is light gray, similar to the color of cement. In addition, the bottom half of the double-core tube (serial no. 2012) contains zones of different color and grain size, including one distinct zone approximately 2.5 cm long that consists primarily of sand-sized and larger particles.

Sieve Analysis

Half the first core-tube sample (serial no. 2013) was removed along the length of the core in three sections defined by two of the transverse cracks described previously. These sections were individually sieved, and the results are presented in figure 10-77; A, B, and C refer to the upper, middle, and lower sections of the core, respectively. The curves are dashed for the material finer than a no. 230 sieve (0.63 mm) because below this size the curves may be considerably in error as a result of very fine particles sticking together in clumps or adhering to larger particles. The sudden break in the distribution makes curve A particularly suspicious below the no. 230 sieve. Unfortunately, it was impossible to go back and check the results because the samples had already been turned over to the biologists who would expose them to the various plants and animals in the LRL.

It is interesting to note that grain size increases with depth in the core sample. The mean grain size (i.e., the sieve opening at which 50 percent of the soil is finer by weight) is plotted versus depth in figure 10-78. It will not be known until more detailed studies are performed whether this increase in grain size is gradual or occurs in discrete steps that would indicate zones in the lunar soil.

Figure 10-79 compares the grain-size distribution of the Apollo 12 core sample with that obtained for the Apollo 11 core samples analyzed by the Lunar Sample Preliminary Examination

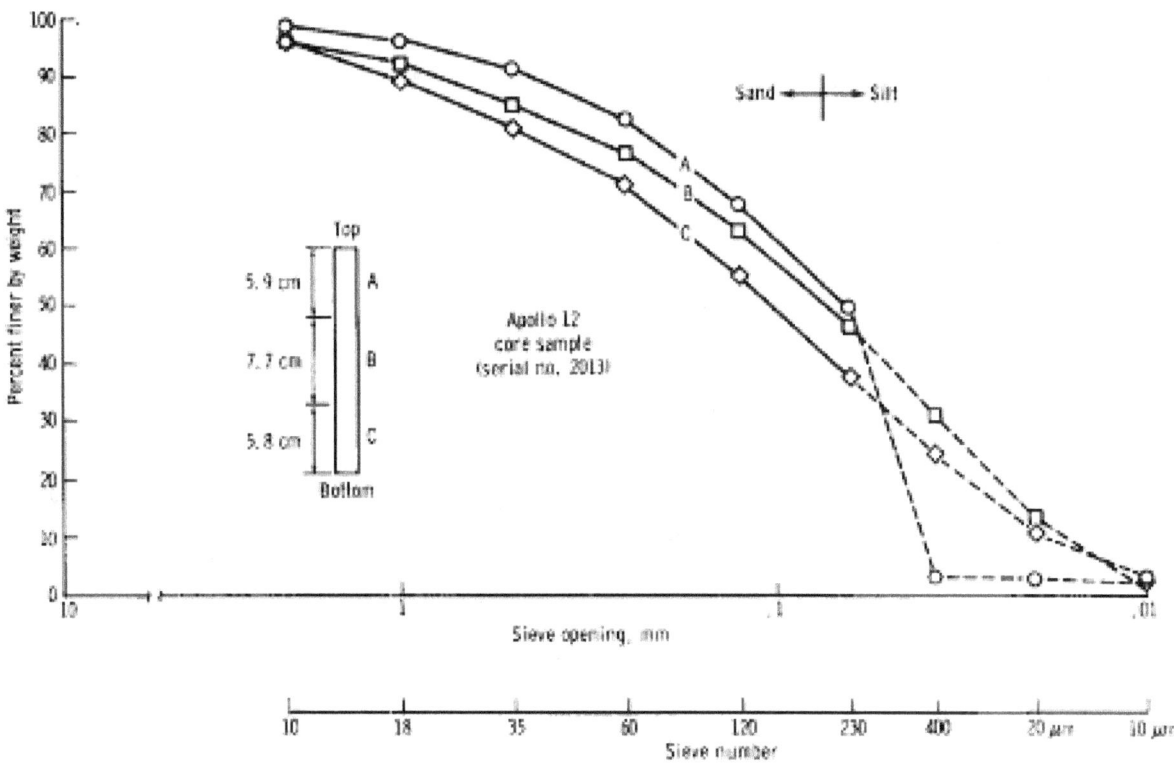

FIGURE 10-77. — Sieve analysis of Apollo 12 core-tube sample (serial no. 2013). The dashed part of the curves is for material that is finer than 0.063 mm (no. 230 sieve).

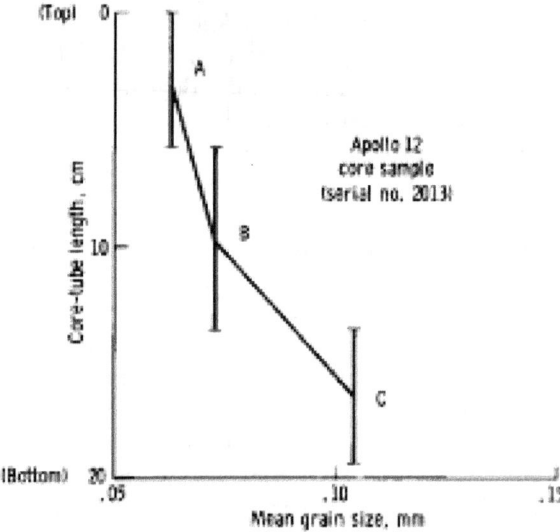

FIGURE 10-78.—Mean grain size versus depth in Apollo 12 core-tube sample (serial no. 2013).

Team (ref. 10-9). Only section B of the Apollo 12 core is shown in figure 10-79, as this section represents roughly the average distribution for the entire sample. It can be seen that the distributions are very nearly the same for grain sizes larger than approximately 0.1 mm. Below this grain size, the distributions differ significantly, with the Apollo 12 analysis indicating a larger proportion of silt-sized particles. It should be pointed out, however, that improvements in the sieving equipment have been made since the Apollo 11 analysis. Thus, the apparent variation is not sufficient evidence to indicate a distinct difference between the two soils. In any case, accurate plots for the grain-size distribution below the no. 230 sieve must await the results of the principal investigators.

Bulk Density

The bulk densities of the Apollo 11 and 12 core-tube samples are presented in table 10-VI. A note of explanation is required concerning the range of diameters shown in the table. A small error in the measured diameter of the sample produces a large error in the calculated sample volume and the bulk density. The Apollo 11 core densities reported previously by the Lunar Sample Preliminary Investigation Team were based on a nominal core-tube bit diameter of 2.00 cm; however, the drawing of the Apollo 11 core bit in figure 10-80(a) indicates that the diameter may be taken to be 1.95 or 1.97 cm. The same is true for the Apollo 12 core bit shown in figure 10-80(b). Thus, the bulk densities shown in table 10-VI have been calculated for diameters of 1.95 and 2.00 cm to indicate the range of uncertainty.

The *in situ* bulk density of the lunar soil has been of great interest for some time (refs. 10-10 and 10-11). The Apollo 11 core samples could not provide an answer, because of the shape of the bit. It can be seen in figure 10-80(a) that the Apollo 11 core-tube bit tapers inward from a diameter of 2.92 to 1.95 cm; these diameters correspond to areas of 6.7 and 3.0 cm^2, respectively. Thus, if the soil were very porous, it could be argued that the bit would compress the *in situ* soil during sampling to as much as double its original density. Conversely, if the soil were densely packed, the shape of the bit would deform the soil and cause the soil to expand to a lower density. Thus, the bulk densities measured in the Apollo 11 core sample could indicate an *in situ* density from 0.75 g/cc to more than 1.75 g/cc.

The Apollo 12 core-tube bit is far from optimal in design, but results in a smaller range of uncertainty. On the other hand, hammering a core into the soil is known to cause more disturbance to the sample than if the core is pushed into the soil at a high, constant speed

TABLE 10-VI. *Core-Tube Bulk Densities*

Core tube serial number	Sample weight, g	Length, cm	Diameter, cm	Bulk density, g/cc
Apollo 11:				
2007	52.0	10.0	1.95 to 2.00	1.66 to 1.75
2008	65.1	13.5	1.95 to 2.00	1.54 to 1.62
Apollo 12:				
[a]2010	56.1	9.3	1.95 to 2.00	1.92 to 2.02
[b]2011	—	—	—	—
[c]2013	189.6	31.5	1.95 to 2.00	1.92 to 2.02
2013	102.9	19.4	1.95 to 2.00	1.69 to 1.77

[a] Upper half of double-core tube.
[b] Not opened; kept in storage at LRL.
[c] Lower half of double-core tube.

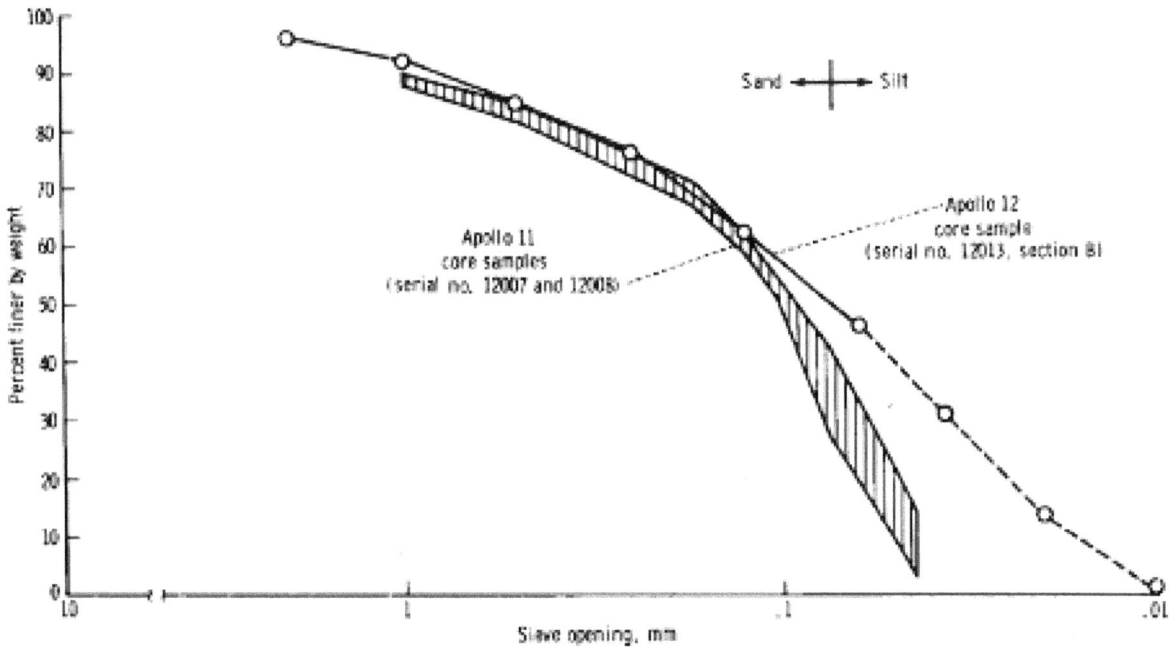

FIGURE 10-79.— Comparison of results of sieve analyses from Apollo 11 core-tube samples (serial nos. 2007 and 2008) and Apollo 12 core-tube sample (serial no. 2013, section B). The dashed part of the Apollo 12 curve is for material finer than 0.063 mm (no. 230 sieve).

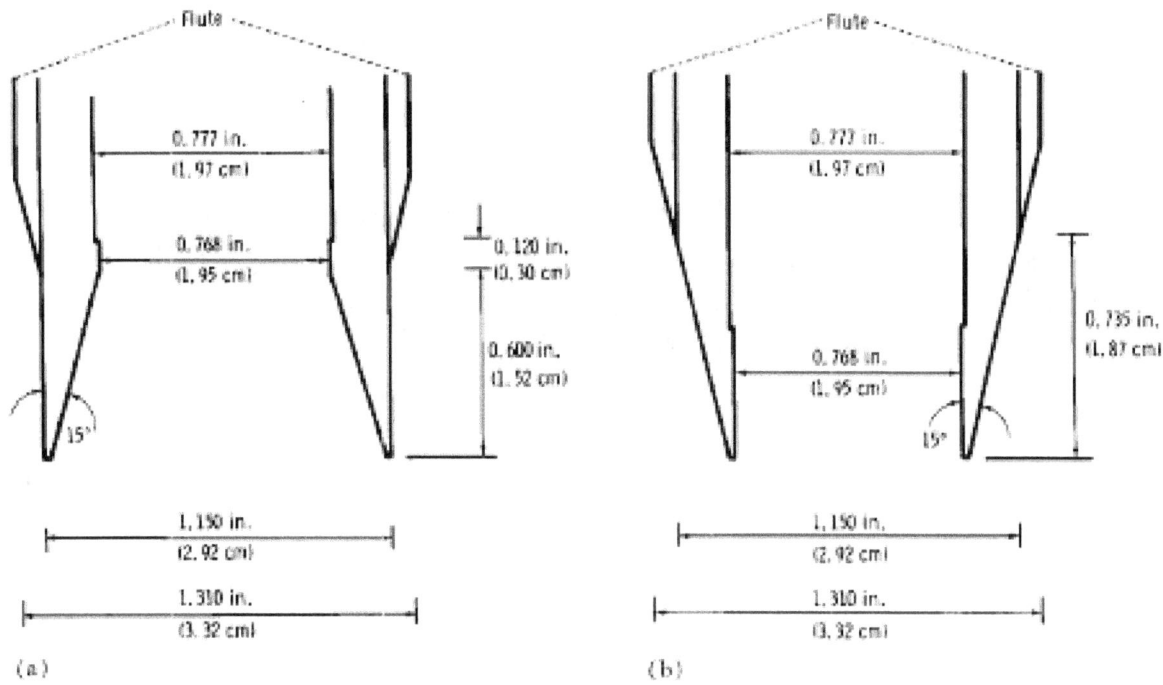

FIGURE 10-80.— Comparison of Apollo 11 and 12 core-tube bits. (a) Apollo 11 core-tube bit. (b) Apollo 12 core-tube bit.

(ref. 10–12). For the time being, the hammering method is unavoidable, and it is necessary to estimate disturbance in terms of the dimensions of the core tube. The degree of sample disturbance has been found to be dependent on the area ratio A_r defined as

$$A_r \text{ (percent)} = 100 \frac{D_e^2 - D_i^2}{D_i^2} \quad (10\text{-}1)$$

where D_e = external diameter
D_i = internal diameter

The smaller the value of A_r, the less disturbed is the sample.

The core-tube bits have raised flutes (shown in fig. 10–80) to facilitate their removal by the astronaut after the sampling operations. By using a weighted average for D_e to include the effect of the flutes, A_r for the Apollo 12 core tubes is calculated as follows:

$$A_r = 100 \frac{(3.06)^2 - (1.97)^2}{(1.97)^2} = 141 \text{ percent}$$

In terrestrial terms, this would be considered very poor, as a standard 2-in.-diameter sampler has an area ratio of only 14 percent. Still, it is a considerable improvement over the reverse-flare core-tube bit of Apollo 11.

It is also important that the area of the bit at the cutting edge be slightly less than the area inside the tube to reduce the friction between the core sample and the inside walls of the core tube. For the Apollo 12 bit, the difference between the two areas was 2 percent.

Taking into account the previously mentioned considerations, it is felt that a rough estimate can be made of the average bulk density of the top 30 cm of the lunar surface at the Apollo 12 landing site. Based on the measured bulk densities in the core tubes, the in situ bulk density is approximately 1.8 ± 0.2 g/cc.

References

10-1. SHOEMAKER, E. M.; BATSON, R. M.; HOLT, H. E.; et al.: Television Observations from Surveyor III. Surveyor III Mission Report. Part II, Scientific Results. JPL Tech. Rept. 32-1177, June 1, 1967, pp. 9-68.

10-2. SHOEMAKER, E. M.; BAILEY, N. G.; BATSON, R. M.; et al.: Geologic Setting of the Lunar Samples Returned by the Apollo 11 Mission. Apollo 11 Preliminary Science Report. NASA SP-214, Oct. 31, 1969, pp. 41-83.

10-3. SHOEMAKER, E. M.; HAIT, M. H.; SWANN, G. A.; et al.: Lunar Regolith at Tranquility Base. Science, vol. 167, no. 3918, Jan. 30, 1970, pp. 452-455.

10-4. TRASK, N. J.: Size and Spatial Distribution of Craters Estimated From Ranger Photographs. Ranger VIII and IX. Part II, Experimenters' Analyses and Interpretations. Tech. Rept. 32-800, Mar. 15, 1966, pp. 252-263.

10-5. SHOEMAKER, E. M.; MORRIS, E. C.; BATSON, R. M.; et al.: Television Observations from Surveyor. Surveyor Project Final Report. Part II, Science Results. JPL Tech. Rept. 32-1265, June 15, 1968, pp. 21-136.

10-6. WILLINGHAM, D.: The Lunar Reflectivity Model for Ranger Block III Analysis. JPL Tech. Rept. 32-664, Nov. 2, 1964.

10-7. COSTES, N. C.; CARRIER, W. D.; MITCHELL, J. K.; and SCOTT, R. F.: Apollo 11 Soil Mechanics Investigation. Apollo 11 Preliminary Science Report. NASA SP-214, Oct. 31, 1969, pp. 85-122.

10-8. SCOTT, R. F.; and KO, H. Y.: Transient Rocket-Engine Gas Flow in Soil. AIAA J., vol. 6, no. 2, Feb. 1968, pp. 258-266.

10-9. The Lunar Sample Preliminary Examination Team: Preliminary Examination of Lunar Samples from Apollo 11. Science, vol. 165, no. 3899, Sept. 19, 1969, pp. 1211-1227.

10-10. SCOTT, R. F.: The Density of the Lunar Surface Soil. J. Geophys. Res., vol. 73, no. 16, Aug. 15, 1968, pp. 5469-5471.

10-11. JAFFE, L. D.: Lunar Surface Material: Spacecraft Measurements of Density and Strength. Science, vol. 164, no. 3887, June 27, 1969, pp. 1515-1516.

10-12. TERZAGHI, K.; and PECK, R. B.: Soil Mechanics in Engineering Practice. John Wiley & Sons, Inc., 1948.

10-13. ANON.: Surveyor III Site 1:2,000. First ed., prepared for NASA under the direction of the Department of Defense by the Army Map Service, Corps of Engineers, Jan. 1968.

11. Lunar Surface Closeup Stereoscopic Photography

T. Gold,[a][†] F. Pearce,[b] and R. Jones[b]

Fifteen stereoscopic pairs were taken with the Apollo lunar surface closeup camera. The camera is similar in all essential respects to the one used on the Apollo 11 mission. A description of the camera is contained in the Apollo 11 Preliminary Science Report (ref. 11-1). All 15 pictures are of excellent quality, and except for a minor malfunction of an exposure counter, the camera functioned perfectly.

All pictures were taken in the close vicinity of the lunar module, and the camera was deployed only for a brief interval at the end of the second extravehicular activity period. Thus, the subject matter is restricted to some targets of interest that were accessible at that time and does not include the many features that were reported during the extended excursions. Each picture is of an area measuring 76 by 82.8 mm. Figure 11-1 is the best available estimate of the locations of the pictures.

Analysis of Pictures

The pictures fall into the following categories:
(1) Dust, disturbed in varying degrees by the descent rocket (figs. 11-2 to 11-11)
(2) Footprints (figs. 11-12 and 11-13)
(3) Rocks (figs. 11-14 to 11-16)

The views in figures 11-2 to 11-16 are included in this report to make the availability of the photographs known. These photographs are available upon request from the National Space Science Data Center, Goddard Space Flight Center, Greenbelt, Md. The following is a list of the 15 photograph numbers keyed to the figure numbers used in this chapter.

[a] Cornell University.
[b] NASA Manned Spacecraft Center.
[†] Principal investigator.

Figure	Photograph
11-2	AS12-57-8441
11-3	AS12-57-8442
11-4	AS12-57-8443
11-5	AS12-57-8444
11-6	AS12-57-8445
11-7	AS12-57-8449
11-8	AS12-57-8451
11-9	AS12-57-8453
11-10	AS12-57-8454
11-11	AS12-57-8455
11-12	AS12-57-8447
11-13	AS12-57-8448
11-14	AS12-57-8446
11-15	AS12-57-8450
11-16	AS12-57-8452

Dust

Figure 11-2 shows patches of coarse material concentrated in shallow depressions, presumably by the action of the descent rocket blowing away the fines. Figure 11-3 shows a pattern of striations presumed to be caused by coarse particles or clumps being moved over the surface by the rocket exhaust. Figure 11-4 shows the effect of molding by the lunar module footpad. Surfaces so molded tend to show bright specks much more than the general surface. These specks are mostly particles of glass. The effect of the mechanical disturbance is to dislodge the coating of fine dust that particles of glass would normally possess. Figure 11-5, which was photographed close to the lunar module descent engine, shows a striation pattern substantially different from the scouring patterns visible in figures 11-2, 11-3, and 11-6. In addition, figure 11-5 shows an unusual hole in the ground that appears to be of natural origin. Such steep-sided holes are probably the consequence of secondary impacts in the speed range of 100 m/sec to 2 km/sec, wherein most of the incoming kinetic energy goes

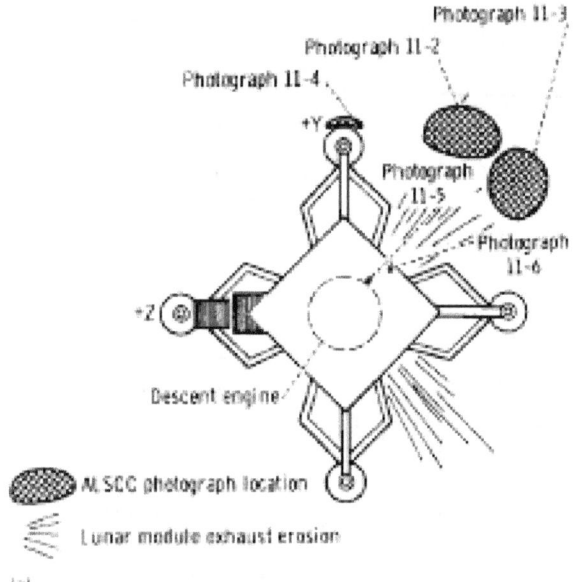
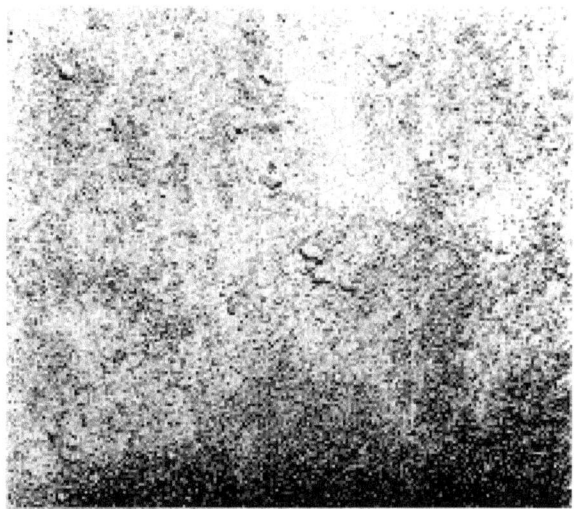

FIGURE 11-2.— Stereoscopic view (AS12-57-8441).

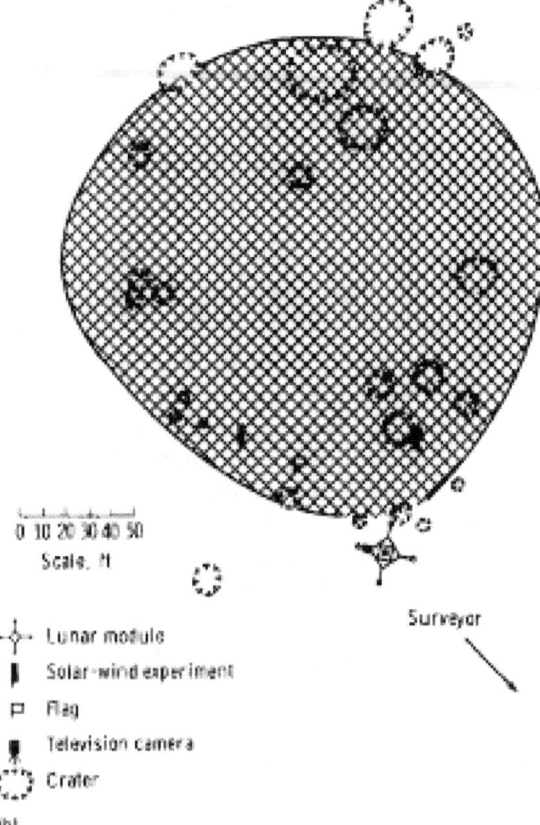

FIGURE 11-3.— Stereoscopic view (AS12-57-8442).

FIGURE 11-1.— Locations of Apollo lunar surface close-up camera (ALSCC) photographs. (a) Locations of figures 11-2 to 11-6. (b) Area in which figures 11-7 to 11-16 were taken.

toward mere displacement of material. Figure 11-6 also was photographed close to the lunar module descent engine. Discoloration and shadowing give a clear indication of the direction of the scouring flow. The cracks are suggestive of a crust, but in fact, a cohesive powder is known to exhibit such crack patterns when disturbed, without any stronger crust being present. Figure 11-7 shows some striation patterns of unknown origin. Figures 11-8 and 11-9 show the consistency of the powdery soil as thrown up by the boots of the astronauts (footspray). This dis-

FIGURE 11-4. — Stereoscopic view (AS12-57-8443).

FIGURE 11-6. — Stereoscopic view (AS12-57-8445).

FIGURE 11-5. — Stereoscopic view (AS12-57-8444).

FIGURE 11-7. — Stereoscopic view (AS12-57-8449).

turbed soil is darker and more lumpy than the undisturbed powder. Some light-colored spray is visible that was presumably caused by a loose agglomerate of a different material having been scattered also by the astronauts' boots. Figure 11-10, which was taken in a small crater, and figure 11-11, which was taken in an open area, show a natural surface of a type not recorded heretofore. The clumps of powder possess a variety of shapes, possibly the consequence of a clumpy ejecta field having been subject to surface erosion processes.

Footprints

By showing the details of the astronauts' footprints, figures 11-12 and 11-13 demonstrate the precision molding of which the soil is capable. The very high cohesion of the soil is quite evident.

Rocks

Figure 11-14 shows the upper surface of a rock. It has a mottled, light appearance and many impact holes in the size range of 0.5 to 3 mm. These holes are mostly glazed and possess

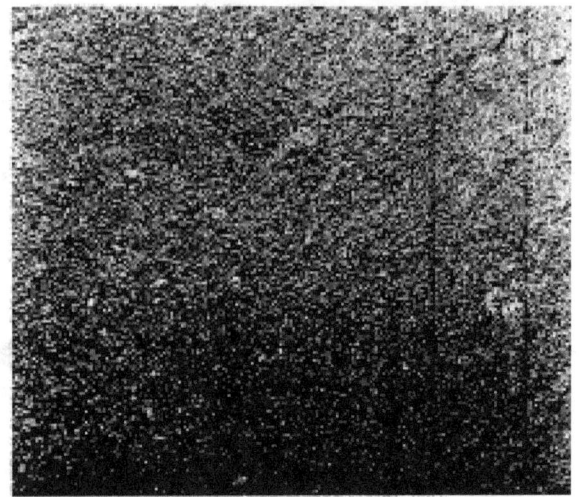

FIGURE 11-8. — Stereoscopic view (AS12-57-8451).

FIGURE 11-10. — Stereoscopic view (AS12-57-8454).

FIGURE 11-9. — Stereoscopic view (AS12-57-8453).

FIGURE 11-11. — Stereoscopic view (AS12-57-8455).

a raised rim, which leaves no doubt that they were caused by hypervelocity impacts.

The remarkable feature seen in this photograph, as on several similar pictures taken during the Apollo 11 mission, is the almost complete absence of dust on the surfaces of such rocks. This cannot be attributed to any cleaning effect by the descent rocket because shadowing would then have to be evident, and shadowing does not occur in any of the cases. In the interval during which all the many impact holes were generated on the surface of the rock, a similar number density of impacts on the neighboring powdery ground must have scattered much powder; and the average situation, if impacts were the only process, would have to be a substantial blanket of dust so that the loss, by impacts, from the dust blanket equaled, in the long run, the gain from scattered material from nearby impacts. The almost complete absence of dust requires an explanation other than such an equilibrium. It must be assumed that there is either a general removal of dust from the lunar surface that dominates all other processes that distribute dust,

FIGURE 11-12. — Stereoscopic view (AS12-57-8447).

FIGURE 11-14. — Stereoscopic view (AS12-57-8446).

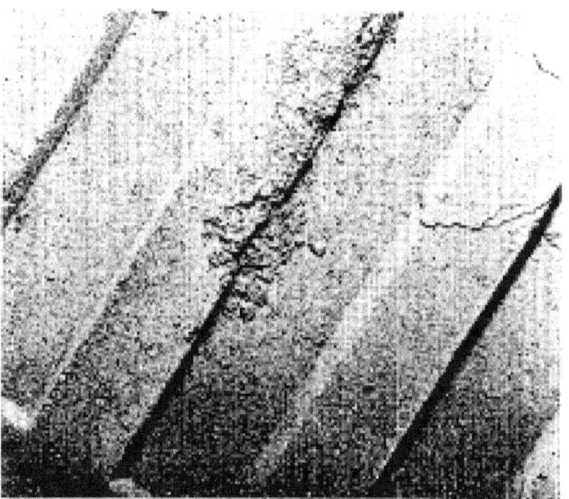

FIGURE 11-13. — Stereoscopic view (AS12-57-8448).

FIGURE 11-15. — Stereoscopic view (AS12-57-8450).

or that there is a dust-transportation process over the lunar surface that has a strong tendency for downhill flow and in which the particles are generally not lifted as high (i.e., more than 5 or 10 cm) as the surfaces of the rocks that exhibit the clean areas. The latter possibility is more in accord with other observations, such as the scarcity of trenches adjoining rocks whose distribution clearly indicates that they fell to their present positions. The trench and pileup that must have been common in the soft soil surrounding a fallen rock must thus be eradicated, yet at the same time no significant amount of material must be deposited on the tops of the rocks. This is a strong indication for a process of surface creep that may be a major process in the long-term evolution of the lunar surface. Figure 11-15 shows a dust-covered rock of rounded shape with a dust-free protrusion. The crack around the rock attracted the astronaut's attention and led him to take this picture. Figure 11-16, taken in a small crater, is a photograph of a rock that has a glassy appearance. Much of the shape of the rock shows patterns of viscous

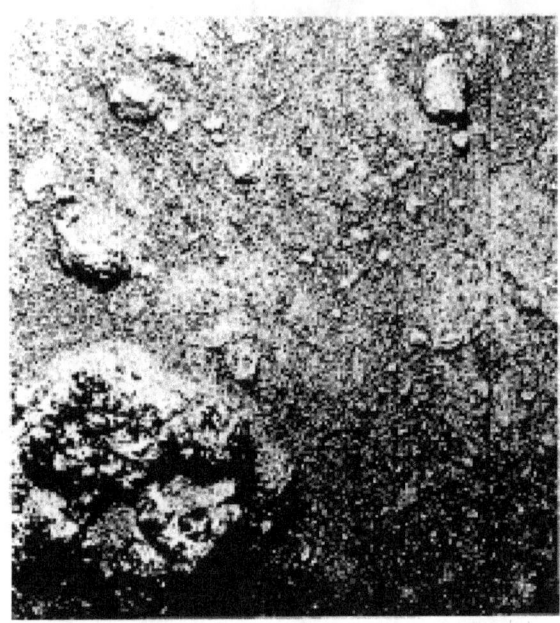

FIGURE 11-16.— Stereoscopic view (AS12-57-8452).

flow. A nearby piece of rock has similar features and may, in fact, be part of the same object. In this case, unlike the case of several Apollo 11 pictures, there is no particular evidence that indicates a surface-glazing process *in situ*. The astronauts reported surface-glazing effects in the centers of many craters, but figure 11–16 shows the only such feature photographed on the Apollo 12 mission.

Conclusions

The pictures give new evidence for the surface processes that appear to be taking place on the Moon. Rocket exhaust features are also seen and can be further analyzed.

Reference

11-1. ANON.: Lunar Surface Closeup Stereoscopic Photography. Sec. 9 of Apollo 11 Preliminary Science Report. NASA SP-214, 1969.

12. Preliminary Examination of Lunar Samples

The Lunar Sample Preliminary Examination Team[a]

This is the first scientific report on the examination of the samples returned from the Apollo 12 mission. Most of these lunar samples have been undergoing physical, chemical, mineralogical, and biological analysis in the Lunar Receiving Laboratory (LRL) at the NASA Manned Spacecraft Center in Houston, Tex., since November 25, 1969.

The Apollo 12 samples may be contrasted with those from Apollo 11 as follows:

(1) While still old by terrestrial standards, the Apollo 12 rocks are approximately 1 billion years younger than those from Apollo 11.

(2) Whereas the Apollo 11 collection contained approximately half microbreccias, there are only two breccias out of 45 rocks in the Apollo 12 return.

(3) The regolith at the Apollo 12 site is approximately half as thick as that of the Apollo 11 site.

(4) The amount of solar-wind material in the Apollo 12 fines is considerably lower than that in the Apollo 11 fines.

(5) The crystalline rocks in the Apollo 12 collection, in contrast to those from Apollo 11, display a wide range in both modal mineralogy and primary texture.

(6) Chemically, the "nonearthly" character of the Apollo 11 samples (high refractory element concentration and low volatile element concentration) is also noted in the Apollo 12 samples, but to a lesser degree.

(7) The chemical composition of the Apollo 12 fine material equals that of the breccias, but does not equal that of the crystalline rocks; this was not observed in the Apollo 11 collection.

The report on the preliminary examination of the lunar samples from Apollo 11 (ref. 12-1) briefly describes the functions of the preliminary examination and the procedures within the LRL, gives the reasons for the sample quarantine, and describes the tools used by the astronauts.

Apollo 12 Samples

Four separate groups of samples (contingency sample, selected sample, documented sample, and totebag sample) were collected on the lunar surface and were returned to Earth. The sample inventory is listed in table 12-I.

The returned lunar material may be divided into the following four groups:

(1) Type A—fine-grained, crystalline igneous rock
(2) Type B—medium-grained, crystalline igneous rock
(3) Type C—breccias
(4) Type D—fines

Contingency sample.—Astronaut Charles Conrad, Jr., collected the contingency sample (table 12-II) 10 m northwest of the lunar module (LM). Using a Teflon bag, Conrad collected the sample early in the first extravehicular activity (EVA) period (EVA-1) to assure that some lunar material would be returned in the event that the EVA periods would be aborted. Approximately five scoops of material were required to obtain 2 kg of material, which included four rocks.

Selected sample.—The selected sample (table 12-III), which replaced the bulk sample of Apollo 11 (ref. 12-1), was collected northwest of the LM (up to 300 m away) during the final hour of EVA-1. Seventeen rocks were collected and placed in one large Teflon bag. Three other large rocks, weighing approximately 2 kg each, were collected and placed in a second large Teflon bag, which was then filled with fine-grained material by using the scoop. Finally, a

---
[a] See "Acknowledgments," p. 215.

TABLE 12–I. *Apollo 12 Sample Inventory*

Material	Mass, g	Container	Gas pressure in container on receipt at LRL
Contingency sample:			
Fines[a]	1 012		
Chips[b]	9		
Rocks (4)	821		
Total	1 932	Teflon bag	1 atm
Selected sample:			
Fines	2 716		
Chips	50		
Rocks (20)	11 940		
Core tube (19 cm)	101		
Total	14 807	1st ALSRC[c]	2×10^{-2} to 4×10^{-2} torr
Documented sample:			
Fines and chips	650		
Rocks (6)	6 124		
Documented bags (13):			
Fines and samples (7)	1 353		
Rock samples (11)	2 288		
Core tube:			
Double-core tube (40 cm)	246		
Unopened	80		
Lunar environment sample container	269		
Gas analysis sample container	101		
Total	11 111	2d ALSRC	0.5 atm
Totebag sample:			
Fines	21		
Chips	10		
Rocks (4)	6 488		
Total	6 519	Teflon bag	1 atm
Summary:			
Fines and chips	5 911		
Rocks	27 661		
Special samples	797		
Total	34 369		

[a] Fines are less than 1 cm.
[b] Chips are between 1 and 4 cm.
[c] Apollo lunar sample return container.

core-tube sample was taken during EVA-1. The two Teflon bags and the core tube were sealed on the lunar surface in the first Apollo lunar sample return container (ALSRC). This ALSRC contained between 40 and 80 μm of gas pressure when it reached the LRL.

Documented sample.—The documented sample (table 12–IV) was collected during the second EVA period, EVA-2, while the astronauts were on their 1.5-km geological traverse (fig. 10–1). The samples were documented by photography using the gnomon. On this geological traverse, the astronauts filled 13 individual sample bags with lunar material, including 13 rocks and seven samples of fine-grained material. Ten additional rocks were collected with the tongs. Two core

TABLE 12-II. *Contingency Sample Inventory*[a]

Sample	Mass, g	Type code	Comments
12070	1102.0	D	Fines
12071	9.16	E	Chips
12072	103.6	A	Basalt
12073	407.65	C	Breccia; originally samples 12073 and 12074 (361.0 + 46.65 g)
12075	232.5	A	Olivine basalt
12076	54.55	A	Basalt
12077	22.63	A	Basalt

[a] Studied in dry-nitrogen cabinets.

TABLE 12-III. *Selected Sample Inventory*[a]

Sample[b]	Mass, g	Type code	Comments
12001	2216.0	D	Less than 1-cm fines; vacuum
12002	1529.5	B	Olivine dolerite;[c] vacuum
12003	300.0	D	Fines and chips; nitrogen
12004	585.0	A	Olivine basalt; vacuum
12005	482.0	A	Olivine basalt; vacuum
12006	206.4	AB	Olivine basalt with radiating feldspar laths; nitrogen
12007	65.2	A	Basalt; nitrogen
12008	58.4	AB	Cumulate (ilmenite); nitrogen
12009	468.2	A	Porphyritic olivine (feldspar), basalt large depression; vacuum
12010	360.0	A	Basalt; vacuum
12011	193.0	A	Olivine basalt; vacuum
12012	176.2	AB	Olivine basalt; vacuum
12013	82.3	A	Igneous breccia; vacuum
12014	159.4	B	Olivine dolerite;[c] vacuum
12015	191.2	A	Porphyritic olivine basalt, large depression; vacuum
12016	2028.3	AB	Basalt; vacuum
12017	53.0	AB	Glass-coated basalt; vacuum
12018	787.0	B	Olivine dolerite;[c] vacuum
12019	462.4	A	Basalt; vacuum
12020	312.0	A	Olivine basalt; vacuum
12021	1876.6	B	Pigeonite dolerite,[c] pegmatite; vacuum
12022	1864.3	B	Olivine dolerite;[c] vacuum

[a] The ALSRC contained 40 to 60 μm of gas pressure when returned to the LRL.
[b] Samples 12002 to 12022 labeled "vacuum" have been in 1 atm of nitrogen for at least 2 hr.
[c] Dolerite is a rock of basaltic composition with an intermediate crystal size.

TABLE 12–IV. *Documented Sample Inventory*[a]

Sample	Mass, g	Type code	Comments
12030	75.0	D	Bag 1–D; fines
12031	185.0	B	Bag 3–D; olivine dolerite
12032	310.5	D	Bag 4–D; fines
12033	450.0	D	Bag 5–D; fines
12034	155.0	C	Bag 6–D; crystal breccia with glass
12035	71.0	B	Bag 7–D; troctolite
12036	75.0	B	Bag 8–D; olivine dolerite
12037	145.0	D	Bag 8–D; fines
12038	746.0	A	Bag 9–D; basalt
12039	255.0	B	Bag 10–D; olivine dolerite
12040	319.0	B	Bag 10–D; olivine dolerite
12041	24.8	D	Bag 11–D; fines
12042	255.0	D	Bag 12–D; fines
12043	60.0	A	Bag 14–D; basalt
12044	92.0	D	Bag 14–D; fines
12045	63.0	A	Bag 15–D; basalt
12046	166.0	A	Bag 15–D; basalt
12047	193.0	A	Bag 15–D; basalt
12048	136.0	D	Bag 7–D; fines
12050	1.0	E	Chip for organic analysis
12051	1660.0	AB	Olivine basalt
12052	1866.0	A	Olivine basalt
12053	879.0	A	Olivine basalt
12054	687.0	B	Shatter cone with glass splash-dolerite
12055	912.0	B	Basalt
12056	121.0	AB	Basalt
12057	650.0	D	Fines and chips from bottom of ALSRC

[a] The ALSRC contained approximately 0.5 atm when returned to the LRL. All of these samples were processed in dry nitrogen.

tubes were driven at two locations; one was a double-core tube. Two special samples, the gas analysis sample container (GASC) and the lunar environment sample container (LESC), which were designed to be sealed individually, were collected (table 12–V). All these samples, except the four largest rocks, were sealed on the lunar surface in the second ALSRC. This second ALSRC contained approximately 0.5 atm of gas pressure when it reached the LRL. The container showed evidence of leakage in three environments: the command module (CM) cabin, air, and dry nitrogen test cabinets.

Totebag sample. — The totebag sample (table 12–VI) consisted of the four rocks that were not sealed in the second ALSRC. The rocks were placed in a large Teflon bag for return to Earth.

All four sample containers were placed in the LM and bagged; then, after rendezvous, the containers were transferred into the CM. The two ALSRC were returned to the LRL by means of the mobile quarantine facility on November 30, 1969.

Lunar Receiving Laboratory operations. — The configuration and operation of the biological barriers in the LRL remained essentially the same as that described for Apollo 11 (ref. 12–1). The selected sample was opened and studied in

TABLE 12-V. *Special Samples*

Sample	Mass, g	Type code	Comments
12023	269.3	D	LESC
12024	101.4	D	GASC
12025	58.1	D	Core 2010 (EVA-2—top of double-core tube)
12026	101.4	D	Core 2013 (EVA-1)
12027	80.0	D	Core 2011 (EVA-2—unopened as of Jan. 1, 1970)
12028	189.6	D	Core 2012 (EVA-2—bottom of double-core tube)

TABLE 12-VI. *Totebag Sample Inventory*[a]

Sample	Mass, g	Type code	Comments
12060	20.7	D	Fines
12061	9.5	E	10 chips from totebag
12062	738.7	AB	Basalt; has depression with raised cone with radial cracks
12063	2426.0	A	Olivine basalt
12064	1214.3	B	Dolerite with cristobalite
12065	2109.0	AB	Pigeonite porphyry, consists of plagioclase, pigeonite, and ilmenite

[a] Studied in dry nitrogen cabinets; handled by people in LRL crew reception area for 20 min.

vacuum (generally on the order of 10^{-6} torr). The documented sample, contingency sample, totebag sample, and core tubes were opened and studied in dry-nitrogen glove cabinets. Approximately 50 g of core-tube material and 450 g of fines and rock chips from the documented and selected samples were committed for biological testing.

Mineralogy and Petrology

The majority of the large rock samples returned by the Apollo 12 crew are holocrystalline, with a range of textures and mineralogical compositions that are characteristic of igneous origin. Two breccias were also returned. The crystalline rocks are similar to the microgabbros and basaltic rocks returned from the Apollo 11 mission in that they consist essentially of clinopyroxene, calcic plagioclase, olivine, and ilmenite. However, the Apollo 12 collection return contrasts to the Apollo 11 collection return in that the Apollo 12 rocks exhibit a wide range in modal mineralogy, grain size, and texture.

Igneous Rocks

Approximately half of the igneous rocks have vesicles present, and all have vugs. The vesicles range in diameter from 0.1 to 40 mm (fig. 12-1) and are commonly lined by tangentially or subparallel-oriented crystals of plagioclase, pyroxene, or olivine. The vugs contain euhedral crystals of pyroxene (fig. 12-2) and olivine and less-well-formed crystals of plagioclase, ilmenite, and spinel. The volume occupied by vugs and vesicles in any rock is generally less than that for the

FIGURE 12-1. — Views of large (up to 8-cm) egg-shaped, smooth-walled vesicles. The walls of these vesicles consist of a matte of lath-shaped plagioclase crystals. Both rocks are highly shocked and consist of plagioclase crystals in a glass matrix. (a) Sample 12009. (NASA S-69-62300) (b) Sample 12015. (NASA S-69-63391)

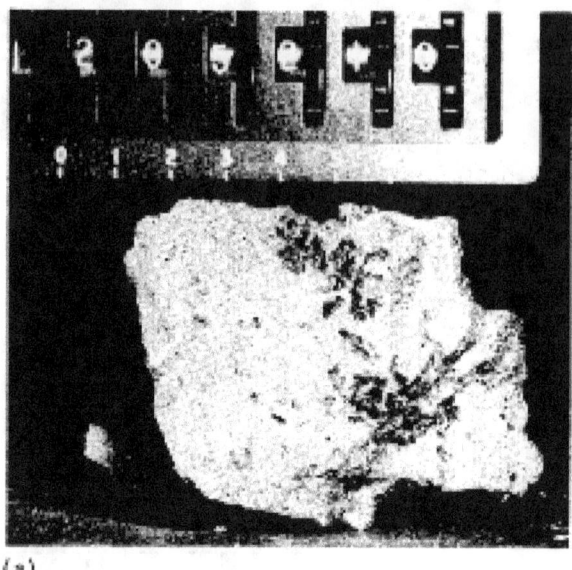

FIGURE 12-2. — A typical equigranular crystalline rock. (a) Sample 12052. (NASA S-69-61241) (b) Closeup of sample 12052. Note the concentration of vugs, perhaps forming a stratigraphic layer. Vugs contain large euhedral crystals of pyroxene. (NASA S-70-21320)

rocks from Tranquility Base. The vugs are irregular, and, in the coarser rocks, occur at the termination of sheaflike aggregates of pyroxene and plagioclase. In one sample, pyroxene crystals appear in raised relief along a joint surface. Crystals in vugs and along joints are considerably coarser in grain size than the groundmass minerals. Variations in cooling rates can explain the variety of grain sizes observed.

The grain size of the igneous rocks ranges from 0.05 to 35 mm. The textures show remarkable variations (fig. 12-3), many of which are common to volcanic and plutonic rocks on Earth. Many of the rocks are equigranular gabbros, some are ophitic to subophitic diabases, and others are variolitic basalts. Feathering sheaves of pyroxene-plagioclase intergrowths are found in the groundmass of porphyritic rocks containing phenocrysts of pigeonite. The olivine crystals in most rocks are equant euhedral grains that are somewhat coarser than the groundmass.

The mineralogical composition of the Apollo 12 rocks reflects high iron oxide (FeO) content. The lower titanium dioxide (TiO_2) content, as compared to that of the Tranquility Base rocks, is reflected in the smaller amounts of ilmenite. The textural and mineralogical variations can be

PRELIMINARY EXAMINATION OF LUNAR SAMPLES

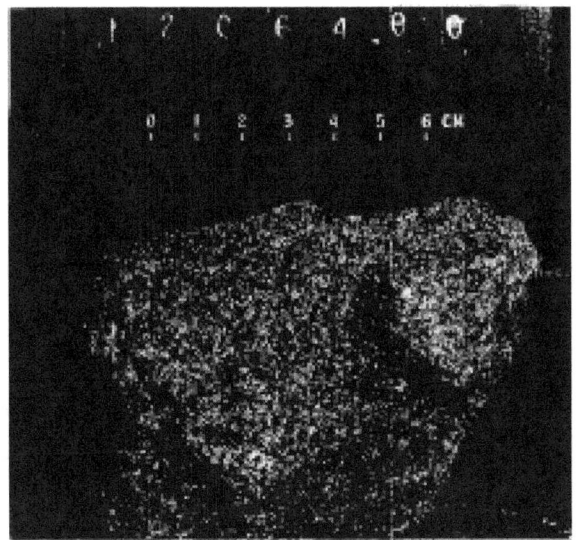

FIGURE 12-3. — Views of rocks and thin sections illustrating the various textures displayed by the Apollo 12 crystalline rocks. (a) Sample 12064, an equigranular holocrystalline rock. (NASA S-69-60899)

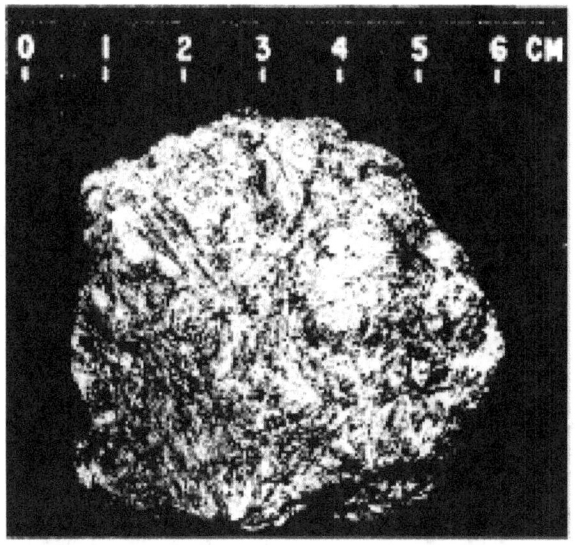

FIGURE 12-3 (continued). — (c) Sample 12031, rock with the porphyritic variolitic texture. (NASA S-69-63651)

FIGURE 12-3 (continued). — (b) Chip of sample 12040, an equigranular holocrystalline rock coarser than sample 12064. (NASA S-69-64828)

FIGURE 12-3 (continued). — (e) Photomicrograph of sample 12021, porphyritic gabbro with variolitic texture. Phenocrysts are pigeonite. Radiating laths are an intergrowth of pyroxene and plagioclase. (NASA S-70-20749)

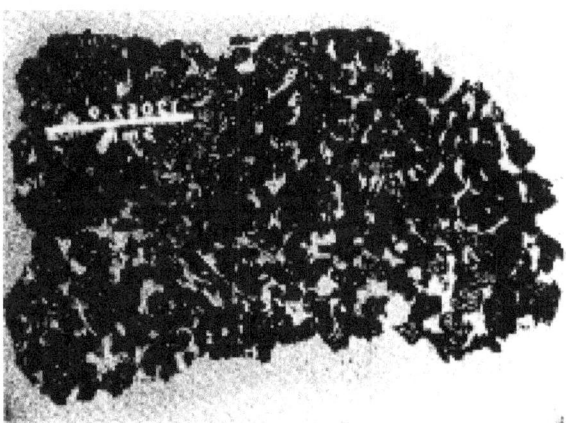

FIGURE 12-3 (continued). — (d) Photomicrograph of olivine basalt from a chip of sample 12057. (NASA-S-69-63409)

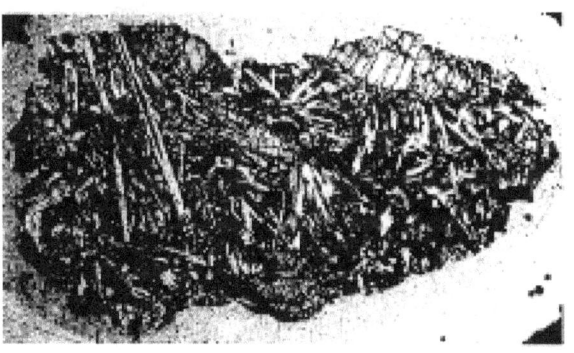

FIGURE 12-3 (continued). — (f) Photomicrograph of sample 12065 illustrating a mafic rock. The texture is variolitic. Plagioclase is approximately 15 percent in this sample. The lath-shaped crystals are pyroxene; the equant crystals are olivine. (NASA S-69-63405)

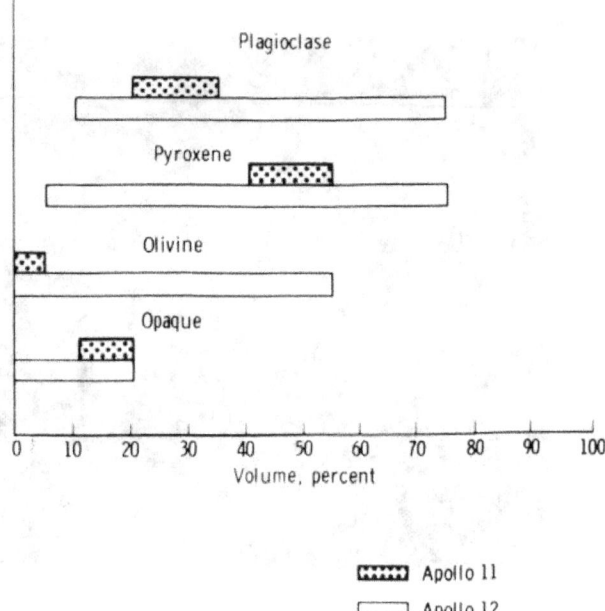

FIGURE 12-4. — Comparison of range of modal mineralogy of Apollo 11 and 12 rocks.

FIGURE 12-3 (continued). — (g) Photomicrograph of sample 12022 illustrating an equant texture in a mafic rock. The large equant crystals are olivine. (NASA S-70-20740)

explained readily by fractional crystallization and mineral accumulation during the cooling of basaltic magmas.

The modal mineralogy shows a wide variation (fig. 12-4), especially when contrasted with the Apollo 11 modes. The Apollo 12 modes range from peridotites (sample 12075: 50 percent pyroxene, 40 percent olivine, and 10 percent plagioclase), olivine gabbros (sample 12036: 25 percent pyroxene, 40 percent olivine, 25 percent plagioclase, and 10 percent ilmenite), and gabbros (like the Apollo 11 rocks) to troctolites (sample 12035: 15 percent pyroxene, 40 percent olivine, and 45 percent plagioclase). Rock sample 12013, which consists largely of plagioclase and sanidine, appears to be a late-stage differentiate on the basis of its modal mineralogy and trace-element content. This sample displays a volcanic breccia or tufflike texture (fig. 12-5).

Some rocks show evidence of planar features, such as fractures and lines of vugs, but generally the rocks are without marked foliation or lineation. All of the rocks are fresh and show no evidence of the hydration or oxidation reactions common during late-stage terrestrial magmatic processes.

Mineralogy

Mineral species identified in the Apollo 12 samples are similar to those observed in the Apollo 11 samples. Glass, plagioclase, pyroxene, olivine, low cristobalite, ilmenite, sanidine, troilite, and metallic iron have been positively identified. Tridymite, spinel, metallic copper, and the

FIGURE 12-5. — Photomicrograph illustrating the volcanic, breccialike structure of sample 12013. (NASA S-69-24223)

FIGURE 12-6. — Photomicrograph illustrating the hourglass structure in a pigeonite crystal from sample 12021. (NASA S-69-24205)

iron analog of pyroxmangite were tentatively identified.

Plagioclase is present in every rock sample, constituting from approximately 5 to 10 percent (in sample 12075) to 70 percent (in sample 12013). Estimates of plagioclase composition in terms of the anorthite (An) component, based on extinction-angle measurements and indices of refraction, range from An_{50} to An_{90}, with the median value falling near An_{80}. Some plagioclase is zoned, and most is twinned. Lath-shaped crystals prevail, and plagioclase-pyroxene intergrowths are very common.

Pyroxenes were identified by X-ray diffraction and optical properties. Pigeonite and subcalcic augites are the most common, as determined optically. Refractive indices indicate that the $Fe/(Fe+Mg)$ ratio is approximately 0.5. Zoned pigeonite phenocrysts occur in porphyritic and in coarse-grained rocks (fig. 12-6). In some crystals, pigeonite occupies the core, and subcalcic augite forms the rim. The subcalcic augite has a deeper brown color than the pigeonite. The groundmass pyroxenes are fine grained, darker in color than phenocrysts, and not zoned. Pyroxene-plagioclase intergrowths are common and range in size from centimeter-long crystals to aggregates with individual crystals only several μm across. Pyroxenes are the dominant minerals in all but two samples.

Olivine was identified by X-ray diffraction and optical properties. Estimates of composition based on optical properties disagree with those based on X-ray data. The discrepancy was shown by spectrographic measurements to be due to a fairly high calcium content. The samples for which X-ray, optical, and spectrographic data are available indicate a composition, in terms of mole percent of end-member pyroxenes, of approximately $Wo_{0.05}$, $En_{0.60}$, $Fs_{0.35}$. Olivine commonly contains inclusions filled with devitrified glass or aggregates of plagioclase, pyroxene, and ilmenite. Some grains contain spherical glass inclusions. Fayalite grains were tentatively identified by their optical properties. In contrast to the Apollo 11 rocks, olivine occurs in almost all Apollo 12 samples.

Low cristobalite occurs as interstitial aggregates and as euhedral to subhedral crystals. Ilmenite was identified by X-ray diffraction, morphology, and optical properties. Ilmenite abundance ranges from less than 1 percent to 25 percent. In reflected light, some grains showed intergrowths with another unidentified oxide phase. In some samples, the ilmenite displays an oriented skeletal pattern (fig. 12-7).

The presence of spinel is suggested by the octahedral forms of opaque minerals of unknown composition that occur within olivine grains and in vugs. At least three other unidentified opaque

FIGURE 12-7.—Photomicrograph illustrating the gross texture of the skeletal ilmenite crystals in sample 12022. (NASA S-70-24742)

phases are also present. Troilite is a ubiquitous phase in all polished thin sections.

Metallic iron occurs as interstitial blobs that are commonly not associated with troilite grains, rather than as blobs in troilite, as noted in Apollo 11. Metallic iron is more commonly associated with ilmenite than with any other phase. Metallic copper, tridymite, and the iron analog of pyroxmangite were tentatively identified by optical methods. Glass occurs as minute interstitial material in some crystalline rocks, as beads and groundmass in clastic rocks, and as a thin coating on several rocks.

Breccias

One rock is a fragmental breccia similar to the Tranquility Base breccias (ref. 12–1). Two breccia chips were also collected. The dominant composition of the breccias appears to be pyroxene and plagioclase, with accessory olivine and glass. Lithic fragments are also present (fig. 12–8). The average mineral composition of the breccias appears to be less olivine-rich than the majority of rock types collected. The rock appears to have a foliation developed in which both lithic and mineral fragments are subparallel. The lithic fragments are as large as 20 by 10 mm and are both igneous and fragmental rocks, an indication of several periods of fragmentation and consolidation.

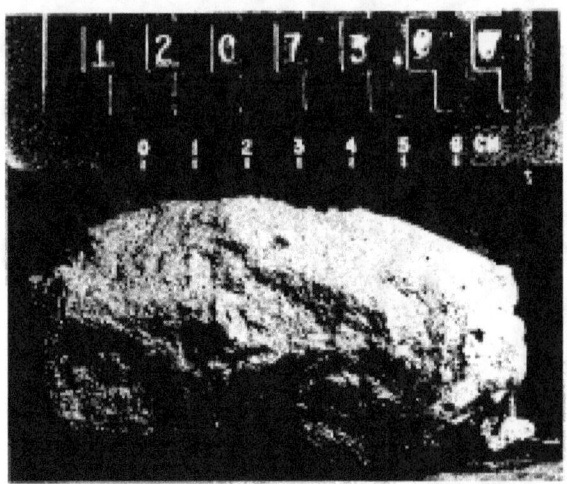

(a)

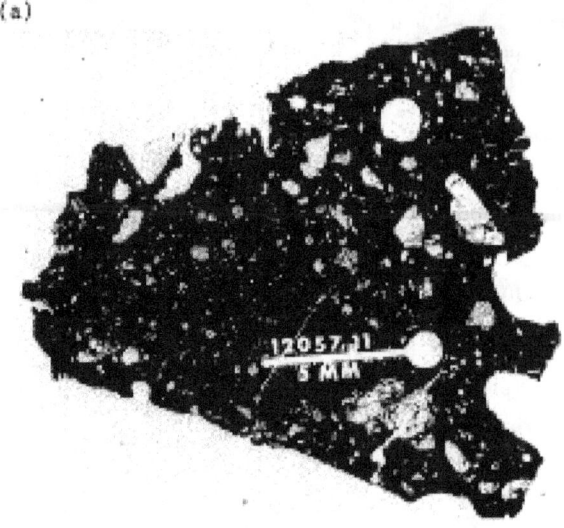

(b)

FIGURE 12-8.—Views illustrating breccias. (a) Sample 12073. Note the glass splash covering the lower right part of the rock. (NASA S-69-61066) (b) Photomicrograph of a chip of sample 12057. Note spherules in matrix. (NASA S-69-63407)

Rock Surface Features

Most of the larger Apollo 12 crystalline rocks are similar to the Apollo 11 rocks in that they are rounded on one surface and exhibit glass-lined pits. Angular fractured surfaces also occur.

Small pits, similar to those previously described (ref. 12–1), occur on rock surfaces (fig. 12–9). The density of the small pits ranges from 1 to 30 pits per square centimeter; however,

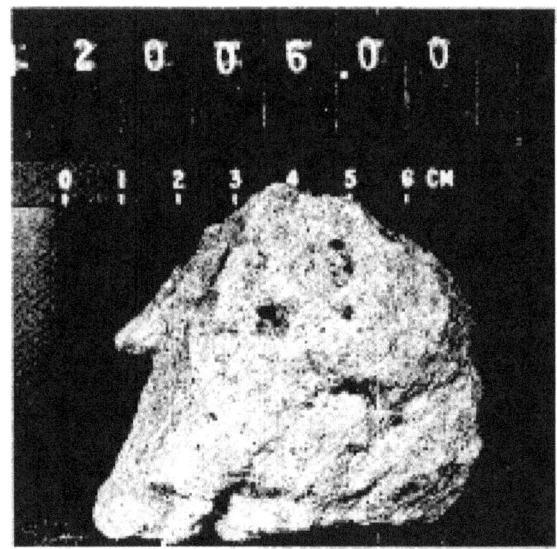

FIGURE 12-9.—View of sample 12006 illustrating a typical surface of crystalline rocks. Note the vugs, glass-lined pits, and white blotchy areas caused by crushed feldspar. (NASA S-69-62341)

angular bottom surfaces have few or no pits. Pits visible under a binocular microscope are 0.1 mm to 1 cm in diameter; depth-to-width ratios are approximately 1:5. Most of the pits are circular; the exceptions are (1) oval pits with their long axis parallel to elongate feldspar or (2) pyroxene crystals in coarse crystalline rocks. Glass lining the pits varies greatly in thickness and vesicularity. Pulverized minerals form white halos, 0.5 to 1.0 crater widths wide, around pits on crystalline rock surfaces. Where pit density is high, a crust (1 to 2 mm thick) of pulverized minerals is formed on the rock surface. Impact pits in glass coatings are surrounded by radiating fractures. On many of the medium- to coarse-grained crystalline rocks, the glass pit linings are raised slightly above the rock surface. These glass-topped pedestals appear to be more resistant to erosion than is the rock.

Irregular patches of glass have been spattered on rock surfaces. Two types that grade into each other can be distinguished: thin films and thick, highly vesicular coatings. The thin (less than 0.1 mm thick) films are brownish black in reflected light and brown in transmitted light, cover 1 to 16 cm², are usually slightly vesicular, and adhere tightly to the rock surface. The thick (0.1 mm to 1 cm) layers found on samples from the bottom of a 3-ft-diameter crater are light to dark brown, have vesicle sizes increasing from 0.1 mm to several millimeters from the base to the outer surface, and have smooth botryoidal surfaces (fig. 12-10). These layers coat several highly fractured breccias and fine-grained crystalline rocks, filling fractures in the rocks to a depth of several centimeters. Contacts between glass and rock are sharp. The glass coatings locally contain a large number of angular, highly shocked rock chips (fig. 12-11). Some of the vesicular glass has been partly devitrified. A light-gray glass of uncertain origin with pronounced flow structure was also collected.

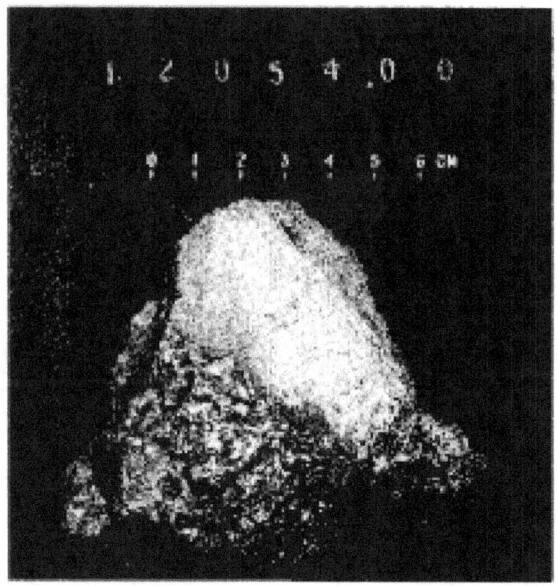

FIGURE 12-10.—View of sample 12054. This photograph illustrates a 1-mm-thick black glass coating that is sealing a fracture (on the left-hand side). The part of the sample that is not glass coated is striated. (NASA S-69-80072)

Fines

The Apollo 12 lunar fines are contrasted with the Apollo 11 fines in that the Apollo 12 fines consist of different proportions of mineral phases and, as a result, are lighter in color. The major constituents, in decreasing order of abundance, are pyroxene, plagioclase, glass, and olivine. The minor constituents, totaling only a few percent,

(a)

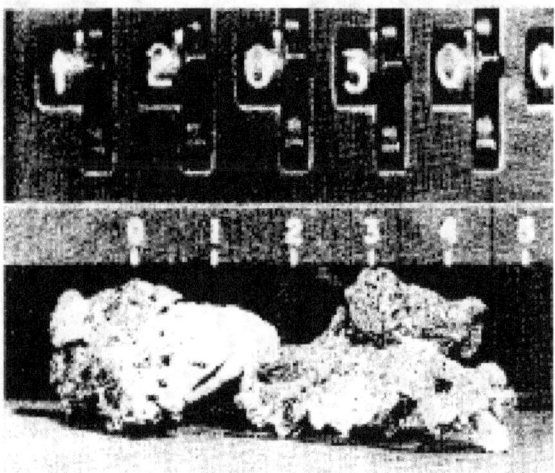

(b)

FIGURE 12-11. — Two views of several chips from sample 12030 that are cemented together with glass. (a) NASA S-69-23380. (b) NASA S-69-23384.

are ilmenite, tridymite, cristobalite, nickel-iron, and several unidentified phases. The iron analog of pyroxmangite was tentatively identified.

Glass totals approximately 20 percent of the Apollo 12 fines. Included in the glass are spheroidal and dumbbell-shaped objects and angular fragments. The glass ranges in color from colorless through pale yellow-brown to dark brown; the index of refraction is generally from 1.55 to 1.75. Bubbles and solid inclusions are common in the colored glasses. In contrast to the Apollo 11 fines, feldspar glass, dark-colored to nearly opaque spheroidal glass, and fragments of dark scoriaceous glass are relatively rare.

The pyroxenes, mostly pale yellow to tan and brown in color, range widely in composition and include augite, subcalcic augite, and pigeonite. The pyroxenes constitute approximately 40 percent of the whole fines. The range of indices of refraction is somewhat greater than that of the pyroxenes of the Apollo 11 fines. Plagioclase has indices of refraction mostly in the bytownite-anorthite range. A small amount of more sodic plagioclase is also present. Olivine totals approximately 5 to 10 percent and is more abundant than in the Apollo 11 fines. In terms of forsterite (Fo), the few carefully measured grains of olivine fell in the compositional range Fo_{60-70}. Many lithic fragments are present. Low tridymite is in the form of anhedral grains, and low cristobalite occurs as microgranular aggregates.

There are two fine samples of notably different character: one is the light layers in the double-core tube, and the other is sample 12033, a documented fine sample taken in a trench dug near the northwest rim of Head Crater. The color of these samples is light gray. Sample 12033 consists

FIGURE 12-12. — View of the 1-mm to 1-cm fraction of lunar fines, sample 12057. The particles are resting on a 1-mm sieve. (NASA S-69-60961)

of clear, angular grains of feldspar with some olivine, pyroxene, and abundant basaltic glass. The coarser (1-mm diameter) glass fragments are pumiceous, exhibiting well-developed flow structure and stretched vesicles; finer fragments are angular and somewhat vesicular. The finer shards also contain flow structure that consists of oriented microlites. Sample 12033 is tentatively considered to be a crystal-vitric ash.

A significant feature of the fines is the presence of a considerable number of very well-rounded grains (fig. 12-12) with minutely chipped surfaces that resemble grains in terrestrial detrital sands. Many of these well-rounded grains are slightly elongate, oblate bodies with a beanlike shape. They range in size to considerably less than 0.1 mm. Presumably the result of mechanical abrasion, the grains are chiefly glass, but some are composed of pyroxene, plagioclase, or intergrowths of these minerals.

Impact Metamorphism

Impact metamorphism in Apollo 12 samples is similar to and as common as in Apollo 11 samples. Shock-vitrified mineral fragments and impact-fused glass spherules and beads are present both in the fines and in the few microbreccias returned from the Apollo 12 mission. Most of the large Apollo 12 crystalline rocks are apparently unshocked or only weakly shocked.

Several smaller crystalline rocks show moderate to strong shock. A thin section of one of these rocks showed extensive fracturing of the plagioclase and partial vitrification and development of lamellar microstructures. The coexisting clinopyroxenes show one to two sets of closely spaced lamellar twinning that are absent in clinopyroxene of similar unshocked crystalline rocks. Another rock shows extensive shock vitrification.

Shocked microbreccias are also present. Many small, fractured microbreccia fragments are held together by glass spatters. Breccias within breccias, indicating a history of multiple shock, are also present.

Small pebbles, which are either glassy or aphanitic, were collected and are highly vesicular or vuggy. They are presumably caused by melting and fragmentation and are probably the result of impact metamorphism. Some of the vesicles are as large as 10 mm in diameter and occupy up to 50 percent of the preserved fragment.

Drive-Tube Core Samples

The Apollo 12 core samples differ from those collected at Mare Tranquillitatis on the Apollo 11 mission in that the Apollo 12 core samples have easily recognizable stratigraphy and two coherent, crustlike layers; otherwise, the Apollo 12 core samples resemble the Apollo 11 core samples in their dominantly fine-grained textures and loose consistence; in their restricted range of medium-gray colors and fresh, unoxidized appearance; and in their abundance of glass, including some spherules. As in the case of sediment from Apollo 11 core samples, dissection of the Apollo 12 core samples produces weakly coherent ephemeral structures, ranging from fine subrounded crumblike units 1 to 2 mm in diameter to subangular blocky, or occasional angular units with maximum dimensions of 5 mm. Coherence was adequate throughout most of the sediment to permit dissection (with care) of small vertical faces that were 1 cm high before slumping occurred.

All core samples opened were broken by fine fracture planes, usually transverse to the core tube. Where such fractures coincide with changes in character of the sediment, they are interpreted as bedding planes. Other, more complex fracture zones, not coinciding with morphologic changes, may be shear fractures produced when the drive tubes were rotated as the core samples were collected.

The core sample collected during EVA-1 is 19.3 cm long and is uniformly medium to dark gray. Stratification shows clearly in the abrupt change in abundance of rock fragments and glass particles coarser than 1 mm below a transverse fracture at a depth of 5.9 cm. This stratification is also reflected by changes in mean grain size with depth (fig. 12-13). Because of coherence of fine particles, it was not possible to obtain reliable mechanical analyses for material finer than 0.062 to 0.031 mm. Material coarser than 2.00 mm could not be analyzed because of the limited sample size.

The three mechanical analyses made of the first Apollo 12 core sample are similar to those

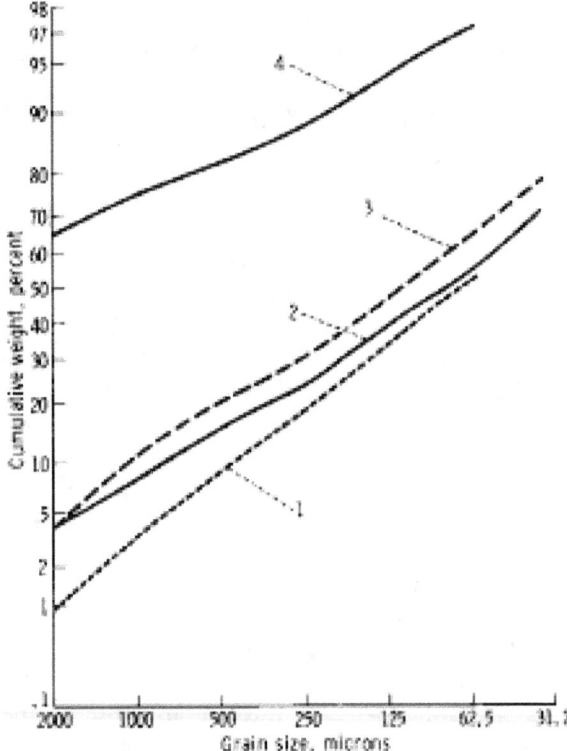

FIGURE 12-13. — Grain-size analysis of core-tube material. Numbers 1, 2, and 3 are top, middle, and bottom, respectively, of core-tube sample 12026; 4 is the coarse layer in core-tube sample 12028.

made of Apollo 11 core samples. The slope of the cumulative curves and hence the sorting is very similar for all three samples. However, successively deeper samples are progressively coarser. The median grain size changes from 0.062 mm in the surface sample material to 0.074 mm for the middle sample material and 0.11 mm for the deepest sample material.

Stratification and morphological change are most evident in the core sample collected in the double-core tube used at Halo Crater (fig. 12-14). The entire lower tube (32 cm) and 9.3 cm of the upper tube were filled with sediment. At least 10 layers or horizons have been recognized. The most distinctive of these units is a coarse layer of angular rock fragments, minerals, and glass, comprised mostly of olivine grains and olivine-rich gabbros (fig. 12-15). The fourth mechanical analysis is of this coarse layer. The slope of the cumulative grain-size curve, and hence

the sorting, is similar to that of the fine material, but the slope displaces markedly towards the coarse end of the graph. Extrapolating to the 50th percentile, the median grain size is approximately 4.9 mm. The sharp contact with the fine material above and below the coarse layer, the lack of fines in the coarse layer, and the log-normal distribution in the coarse layer suggest that the coarse layer is of primary impact origin. The gradual increase in grain size with depth shown by the three analyses from the first core sample suggests also that grain size of the debris has decreased because of reworking, probably by successive impacts.

Other units found in the deep core of the double-core tube include a fine-textured zone of lighter-medium-gray material; a zone of mixed, incoherent light- and medium-gray sediments; and at the base of the core, a layer of much lighter gray material. This lowermost layer is similar in appearance to sample 12033. Both the coarse layer and a medium-gray layer 2 cm thick, just below the surface at Halo Crater, have a friable consistence unlike any core materials observed previously, and particles coarser than 1 mm in both layers may be strongly bonded aggregates indigenous to the layers, rather than admixed fragments.

Chemistry

Chemical analyses of the samples were carried out mainly by optical spectrographic techniques conducted inside the LRL biological barrier. An instrument with a dispersion of 5.2 Å/mm was used.

The procedures were generally similar to those used to analyze the Apollo 11 samples, but modifications were made to cope more effectively with the high concentration of refractory elements. For example, the weight ratio of admixed carbon to sample was increased from 1:1 to 4:1 for the method using palladium as internal standard, and the range of intensity measurements possible was increased by a factor of 3, thus enabling a wider spread in concentrations to be effectively covered (e.g., barium and zirconium).

The overall precision of the determinations is ±5 to ±10 percent of the amount actually present. Accuracy was controlled by use of the inter-

PRELIMINARY EXAMINATION OF LUNAR SAMPLES 203

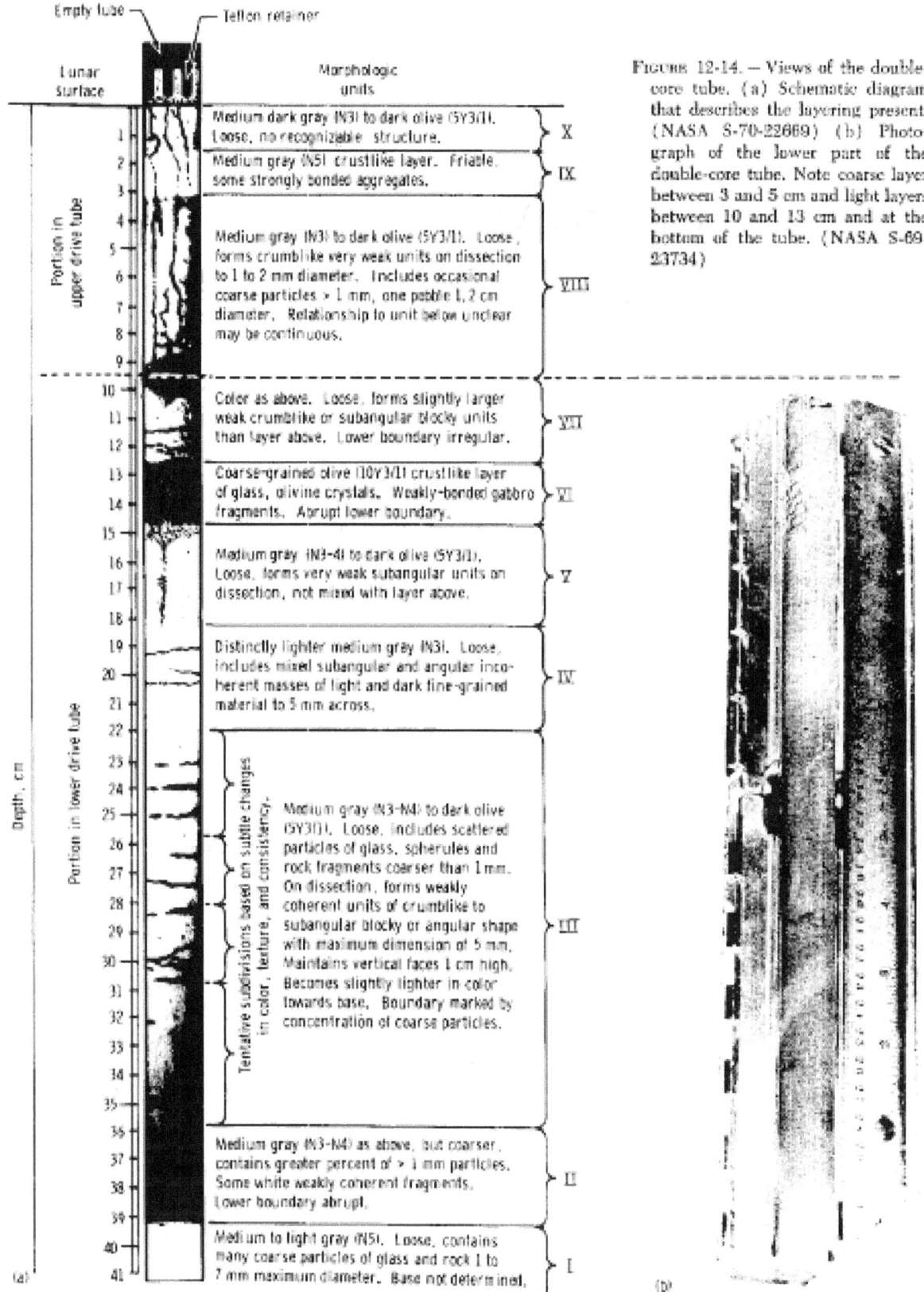

FIGURE 12-14. — Views of the double-core tube. (a) Schematic diagram that describes the layering present. (NASA S-70-22669) (b) Photograph of the lower part of the double-core tube. Note coarse layer between 3 and 5 cm and light layers between 10 and 13 cm and at the bottom of the tube. (NASA S-69-23734)

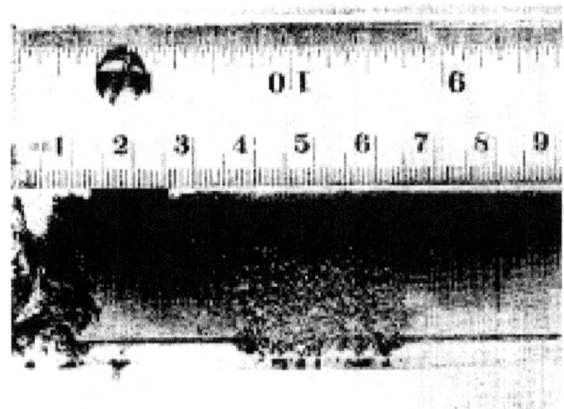

FIGURE 12-15.—View illustrating the coarse layer near the top of the lower part of the double-core tube. (NASA S-69-23404)

national rock standard samples (G–1, W–1, SY–1, BCR–1, AGV–1, GSP–1, G–2, PCC–1, and DTS–1) for calibration.

The spectrographic plates were examined to establish the presence or absence of all elements that have spectral lines in the wavelength regions covered (2450 to 4950 Å and 6100 to 8600 Å). Line interferences were checked for all lines. Several samples were brought from behind the biological barrier and analyzed by atomic absorption procedures for aluminum, calcium, magnesium, iron, titanium, sodium, and potassium and by chemical colorimetric procedures for silicon. The data reported for these elements in sample 12013 were obtained by these methods. A total of 32 samples were processed, and a representative set of 14 analyses is given in table 12–VII. Sample weights provided for analysis were typically 100 to 150 mg.

The samples appear to be free from inorganic contamination from either the rock box or the LM. Niobium, present at 88 percent in the skirt of the LM descent engine, was detected in only two samples—sample 12013 with 170 ppm and sample 12033 with 44 ppm. These amounts are almost certainly indigenous to the rocks, and the presence of niobium is consistent on geochemical grounds with the high abundance of geochemically associated elements zirconium and yttrium. Indium, present in the seal of the rock boxes, was not detected. Sporadic copper contamination of up to several hundred parts per million was encountered.

The major constituents of the samples are, in general order of decreasing abundance, silicon, iron, magnesium, calcium, aluminum, and titanium. The major silicate and oxide mineral phases present in the samples allow the deduction that oxygen comprises the major anion. Chromium, sodium, manganese, and potassium are minor constituents, with concentrations ranging from 0.05 to 0.6 percent. Occasionally, barium and zirconium reach these concentration levels. The other constituents are present mostly at less than 200 ppm (0.02 percent). The volatile elements (lead, boron, bismuth, thallium, etc.) are generally below the limits of detection of the spectrographic methods employed, although lead (≈ 30 ppm) and boron (≈ 20 ppm) were detected in sample 12013. Gold, silver, and the platinum group elements were not detected in any samples.

The chemistry of the crystalline rocks is distinct from that of the fine material and the breccias. The rocks are lower in rubidium, potassium, barium, yttrium, zirconium, and lithium. Iron and chromium contents are higher in the rocks. Several critical-element ratios are likewise distinct: K/Rb ratios average 850 in the rocks as compared to 450 in the fines. The Fe/Ni ratio in the rocks (average: 3000; range: 2000 to 11 000) is much higher than in the fine material (with ratios of approximately 600). The Rb/Sr ratios, which are very low in the rocks (0.005), are higher (0.02) in the fines.

The fine material and the breccias are generally quite similar in composition and could not have formed directly from the large crystalline rock samples. The nickel content of the breccias and the fine material places an upper limit on the amount of meteoritic material contributed to the lunar surface regolith. By using an average meteoritic nickel content of 1.5 percent, the nickel content of the fine material represents a meteoritic contribution of approximately 1 percent, if all the nickel were extralunar.

The crystalline rocks show minor, but significant, internal variations in chemistry (table 12–VII). The rock samples are tabulated in order of decreasing magnesium content, which appears to be the most significant parameter among the major constituents. A number of interesting geo-

PRELIMINARY EXAMINATION OF LUNAR SAMPLES 205

TABLE 12-VII. *Composition of Apollo 12 Samples*

Component	Crystalline rocks									Sample 12070 (a)	Sample 12073 (b)	Sample 12010 (c)	Sample 12023 (a)	Sample 12013 (c)
	Sample 12002	Sample 12004	Sample 12015	Sample 12022	Sample 12009	Sample 12065	Sample 12052	Sample 12064	Sample 12038	Average					
Elements:															
Rubidium (Rb), ppm	0.64	0.47	1.0	0.17	0.57	0.72	0.80	0.76	0.70	0.64	3.2	4.9	2.0	7.5	33
Barium (Ba), ppm	38	60	44	38	65	70	50	55	230	72	420	510	180	720	2150
Potassium (K), ppm	460	480	510	580	520	600	570	700	470	540	1500	2100	1300	3240	1.66
Strontium (Sr), ppm	110	145	115	160	110	135	135	165	230	145	170	230	145	260	150
Calcium (Ca), percent	6.6	7.1	7.0	7.9	7.1	9.0	7.9	8.6	7.9	7.6	7.1	8.2	7.0	8.2	4.5
Sodium (Na), percent	.38	.36	.27	.27	.38	.29	.33	.31	.45	.33	.30	.29	.39	.40	.51
Ytterbium (Yb), ppm	—	—	—	—	—	—	—	—	—	—	—	—	—	12	20
Yttrium (Y), ppm	40	52	46	62	48	48	42	55	68	51	130	180	57	260	240
Zirconium (Zr), ppm	120	170	160	160	150	180	170	170	260	170	670	1200	380	950	2200
Chromium (Cr), ppm	3900	5800	3900	2850	5200	3500	3700	3000	2200	3750	2800	2800	3050	2100	1050
Vanadium (V), ppm	65	85	95	65	77	135	105	100	70	88	64	50	82	37	13
Scandium (Sc), ppm	36	45	44	52	42	60	52	60	55	50	47	42	50	33	21
Titanium (Ti), percent	1.9	2.0	1.9	3.4	2.0	2.3	2.2	2.9	1.9	2.3	1.9	1.9	2.2	1.6	.72
Nickel (Ni), ppm	135	90	70	40	57	35	32	15	14	54	200	350	80	140	105
Cobalt (Co), ppm	48	50	47	36	46	34	42	40	23	40	42	30	39	34	13
Iron (Fe), percent	17.9	17.9	17.1	17.1	15.5	17.1	16.3	17.0	13.2	16.6	13.2	13.0	15.2	12.4	7.8
Manganese (Mn), ppm	1800	1750	2550	1350	1450	3200	2400	2500	2000	2050	1900	1500	1400	1800	950
Magnesium (Mg), percent	10.6	9.0	8.4	7.8	7.5	6.6	6.0	4.8	3.9	7.2	7.2	6.6	6.6	6.5	3.6
Lithium (Li), ppm	3.9	4.2	10	3.1	5.5	6.0	4.5	6.7	5.5	5.5	11	25	7	15	100
Aluminum (Al), percent	5.7	5.6	5.8	5.8	5.7	6.4	5.8	6.3	6.4	5.9	7.4	7.9	6.1	8.5	6.3
Silicon (Si), percent	18.4	17.3	17.9	16.8	19.2	18.2	19.6	18.7	22.9	18.5	19.6	19.1	20	19.2	28.5
Oxides:															
Silicon dioxide (SiO$_2$), weight percent	35	37	38	36	41	39	42	40	49	40	42	41	43	41	61
Titanium dioxide (TiO$_2$), weight percent	3.1	3.4	3.2	5.1	3.3	3.8	3.6	4.9	3.2	3.7	3.1	3.1	3.7	2.6	1.2
Aluminum oxide (Al$_2$O$_3$), weight percent	11	10.5	11	11	11	12	11	12	12	11.2	14	15	11.5	16	12
Iron oxide (FeO), weight percent	23	23	22	22	20	22	21	22	17	21.3	17	17	19.5	16	10
Magnesium oxide (MgO), weight percent	17.5	15	14	13	12.5	9	10	8	6.5	11.7	12	11	11	11	6.0
Calcium oxide (CaO), weight percent	9.3	10	9.8	11	10	12.6	11	12	11	10.7	10	11.5	10	11.5	6.3
Sodium oxide (Na$_2$O), weight percent	.53	.48	.37	.36	.51	.39	.45	.42	.60	.45	.40	.50	.53	.54	.69
Potassium oxide (K$_2$O), weight percent	.055	.058	.062	.068	.063	.072	.069	.084	.057	.065	.18	.25	.16	.39	2.0
Manganese oxide (MnO), weight percent	.17	.23	.33	.17	.19	.41	.31	.32	.26	.26	.25	.19	.18	.23	.12
Chromium oxide (Cr$_2$O$_3$), weight percent	.57	.85	.57	.39	.76	.51	.54	.44	.32	.55	.41	.41	.45	.31	.15
Zirconium dioxide (ZrO$_2$), weight percent	.016	.023	.022	.022	.020	.024	.023	.23	.035	.023	.09	.16	.05	.13	.30
Nickel monoxide (NiO), weight percent	.017	.011	—	—	—	—	—	—	—	—	.025	.044	—	.018	.013
Σ, weight percent	100.2	101.1	99.4	99.6	99.1	99.8	100.0	100.2	100.0	99.9	99.5	100.2	100.1	99.7	99.8

a Fine material.
b Breccia.
c Breccia (?).

d Light-colored fines; sample 12003 also contains 44 ppm Nb.
e Sample 12013 also contains 30 ppm Pb, 15 ppm B, and 170 ppm Nb.
f In percent.

chemical trends appear when the samples are so arranged (fig. 12-16). Nickel shows a striking decrease (by an order of magnitude) in concentration. Chromium shows a smaller relative decrease in the same direction, and cobalt shows a slight decrease. Silicon increases as magnesium decreases. Similar trends are shown by vanadium, scandium, zirconium, yttrium, potassium, barium, and calcium, although the variations are small. As the amount of magnesium decreases, the critical-element ratios Fe/Ni, Ni/Co, and Cr/V decrease, and V/Ni increases. No significant trends are shown by K/Rb, Rb/Sr, or K/Ba ratios.

Sample 12013 is unique among lunar samples. This rock sample contains the highest concentration of silicon dioxide (SiO_2) yet observed (61 percent), and the amounts of potassium, rubidium, barium, zirconium, yttrium, lithium, and the rare-earth element ytterbium are enriched 10 to 50 times, as compared to the other rocks. These high concentrations are reminiscent of the terrestrial enrichment of elements in residual melts during the operation of fractional crystallization processes. The amounts of magnesium, iron, chromium, manganese, nickel, titanium, scandium, and cobalt in sample 12013, although low in comparison with the other rocks, are not strikingly depleted. Nickel, in particular, is not depleted, as it would be in terrestrial analogs. Sample 12033 from the light-gray fine material shows some analogs in composition to the rock (sample 12013), containing high quantities of ytterbium, niobium, and rubidium.

A comparison of the Apollo 12 samples from Oceanus Procellarum with the Apollo 11 samples from Mare Tranquillitatis shows that the chemistry at the two mare sites is clearly related. Both sites show the distinctive features of high concentrations of refractory elements and low contents of volatile elements; these two features most clearly distinguish lunar material from other material. In detail, there are numerous and interesting differences between the Apollo 12 and 11 rocks, including the following:

(1) There is a lower concentration of titanium both in the rocks and in the fine material of Apollo 12. The range in composition is 0.72 to 3.45 percent titanium (1.2 to 5.1 percent TiO_2), as compared with the range in the Apollo 11

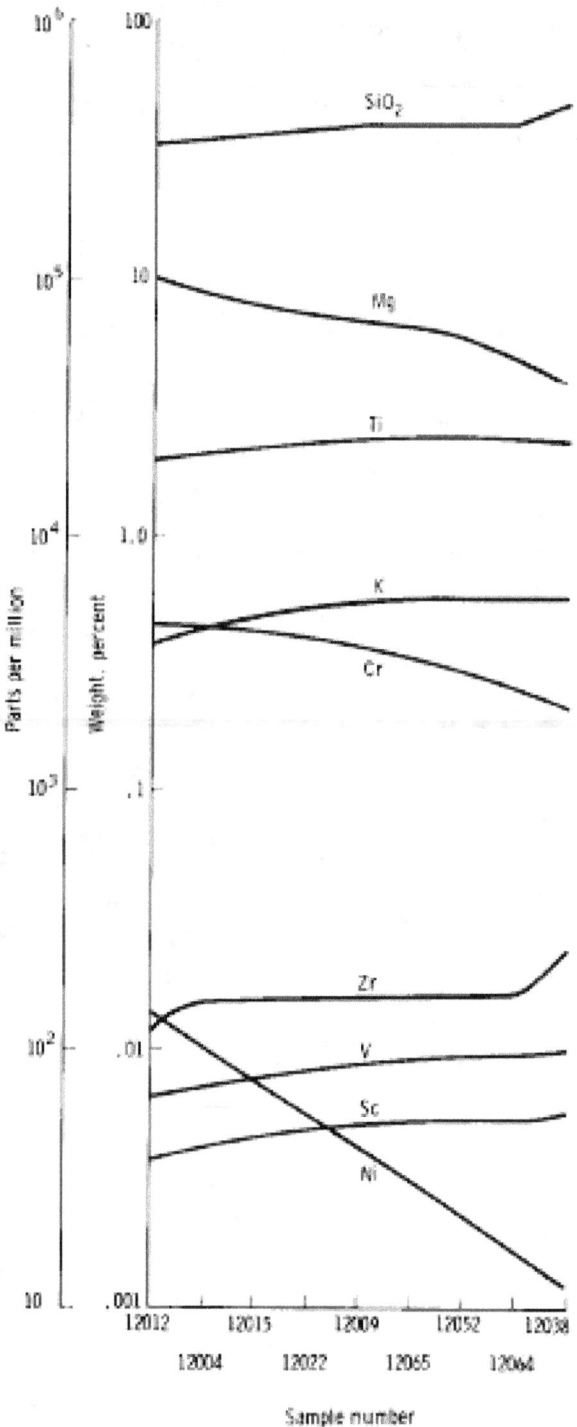

FIGURE 12-16. — Element variation within the crystalline rocks. Samples are arranged in order of decreasing magnesium content. Note the decrease in nickel content by an order of magnitude in this sequence.

rocks of 4.7 to 7.5 percent titanium (7 to 12 percent TiO_2).

(2) There are lower concentrations of potassium, rubidium, zirconium, yttrium, and barium in Apollo 12 rocks.

(3) There are higher concentrations of iron, magnesium, nickel, cobalt, vanadium, and scandium in the crystalline rocks from Apollo 12. These data are consistent with the more mafic character of the Apollo 12 rocks.

(4) The significant variation in the Apollo 12 rocks is among the elements that favor ferromagnesian minerals. The range of abundance was not nearly as great in the Apollo 11 rocks.

(5) The fine material at the Apollo 12 site differs from that at the Apollo 11 site in containing approximately one-half the titanium content; more magnesium; and possibly higher amounts of barium, potassium, rubidium, zirconium, and lithium. The light-gray fine material (sample 12033) is strongly enriched in rubidium, zirconium, ytterbium, and niobium, relative to the other fine material.

The chemistry of the Apollo 12 samples does not resemble that of chondrites; nickel, in particular, is strikingly depleted. The eucrite class of the basaltic achondrites is closest in composition, and sample 12038 shows some similarities in composition (table 12–VII). The felsic rock (sample 12013) does not resemble terrestrial diorites, dacites, anorthosites, or tektites in the abundances of most elements.

The Apollo 12 material is enriched in many elements by 1 to 2 orders of magnitude in comparison with estimates of cosmic abundances.

Noble Gases

Several samples each of fine material, breccias, and crystalline rocks returned by Apollo 12 have been analyzed for noble-gas isotopes. Analyses were performed by mass spectrometry, and the general technique used was the same as for the preliminary examination of Apollo 11 samples (ref. 12–1). As was the case for Apollo 11 material, the Apollo 12 fines and breccias are characterized by large abundances of noble gases of solar-wind origin, as exemplified by the large relative abundances of helium and by the characteristic $^4He/^3He$ and $^{20}Ne/^{22}Ne$ ratios. The crystalline rocks, however, contain much smaller amounts of noble gases that arise mainly from spallation reactions within the samples or from radiogenic decay. Table 12–VIII lists noble-gas isotopic contents for typical samples of all three types of lunar material from Apollo 12. Ratios are believed to be accurate to within ±2 percent (except $^4He/^3He$, which has a larger uncertainty), and abundances are accurate to within ±20 percent (except ^{84}Kr, which is less well known).

In spite of the general similarity in noble-gas content of Apollo 12 and 11 material, there are several real and significant differences between

TABLE 12–VIII. *Noble-Gas Abundances of Apollo 12 Lunar Material*[a]

Type	Sample	^4He	$^4He/^3He$	^{20}Ne	$^{20}Ne/^{22}Ne$	$^{22}Ne/^{21}Ne$	^{40}Ar	$^{40}Ar/^{36}Ar$	$^{38}Ar/^{36}Ar$	^{84}Kr	^{132}Xe
Fines	12070	6 700 000	2 300	120 000	13.3	28	12 000	0.60	5.1	13	2.6
	12060	4 000 000	2 100	73 000	13.2	26	9 800	.80	5.2	4	1.1
Breccia	12034	130 000	940	8 700	12.0	13	3 500	2.1	4.8	.95	1.1
	12071	1 700 000	2 100	33 000	12.3	21	5 300	.90	5.1	4.8	1.2
Typical	12062	9 000	59	63	2.4	1.3	1 400	.57	.98	.07	.021
crystalline	12064	13 000	64	60	1.6	1.2	1 300	50	.83	.033	.012
rocks	12004	15 000	200	46	3.6	1.5	900	87	1.9	.02	.014
Special	12013	360 000	10 000	100	7.8	2.5	820 000	51 000	2.5	.21	.15
cases	12010 (light)	24 000	270	20	5.2	1.8	1 400	72	1.6	.015	.017
	12010 (dark)	2 700 000	2 800	88 000	11.9	27	42 000	3.4	5.2	2.9	1.2

[a] In 10^{-8} cm^3 at standard temperature and pressure per gram every 10^6 yr.

the two. The noble-gas content of the Apollo 12 fines and breccias is lower—the fines by a factor of 2 to 5 and the breccias by an order of magnitude. By using a model of formation of fine material by degradation of surface rock, surface irradiation by constant solar wind, and subsequent burial by additional fine material, the lower gas content of the fines implies a higher accumulation rate of material. The Apollo 12 breccias are approximately a factor of 2 lower in content than Apollo 12 fines. This finding constitutes a reversal of the trend observed with Apollo 11 material and implies that the Apollo 12 breccias were formed from fine material of lower solar-wind gas content. Possibly, the Apollo 12 breccias were formed at some distance or depth from their current lunar location.

For breccia sample 12034, the total noble-gas content is low enough for a spallation component to be quite evident. Sample 12010 is an unusual case of a breccialike rock that was identified on the basis of its noble-gas content. This sample is characterized by a large relative abundance of lithic material, with dark-gray fine-grained material occurring as veins. In table 12-VIII, sample 12010, labeled "dark," contains approximately 70 percent of this fine-grained material, whereas the sample labeled "light" is essentially pure lithic material. The typical breccialike nature of sample 12010 in terms of its noble-gas content is obvious.

The $^{40}Ar/^{36}Ar$ ratio in the Apollo 12 fines and breccias is lower than that for Apollo 11 samples (with the exception of sample 12010), although the ratio still shows larger values for breccias than for fines. Theoretical considerations prohibit $^{40}Ar/^{36}Ar$ ratios even as large as 0.6 for the Sun, making a solar-wind origin of the ^{40}Ar unlikely. The amount of ^{40}Ar in the fines is too large to be generated by in situ decay of potassium; however, for the breccias, this mode of origin may be possible. Excess ^{40}Ar of lunar origin thus appears to have been acquired by the fine material. The amount of ^{40}Ar and the $^{40}Ar/^{36}Ar$ ratio in sample 12010 demonstrate this phenomenon well. While the lithic phase resembles the other crystalline rocks in its argon content, the fine-grained material shows not only large amounts of solar-wind argon, but also large excesses of ^{40}Ar.

The crystalline rock samples reported in this chapter were completely interior chips and contain noble gas in amounts that are orders of magnitude less than that in the fine material. The exceptions are ^3He, ^{21}Ne, ^{40}Ar, and some of the lighter isotopes of krypton and xenon, isotopes whose abundances have been greatly increased by spallation reactions and radioactive decay. By using the potassium concentrations obtained for the rocks, K-Ar ages have been calculated. Several crystalline rocks give ages between 1.7×10^9 and 2.7×10^9 yr, with an average value of 2.3×10^9 yr. Rock sample 12013 has a unique chemistry, characterized in part by much higher abundances of potassium and uranium and, consequently, also of radiogenic ^4He and ^{40}Ar. Sample 12013 also contained excess radiogenic ^{40}Ar, rendering the K-Ar age meaningless. However, as is the case for many of these crystalline rocks, the U-Th-^4He age is consistent, with a value of approximately 2.3×10^9 yr. This age is considerably less than that found for the Apollo 11 crystalline rocks, although the range of ages for the two sites overlap. It appears that at least this portion of Oceanus Procellarum has a more recent crystallization age than the rocks sampled by the Apollo 11 crew in Mare Tranquillitatis, which implies that lunar maria have a formation history of at least 1 billion years.

Cosmic-ray exposure ages (i.e., integrated exposure time at the lunar surface) have been calculated for several rocks on the basis of 2π geometry and a ^3He production rate of 1×10^{-8} cm^3 at standard temperature and pressure per gram every 10^6 yr, with some apparent grouping of ages. The breccias also give radiation ages in this group. These exposure ages resemble those found for the Apollo 11 rocks. Spallation-produced isotopes other than ^3He are consistent with the ^3He ages and with the special chemistry of the lunar material. Because of the high-alkaline earth, yttrium, and zirconium abundances, spallation-produced krypton and xenon are quite obvious in the rocks. The amounts of these gases are also roughly consistent with the chemistry.

Total-Carbon Analyses

The total carbon content of the lunar samples was determined by oxygen combustion followed by gas chromatographic detection of carbon dioxide (CO_2) produced. Samples weighing 50 to 600 mg, together with iron chips and a copper-tin accelerator, were placed in a preburned refractory crucible that was then heated in excess of 1600° C in a flowing oxygen atmosphere. The combustion products were analyzed using previously described procedures (ref. 12-2).

The system was calibrated by using National Bureau of Standards Steel Standard 101e. Samples of this standard, containing from 5 to 50 μg of carbon, were analyzed, using the same conditions as for the lunar samples. To reduce the background, the crucible was burned in air at 1000° C for at least 1 hr. Only crucibles heated to 1000° C in a single batch were used in a sequence of standards and samples. The precision of the method was evaluated by making replicate runs on sample blanks. A typical standard deviation of a series of 10 runs was 2 μg of total carbon. The results for the standard samples were plotted on linear graph paper, and the carbon content in lunar samples was read directly from the standard linear curve.

The results of the analyses are given in table 12-IX. The highest carbon abundances, like

TABLE 12-IX. *Total Carbon Analysis*

Type	Sample	Total carbon, ppm
Fines	12003	180
	12024	115
	12032	25
	12033	23
	12042	130
	12059	200
Breccia	12034	65
	12057	120
Coarse-grained rocks	12040	45
	12040	45
	12044	44
Fine-grained rocks	12052	34
	12052	34
	12063	35
	12065	31

those of the Apollo 11 samples (ref. 12-2), tend to be found in the fines. Exceptions to this generalization are sample 12032 from the ejecta blanket of Bench Crater and sample 12033 from the bottom of a trench at Head Crater. Carbon is more abundant in the breccias than in the igneous rocks. The igneous rocks are consistently low in carbon abundance. A total carbon abundance of approximately 40 ppm appears to be indigenous to lunar rocks. Additional carbon appears to have been added to the fines and subsequent breccias by meteoritic impact, by the solar wind, and possibly by contamination. The total-carbon results give no indication of the specific chemical species present.

Gamma-Ray Spectrometry

Operation of the LRL radiation counting laboratory (RCL) followed the general procedures developed for Apollo 11 studies (ref. 12-1). The first RCL sample was received for measurement on November 30, 1969.

Analysis was performed by use of the NaI (Tl) low-background spectrometer and the online computer data-acquisition system described in reference 12-1. Samples were mounted in stainless-steel cans of 16.2-cm diameter and 0.8-mm wall thickness with bolt-type indium seals. Standard containers of overall heights of 5.6 or 7.6 cm were used for all rock samples. Fines were packaged in a cylindrical container used in searches for magnetic monopoles.

For this preliminary study, calibrations were obtained with a series of radioactive standards that were prepared by dispersing known amounts of radioactivity in quantities of iron powder. Time did not permit recording a library of standard spectra with the standard sources placed inside the steel containers actually used. Empirical corrections for the effects of the containers were applied; these corrections were least important for the ^{40}K data and most serious for the ^{26}Al and ^{22}Na results.

The results are summarized in table 12-X. Because of the preliminary nature of the investigation, rather large errors have been assigned. These errors include, in addition to the statistical errors of counting, estimates of possible systematic errors caused by uncertainties in the detector efficiency calibration.

Although there are many qualitative similarities between the RCL data on samples from Apollo 11 and 12, there are also some notable differences. The potassium concentration of the crystalline rocks is remarkably constant at approximately 0.05 weight percent, and the K/U ratio is approximately 2200. These properties appear significantly lower than the typical crystalline rocks (ref. 12-1) from Apollo 11; however, one of the coarsely crystalline Apollo 11 rocks (sample 10003) did resemble very closely the chemical composition of the crystalline rocks shown in table 12–X. The Th/U ratio, as found for the materials from Tranquility Base, is approximately 4 for all typical materials listed in table 12–X. The concentrations of the radioactive elements potassium, thorium, and uranium in the crystalline rocks listed in table 12–X are all remarkably constant and, on the average, are much lower than comparable Apollo 11 rocks. Because so few samples can be compared, it is not possible to discount biased sampling of the lunar surface material as a factor in the explanation of the differences between the soils and the rocks.

The breccias and fines are quite different from the crystalline rocks in several respects. The K/U ratio is only 1400 to 1500, compared with an average K/U ratio of approximately 2800 for all Apollo 11 materials. Thus, the Apollo 12 samples show even greater differences from terrestrial rocks and meteorites than did the surface material from Tranquility Base. Although the Th/U ratio remains approximately 4 in the Apollo 12 samples, the concentrations of all the radioactive elements are much higher in the breccias than in the crystalline rocks.

In general, the amounts of cosmogenic ^{26}Al and ^{22}Na appear saturated but show variations that may be related to chemical composition or to cosmic-ray exposure. For example, sample 12034 was recovered from a trench dug during the lunar surface activities. When collected, it was buried to a depth of 10 to 15 cm. The saturation activities of ^{26}Al and ^{22}Na are reduced by the approximate factor expected because of attenuation of the irradiation flux in the lunar soil.

Biology

To date, it has been impossible to demonstrate any viable organism in the lunar material, and there is no evidence of previous living or fossil material. Direct observations involved light microscopy using white light, ultraviolet light, and

TABLE 12–X. *Gamma-Ray Analyses of Lunar Samples*

Sample type	Sample	Weight, g	Potassium (K), weight percent	Thorium (Th), ppm	Uranium (U), ppm	^{26}Al, dpm[a]/kg	^{22}Na, dpm[a]/kg	Other radionuclides detected	Remarks
Crystalline rocks	12002	1530	0.044±0.004	0.96±0.1	0.24±0.033	72±14	53±10	^{54}Mn, ^{56}Mn, ^{60}Co, ^{46}Sc, ^{48}V	—
	12004	502	.045± .004	.88± .09	.25± .033	112±22	65±13	^{54}Mn, ^{56}Co, ^{46}Sc, ^{48}V	—
	12039	255	.060± .005	1.20± .12	.31± .040	80±16	45±9	^{56}Co, ^{54}Mn	—
	12053	879	.051± .004	.89± .09	.25± .033	85±17	42±9	^{56}Co, ^{48}V, ^{46}Sc, ^{54}Mn	—
	12054	687	.052± .004	.77± .08	.21± .030	50±11	42±9	^{48}V, ^{56}Co, ^{54}Mn, ^{46}Sc	—
	12062	730	.052± .004	.81± .08	.21± .030	65±13	34±7	^{48}V, ^{56}Co, ^{54}Mn, ^{46}Sc	—
	12064	1205	.053± .004	.88± .09	.24± .035	58±12	44±9	^{56}Co, ^{46}Sc, ^{48}V, ^{54}Mn	—
Miscellaneous samples	12034	154	.44 ± .035	13.2 ±1.3	3.4 ± .4	58±12	27±6	^{54}Mn	Breccia
	12073	405	.278± .022	8.2 ± .8	2.0 ± .3	125±25	60±12	^{56}Co, ^{54}Mn, ^{46}Sc	Breccia
	12070	354	.206± .016	6.0 ± .6	1.5 ± .2	140±25	65±13	^{56}Co, ^{48}V, ^{46}Sc, ^{54}Mn	Fine material
	12013	80	2.02 ± .016	34.3 ±3.4	10.7 ±1.6				Feldspathic differentiate

[a] Disintegrations per minute.

phase-contrast techniques. A wide variety of biological systems are now undergoing tests with lunar material to determine if there is any toxicity, microbial replication, or pathogenicity. Germ-free in vivo systems include mice, plants, plant tissue cultures, and viral assay tissue cultures. Other in vivo tests are being performed on animals that are not germ free, including fish, insects, Japanese quail, oysters, flatworms, protozoans, and shrimp. Histological studies are being made to determine whether there is any evidence of pathogenicity. Other activities involve extensive in vivo study of the early biosample and of the regular lunar samples.

Organic Chemistry

A computer-coupled high-sensitivity mass spectrometer has been used to estimate the abundances of organic matter in the lunar samples. Information on the volatile organic matter or pyrolyzable organic matter, or both, as a function of sample temperature has been obtained from detailed mass-spectrometer data. From these data on the lunar samples and from blanks and controls, an assessment has been made of the relative contributions of terrestrial contaminants as opposed to possible indigenous lunar organic matter.

Samples were sealed in stainless-steel vials with aluminum caps and were heat sterilized at 130° C for 30 hr. Portions of these samples (35 to 500 mg) were transferred to a nickel container for mass-spectral analysis. The inlet system on the mass spectrometer permitted insertion of these capsules into an oven heated to 500° C. This oven is connected to the mass-spectrometer ionization chamber by means of a heated 8-cm quartz tube. The mass spectrometer was operated on line with a Sigma-2 computer and was equipped with a high-efficiency ion source and a high-gain electron multiplier. The online computer permitted continuous control of the scanning circuit and recording of the mass spectra, as well as calibration of the mass scale, spectrum normalization, and visual display of the spectrum during each scan cycle.

One hundred mass spectra were collected by scanning the magnet, beginning with the insertion of the sample into the oven and continuing every 15 sec for 25 min. The data are normally presented in the following three types of plots:

(1) *Summarize plot* — A plot is made of the total ion current of each scan as a function of the scan number. The total ion current of each scan is arrived at by a summation of all peak areas in the scan (except major background peaks) and is directly proportional to the amount of volatilizable material or pyrolyzable material, or both, in the sample. Increasing scan number also represents increasing sample temperature, but the sample temperature was not actually measured. The summarize plot is, then, a plot of intensity versus temperature.

(2) *m/e plot* — The m/e plot is a single-scan plot of peak intensity (area) versus mass number; that is, the mass-to-charge (m/e) ratio for the ion. These plots are usually normalized to the most intense peak above mass 50.

(3) *Summarize-m/e plot* — The summarize-m/e plot is a plot of the ion current for a given mass as a function of scan number. The summarize plot gives information on the thermal stability and amounts of volatile material or pyrolyzable material, or both, while the m/e plot allows qualitative identification of the material. Mass number was calibrated using perfluorokerosene (PFK). Sensitivity calibrations were made by introducing samples of n-tetracosane (C_{24}) on clean quartz at concentrations of 1, 2, and 3 ppm to determine the total ion current produced for given amounts of volatile or pyrolyzable material, or both.

Typical summarize, m/e, and summarize-m/e plots are shown for sample 12026 in figure 12–17. The summarize plot shows two distinct peaks at scan numbers 7 and 21 and a possible peak at scan 53. The summarize-m/e plot for mass 44 (CO_2) shows that the initial peak at scan 7 is mainly CO_2 (possibly adsorbed), but the peak at scan 21 is not CO_2. The CO_2 peak at scan 45 is common to most of the lunar samples and could be caused by thermal decomposition of a carbonate present in ppm quantities. The m/e plots (fig. 12–17(c)) aid in identifying possible organic material coming off at various scans.

Samples representative of all Apollo 12 rock types and core-tube fines were analyzed, as were appropriate blanks and controls. In comparison with the results reported from examination of Apollo 11 samples, the data obtained for the

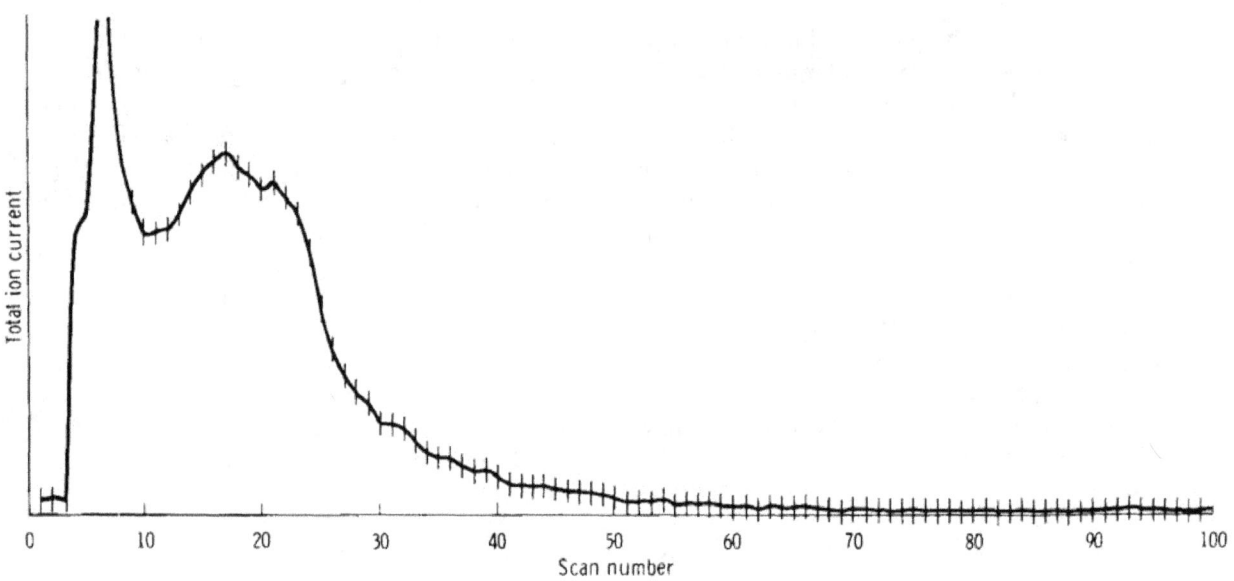

FIGURE 12-17. — Typical organic mass spectra for Apollo 12 samples (representative sample 12026). (a) A summarize plot.

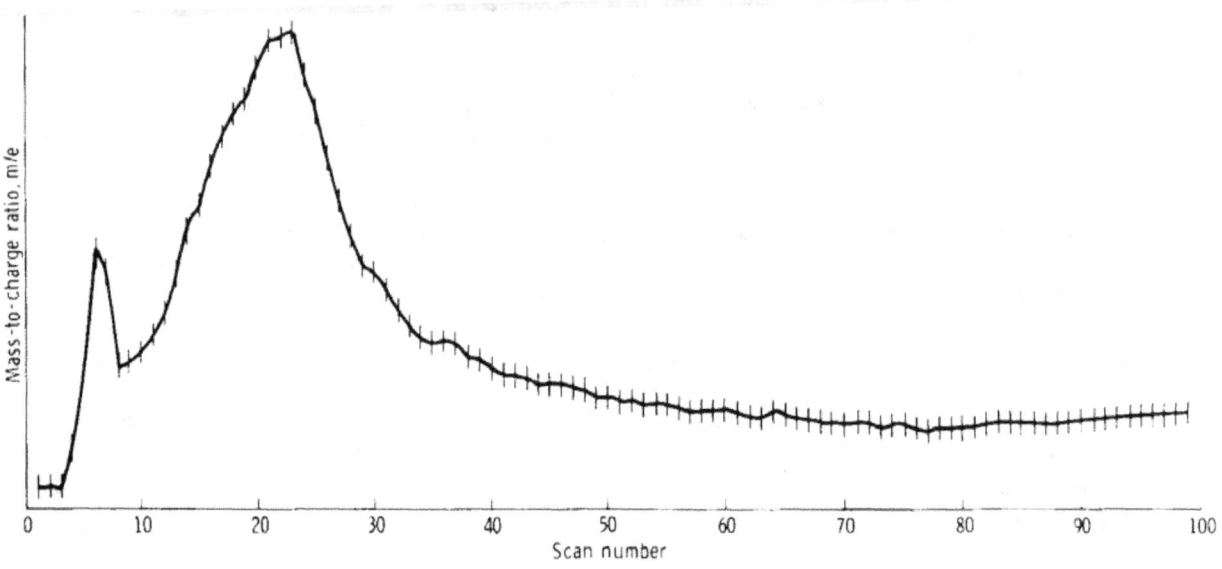

FIGURE 12-17 (continued). — (b) A summarize-m/e plot for mass 44.

Apollo 12 samples, as well as the controls, indicate a significant reduction in the level of ALSRC and LRL organic contamination to values below the 1 ppm level. This situation has alleviated significantly the previous ambiguities in interpretation of data. The levels of organic material in the rock chips were between 0.1 and 0.4 ppm. The levels for the fine material (GASC, core-tube fines, and other fines) ranged from 0.3 to 1.0 ppm. A feature common to all the samples was the high yield of CO_2. The evolution of CO_2 characteristically reaches a maximum rate approximately 10 to 15 min after sample insertion into the mass spectrometer and then decreases, although in many cases CO_2 remains a significant contributor to the total ionization throughout the analysis (approximately 30 min). Mass spectra from the core-tube fines

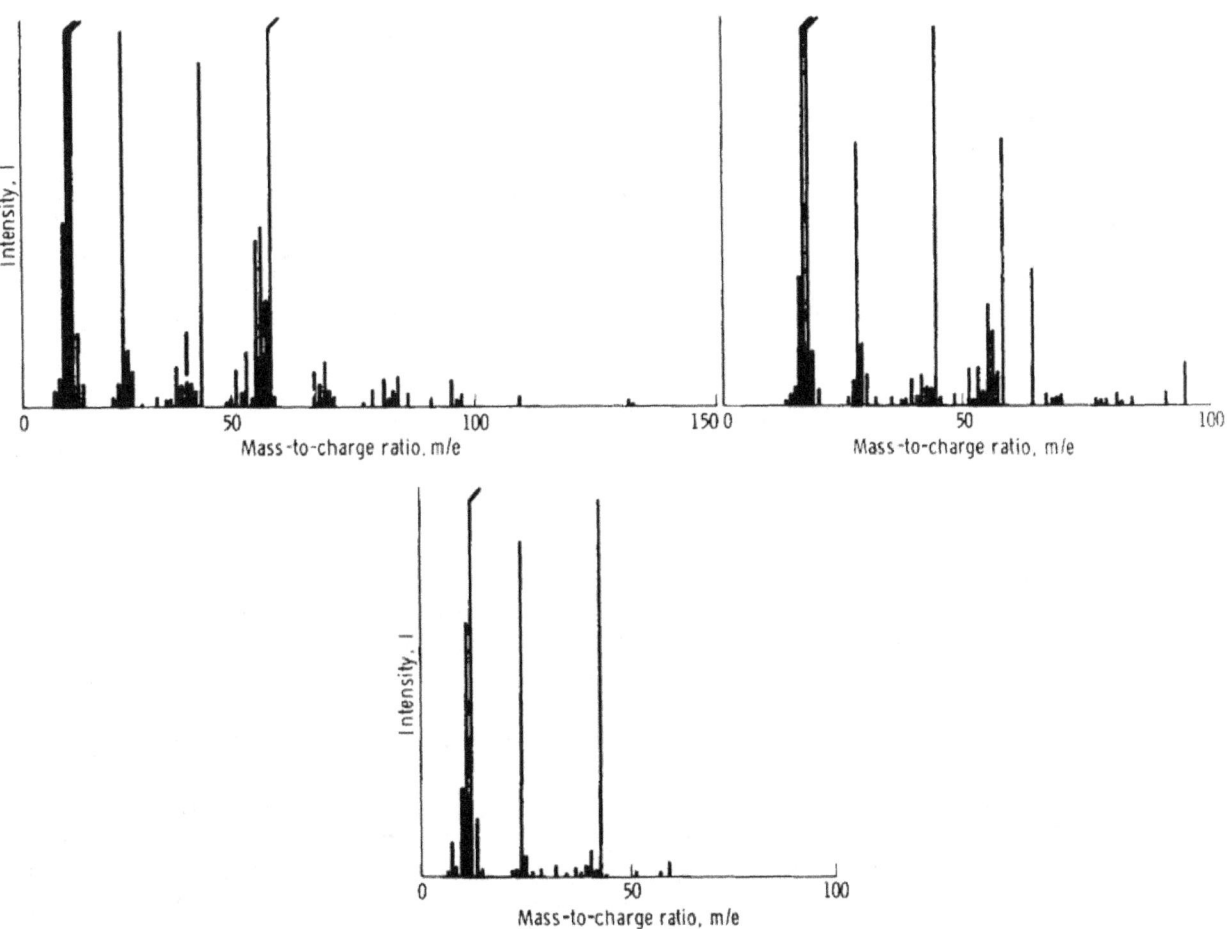

FIGURE 12-17 (continued). — (c) Representative m/e plots.

display ions up to the $m/e=120$ range that, if they were ions of aliphatic hydrocarbons, would be unsaturated but not aromatic.

Comparison of the mass spectra and the volatilization characteristics of the evolved gases from the blanks and controls with the corresponding data on the lunar sample shows that most of the organic matter observed can be attributed to terrestrial contamination. The evolution of sulfur dioxide (SO_2) and relatively large quantities of carbon monoxide (CO) and CO_2 would be consistent with reactions of elemental carbon and sulfur with the mineral matrix.

A somewhat striking contrast is evident between the Apollo 11 and 12 material. There is no indication (based on the observations late in the heating cycle of $m/e=78$ and $m/e=91$) that pyrolysis of any indigenous organic matter is occurring in the Apollo 12 samples.

In conclusion, then, if there is any naturally occurring organic matter present in the Apollo 12 samples, its concentration is extremely low (no more than 10 to 200 ppb).

Conclusions

The major finding of the preliminary examination of the lunar samples is that the conclusions reached concerning the rocks from Tranquility Base (ref. 12–1) also apply to the Apollo 12 rocks, except as follows:

(1) Whereas all the Apollo 11 crystalline rocks displayed essentially one texture (lath-shaped ilmenite and plagioclase with interstitial pyroxene) and similar modes (50 percent pyroxene, 30 percent plagioclase, 20 percent opaque, 0 to 5 percent olivine), the Apollo 12 crystalline rocks show a wide range in both texture and mode.

(2) Most of the igneous rocks fit a fractional crystallization sequence indicating either that they represent parts of a single intrusive sequence or that they are samples of a number of similar sequences.

(3) Breccias are of lower abundance at the Oceanus Procellarum site, as compared to Tranquility Base, presumably because the regolith at the Oceanus Procellarum site is less mature and not as thick as that at Tranquility Base.

(4) Complex stratification, presumably due mainly to the superposition of ejecta blankets, exists in the lunar regolith. At present, the possibility of a layer of volcanic ash cannot be discarded.

(5) The greater carbon content of the breccias and fines as compared to that of the crystalline rocks is presumably due largely to contributions of meteoritic material and the solar wind.

(6) The level of indigenous organic material capable of volatilization or pyrolysis, or both, appears to be extremely low (i.e., no greater than 10 to 200 ppb).

(7) The content of noble gas of solar-wind origin is less in the fines and breccias of the Apollo 12 rocks than in the fines and breccias from Tranquility Base. The breccias contain less solar-wind contribution than the fines, indicating that the breccias were formed from fines lower in solar-wind noble gases than the fines presently at the surface.

(8) The presence of nuclides produced by cosmic rays shows that the rocks have been within 1 m of the surface for 1×10^6 to 200×10^6 yr.

(9) The ^{40}K-^{40}Ar measurements on igneous rocks show that they crystallized 1.7×10^9 to 2.7×10^9 yr ago.

(10) The Apollo 12 breccias and fines are similar chemically and contain only half the titanium content of the Apollo 11 fines. The composition of the crystalline rocks is distinct from that of the fine material in containing less nickel, potassium, rubidium, zirconium, uranium, and thorium.

(11) The Apollo 12 rocks contain less titanium, zirconium, potassium, and rubidium and more iron, magnesium, and nickel than the Apollo 11 samples.

(12) Systematic variations among the magnesium, nickel, and chromium contents occur in the crystalline rocks, but there are only small differences in the potassium and rubidium contents.

(13) Rock sample 12013 has a distinctive composition, similar to that of a late-stage basaltic differentiate. This sample contains higher amounts of silicon, potassium, rubidium, lead, zirconium, yttrium, ytterbium, uranium, thorium, and niobium than previously encountered.

Discussion

Even though the preliminary examination has included only superficial scientific reconnaissance of samples that will be studied for years to come, a number of new, important findings have emerged from the data, especially by comparing the data with those for the Tranquility Base rocks. A comparison of the Apollo 12 samples from Oceanus Procellarum with the Apollo 11 samples from Mare Tranquillitatis shows that the chemistry at the two mare sites is clearly related. Both sites show the distinctive features of high concentrations of some refractory elements and low contents of volatile elements; these two features most clearly distinguish lunar material from other material. This overall similarity indicates that the Apollo 11 sample composition is not unique. Taken in conjunction with the Surveyor 5 and 6 chemical data, this similarity is suggestive of a similar chemistry for mare basin fill. In detail, there are numerous and interesting chemical differences between the Apollo 12 and 11 rocks. It is clear that, although they differ in details, the geochemical problems raised by the Apollo 11 samples are still present for the Apollo 12 material.

Unlike the Tranquility Base samples, the element abundances in the fines in the Apollo 12 samples display a generally more fractionated character than in the rocks. The fine material and the breccias are generally quite similar in composition and could not have formed directly from the large crystalline rock samples. The chemistry of the fine material is not uniform in the different maria.

The overall geochemical behavior of the rocks is consistent with the patterns observed during fractional crystallization in terrestrial igneous rocks involving olivine and pyroxene separation. Thus, the silicate melt is observed to be depleted in elements such as nickel and chromium that preferentially enter olivine and pyroxene, and the residual melt is enriched in elements such as barium and potassium that are excluded from the early crystal fractions. The slight degree of enrichment of the elements such as barium and potassium indicates an early stage of fractional crystallization process. Sample 12038 fits neatly as a late-stage differentiate in such a process. Whether these rocks form a related sequence or are a heterogeneous collection of similar origins cannot be answered from the chemical evidence.

The chemistry of the Apollo 12 samples does not resemble that of any known meteorite, because nickel is, in particular, strikingly depleted. The Apollo 12 sample chemistry has interesting similarities with the eucrites; sample 12038 shows many similarities in composition. The concentrations of titanium, zirconium, strontium, and barium are sufficient to distinguish the eucrites, but it now seems to be a fairly good possibility that rocks of similar chemistry to the eucritic meteorites are present on the Moon, in view of the variation in composition observed between and at the two mare sites. This possibility has previously been suggested (ref. 12–3). Even though the Apollo 12 rocks are more similar to tholeiitic and alkaline basalts than the Apollo 11 rocks, there are still some striking dissimilarities.

The Apollo 12 material is enriched in many elements by 1 to 2 orders of magnitude in comparison with estimates of cosmic abundances, and the mare material is strongly fractionated relative to ideas of the composition of the primitive solar nebula. The Apollo 12 site appears to be less geomorphologically mature than Tranquility Base, with a thinner regolith. The lower amount of solar-wind material in the fines, compared to that in samples from Tranquility Base, also suggests that Oceanus Procellarum mare material is younger than that in Mare Tranquillitatis.

The single most interesting scientific observation is the K-Ar age of the Apollo 12 rocks. The K-Ar age of these rocks reinforces the possibility indicated from the data obtained on the Apollo 11 rocks, that is, that the lunar maria are geologically very old. If the minimum ages established by this method are indicative of the true age of the Apollo 12 rocks, then the mare material from Oceanus Procellarum at the Apollo 12 site is approximately 1 billion years younger than the material from the Apollo 11 site. Although this K-Ar age is subject to various uncertainties, the younger age for the Apollo 12 material is consistent with geological observations. This large age difference indicates a prolonged period of mare filling.

References

12-1. Lunar Sample Preliminary Examination Team: Preliminary Examination of Lunar Samples from Apollo 11. Science, vol. 165, no. 3899, Sept. 19, 1969, pp. 1211-1227.

12-2. MOORE, C. B.; LEWIS, C. F.; GIBSON, E. K., JR.; and NICHIPORUK, W.: Total Carbon and Nitrogen Abundances in Lunar Sample. Science, vol. 167, no. 3918, Jan. 30, 1970, pp. 495-497.

12-3. DUKE, M. B.; and SILVER, L. T.: Petrology of Eucrites, Howardites, and Mesosiderites. Geochim. Cosmochim. Acta, vol. 31, no. 10, 1967, pp. 1637-1665.

ACKNOWLEDGMENTS

The following people contributed directly to obtaining the data and to the preparation of this report: D. H. Anderson, Manned Spacecraft Center (MSC); E. E. Anderson, Brown and Root-Northrop (BRN); P. R. Bell, MSC; Klaus Beimann, Massachusetts Institute of Technology (MIT); D. D. Bogard, MSC; Robin Brett, MSC; A. L. Burlingame, University of California at Berkeley; Patrick Butler, Jr., MSC; A. J. Calio, MSC; E. C. T. Chao, U.S. Geological Survey (USGS); R. S. Clark, MSC; D. H. Dahlem, USGS; J. S. Eldridge, Oak Ridge National Laboratory (ORNL); M. S. Favaro, U.S. Public Health Service; D. A. Flory, MSC; C. D. Forbes, MSC; T. H. Foss, MSC; Clifford Frondel, Harvard University; R. Fryxell, Washington State University; John Funkhouser, State University of New York, Stony Brook; E. K. Gibson, Jr., MSC; W. R. Greenwood, MSC; R. S. Harmon, MSC; J. Hauser, University of California at Berkeley; G. H. Heiken, MSC; Walter Hirsch, BRN; P. H. Johnson, BRN; J. E. Keith, MSC; C. F. Lewis, Arizona State University (ASU); John F. Lindsay, MSC; Gary E. Lofgren, MSC; V. A. McKay, ORNL; N. Mancuso, MIT; J. D. Menzies, U.S. Department of Agriculture; Carleton B. Moore, ASU; D. A. Morrison, MSC; R. Murphy, MIT; G. D. O'Kelley, ORNL; M. A. Reynolds, MSC; R. T. Roseberry, ORNL; O. A. Schaeffer, State University of New York, Stony Brook; Ernest Schonfeld, MSC; J. W. Schopf, University of California at Los Angeles; D. H. Smith, University of California at Berkeley; R. L. Smith, USGS; R. L. Sutton, USGS; S. R. Taylor, Australian National University; Jeff Warner, MSC; Ray E. Wilcox, USGS; D. R. Wones, MIT; and J. Zahringer, Max-Planck-Institut, Heidelberg, Germany.

The members of the Lunar Sample Preliminary Examination Team wish to acknowledge the technical assistance of the BRN staff: John H. Allen, Travis J. Allen, A. Dean Bennett, L. E. Cornitius, J. B. Dorsey, Paul Gilmore, George M. Greene, William R. Hart, D. W. Hutchison, Robert W. Irvin, Carl E. Lee, J. D. Light, E. Allen Locke, J. Roger Martin, David R. Moore, Weldon B. Nance, Albert F. Noonan, David S. Pettus, Clifford M. Polo, W. R. Portenier, M. K. Robbins, Louis A. Simms, and R. B. Wilkin.

13. Preliminary Results from Surveyor 3 Analysis

R. E. Benson,[a] B. G. Cour-Palais,[a] L. E. Giddings, Jr.,[b] Stephen Jacobs,[a] P. H. Johnson,[b] J. R. Martin,[b] F. J. Mitchell,[c] and K. A. Richardson[a]

Surveyor 3 was launched on April 17, 1967, and landed on the lunar surface in the Ocean of Storms on April 20, 1967. The Surveyor 3 mission and scientific results are described in detail in reference 13-1.

The vernier engines remained on through the first two lunar touchdowns of Surveyor 3 at a thrust level equal to approximately 90 percent of the spacecraft lunar weight, which caused the spacecraft to rebound each time from the lunar surface. The vernier engines were shut down by ground command approximately 1 sec before the third touchdown, and the spacecraft came to rest. The failure of the vernier engines to shut down before landing appeared to have caused contamination or pitting of the optical surfaces of the Surveyor 3 television (TV) camera. The Surveyor 3 spacecraft was apparently in good condition when it was secured for the lunar night on May 4, 1967, 14 days after touchdown.

The Apollo 12 lunar module (LM) landed on the northwest rim of Surveyor Crater in the Ocean of Storms, approximately 183 m from the Surveyor 3 spacecraft (fig. 13-1). During the second extravehicular activity (EVA) period, the astronauts removed several pieces of hardware from the Surveyor 3 spacecraft. The hardware removed from Surveyor 3 by the Apollo 12 astronauts and returned to Earth presents a unique opportunity to evaluate the influence of prolonged lunar-environment exposure on the elements that comprise typical spacecraft engineering systems. The principal objectives of the engineering investigations will be to improve the technology base that will be used in the design of future lunar and space vehicles. The Surveyor 3 TV camera contains elements common to most space vehicle systems, including electronics, active mechanical devices, optical elements, detectors, and a wide variety of materials.

Preliminary Results

The engineering investigations have just begun, and it is impossible to draw definitive conclusions as yet. Based on examinations to date, the following preliminary observations can be made.

Cold Welding

During the disassembly of the camera housing for biological sampling, removal torques were measured for the collar nuts, the connector-retaining nuts, and the screws that hold the lower shroud. These removal torques were all within the range specified for installation prior to launch, which indicates the absence of cold welding for these particular elements. The force required to demate the three connectors on the front of the camera was measured, and again, no evidence of cold welding was found.

Difficulty was encountered in retracting one of the connector bodies from the lower shroud. The mating surfaces will be examined when they can be removed to determine if cold welding was a contributing factor to this difficulty.

Discoloration

The discoloration observed on the exterior surfaces of the TV camera includes unusual patterns of relative light- and dark-colored areas that do not appear to correlate with solar illumination and the resulting radiation degradation. Handling during retrieval and return has pro-

---

[a] NASA Manned Spacecraft Center.
[b] Brown & Root-Northrup.
[c] U.S. Air Force.

FIGURE 13-1. — The Apollo 12 LM approximately 183 m from the Surveyor 3 spacecraft on the lunar surface.

duced considerable disturbance of the original discoloration patterns. The discoloration is attributable to at least three possible sources: (1) radiation darkening of the surface coating; (2) lunar debris accumulated on the camera surface; and (3) contamination from various sources, including prelaunch environments, retrograde engines, and spacecraft outgassing. It is not feasible at this time to assign relative importance to these three possible sources, nor is it feasible to preclude the presence of additional contributing factors.

Inspection indicates that some of the unusual patterns of the light- and dark-colored areas can be traced to a lightening mechanism that apparently originated above and behind the camera in the general direction of the LM. While a "sandblasting" mechanism is most likely, other dominant or contributing mechanisms cannot be precluded. For example, if a significant fraction of the camera surface were covered by lunar material during the Surveyor 3 landing, such material would shield the paint underneath from radiation darkening, and a mechanism that simply removes this lunar debris, without removing any of the radiation-darkened surface-coating layer, would produce the light-colored areas observed.

Organic contaminants on the order of a few thousandths of an angstrom thick, which result either from prelaunch contamination or from outgassing of spacecraft components, can produce deep discoloration that is difficult to distinguish from radiation damage to the coating or from accumulated lunar material. Because the camera was stored in a polyethylene bag upon its return to the Lunar Receiving Laboratory (LRL), it may not be possible to clearly eliminate photolyzed organic contaminant as a source of discoloration.

Polished Tube

The unpainted tube that was cut from a support strut adjacent to leg 2 of the Surveyor 3 spacecraft was contaminated on one side approximately one-half to two-thirds of the way around the circumference. The contaminant on this tube was much heavier at one end than at the other. It is possible that some of the contamination was rubbed off during return and handling. The total content of this contaminant has not been determined, but it appears to contain a significant portion of particulate matter that is probably lunar debris. Obviously the discoloration visible on the polished tube and on the bare areas of the camera could not include radiation damage, except to contaminants.

Optics

The mirror surface of the camera is very diffuse because of accumulated lunar dust but is substantially less diffuse than the photographs taken on the Moon indicate. Part of the Surveyor 3 investigation will include attempts to compare the current dust coverage with that which was present when the camera was photographed by the Apollo 12 astronauts on the Moon and with the dust coverage present at the time of the Surveyor 3 mission. No pitting of the mirror has been observed, but an adequate examination will be difficult until dust has been removed from selected areas.

There is a considerable amount of dust on the filters immediately below the mirror and on those areas that are partially protected by the filter-wheel drive mechanism. Until the filter wheel is actuated or removed, it will be impossible to determine whether the partially shielded areas are as thoroughly covered with dust as are the exposed areas.

Lunar Dust

Some fine debris estimated to be a few milligrams in weight was found to be clinging to the camera surface in the recessed area under the support collar. The sample of this material was analyzed by emission spectroscopy, as reported elsewhere in this section. The remaining debris has, at this time, received only cursory low-power microscopic examination. This remaining debris appears to be lunar fines and contains various minerals with a wide range of particle sizes (up to approximately 150 μm). This debris probably entered the recess under the collar through the inspection hole, either during the original Surveyor 3 landing or, more likely, during the Apollo 12 LM landing. Entrance during LM landing would account for the larger particles in this area and not those in the open front of the camera. The inspection hole could "see" the LM landing site, whereas the front of the camera could not. It may be possible to determine some parameters of the disturbance caused by the LM descent from debris particle size and from acceptance-angle geometry of the inspection hole.

One of the screws and its matching washer from the lower shroud that was removed during operations at the LRL have been examined in the scanning electron microscope. The surface has considerable particulate debris that covers an estimated 15 to 20 percent of the surface area. The material contains a substantial number of spheres and angular particles that range in size from a fraction of a micron to approximately 4 μm. The absence of particles larger than approximately 4 μm may be the result of their absence in the original source of the dust, their failure to adhere on the Moon, or mass-to-adhesion characteristics which are such that the larger particles fell off during or after the return to Earth. Preliminary examination indicates approximately uniform distribution of particles over all surfaces of the screwhead.

It may be possible to determine the source and transport mechanism of the dust by examination of the screws from various locations on the camera. Because approximately 500-power magnification is required to clearly resolve the particulate material, such an examination will be a Herculean task and can only be undertaken on a statistical basis.

Preliminary Scientific Investigation

A large number of scientific tasks have been approved for the returned Surveyor items, and these tasks are outlined in the official flow plans. Proposed work on radioisotope production by energetic charged particles, cosmic-ray tracks, soil mechanical properties, albedo and particle size, biological assay, optical and scanning electron microscopy and electron microprobe of meteoroid impacts, and alpha- and gamma-particle activity of returned parts has either been underway or will soon commence.

The preliminary reports contained in this section consist of work performed by scientists at the NASA Manned Spacecraft Center (MSC) and were begun when the returned Surveyor parts were still at the LRL. These preliminary studies report on the radioactivity analysis of the Surveyor 3 TV camera, the analysis of the dust from the Surveyor 3 TV camera, the examination of Surveyor 3 components for meteoroid damage, and alpha-particle activity of the Surveyor 3 spacecraft.

Radioactivity Analysis of the Surveyor 3 TV Camera

The Surveyor 3 TV camera was examined by gamma spectrometry on January 7, 1970. The purpose of this examination was to measure the radioactivity present and, more particularly, to determine if there was induced radioactivity present that could be attributed to the exposure of the camera to solar and galactic cosmic radiations while the camera was located on the lunar surface.

The camera was counted for 50 000 sec in the gamma spectrometry system located in the MSC Radiation Counting Laboratory. This system, which is located in an exceptionally low-background counting room, includes an array of six 5- by 4-in. NaI (Tl) thallium-activated crystals and associated electronics with a 4096-channel analyzer.

Qualitative analysis indicated the presence of ^{40}K, ^{56}Co, and ^{22}Na in the Surveyor 3 camera. These tentative results indicate that some induced radioactivity was present. The total amount of radioactivity, for all radionuclides present, was very small and amounted to approximately 0.003 μCi/kg. These results are considered to be preliminary because the data will be subjected to further analysis. Arrangements have been made to obtain an identical nonflight model of the Surveyor 3 camera for use as a counting standard. Following completion of the gamma-spectrum analysis of the nonflight model camera, analysis of the Surveyor 3 camera data can be completed, and a report of final results will become available.

Analysis of Dust From the Surveyor 3 TV Camera

Dust from the Surveyor 3 TV camera was analyzed by quantitative emission spectroscopy. The sample came from a fine line of the dust that lay along the lower edge of the positioning bearing attached to the outer cover of the camera. It is assumed that the sample is composed of bearing material, lunar surface fines material, or dust from another source.

The sample weighed slightly less than 0.5 mg, which is an insufficient amount for the normal 1:1:4 ratio of sample to strontium carbonate to graphite. The total sample was mixed with approximately 4 g of 1:4 strontium carbonate-graphite mix. All this material was loaded into an electrode and was excited by an 8-A dc arc. The atmosphere was 80 percent argon and 20 percent oxygen, and the sample was burned to completion. From the results of the analysis, it appears that the Surveyor 3 dust consists of lunar surface fines material.

Examination of Components for Meteoroid Damage

The external surfaces of the Surveyor 3 TV camera and of the polished aluminum tube were microscopically examined for evidence of meteoroid impact. Approximately 60 percent of the TV camera surface area of nearly 1900 cm^2 was scanned at 25-power magnification. Every suspected impact crater on selected areas of the flat surfaces was recorded. The remainder of the camera surface was scanned at lower magnifications to insure that no significant meteoroid damage had occurred. The polished tube, 19.7 cm long and 1.3 cm in diameter, was carefully scanned at a general level of 40-power magnification with a stereozoom microscope. Local areas of interest were examined at much higher magnifications. Typical surface effects and suspected impact craters were photographed (figs. 13-2 and 13-3).

Two 2.5-cm-long sections of the polished tube have been under detailed examination since the preliminary examination at the LRL. These sec-

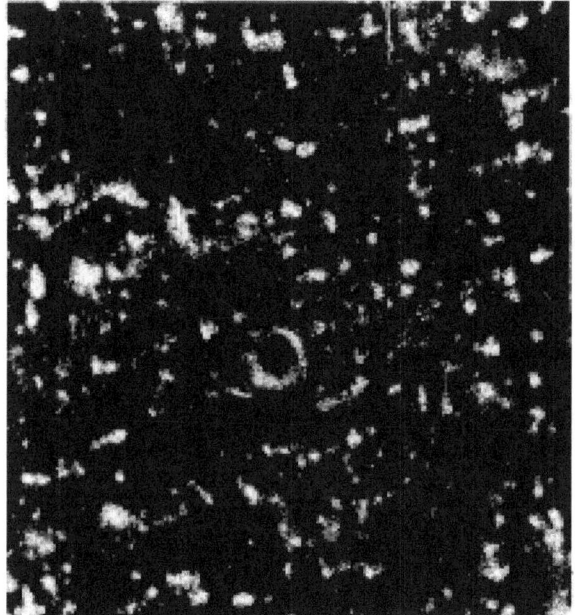

FIGURE 13-2. — Possible 133-μm-diameter crater near the top right-hand screw on the gear housing.

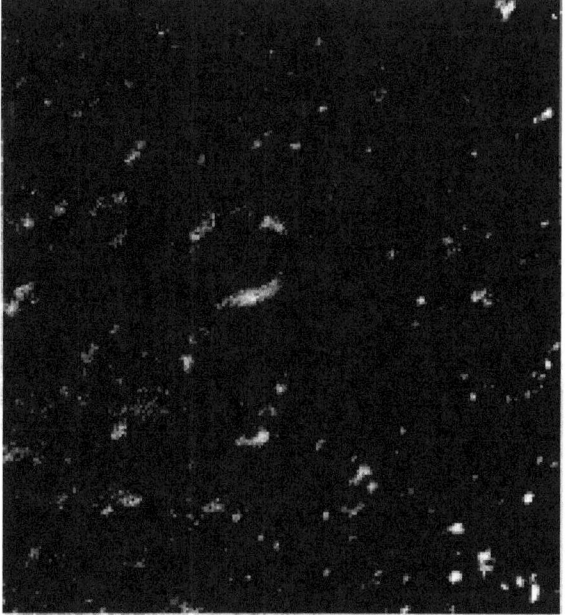

FIGURE 13-3. — Possible 250-μm-diameter crater near the bottom left-hand screw on the gear housing.

tions have been optically scanned at up to 400-power magnification. The specimens were mounted on a special jig that allowed 360° rotation of the tube. The jig was then indexed to the microscope base, which had the necessary translational capability. Typical samples of the polished tubing and the painted surface of the camera were supplied by the prime contractor for Surveyor 3. These samples were examined with the same microscope to determine typical surface background.

A preliminary assessment of the meteoroid examination of the TV camera indicates that there is no damage of any consequence by primary or secondary meteoroids. After approximately 950 days of exposure to the meteoroid environment, there were no impacts that completely penetrated the protective paint or that damaged the surface of the mirror. There are five craters on the camera surface that are possibly meteoritic in origin. These craters range between 0.025 and 0.25 mm in diameter. However, this is compatible with the MSC flux estimates used to calculate the meteoroid reliability for the Apollo lunar missions. This suggests that there is a greater confidence in these flux calculations, particularly for the calculations that pertain to EVA periods.

In addition to the possible meteoroid impacts, numerous surface effects of probable low-velocity origin were noted on the TV camera surface. In general, these low-velocity effects were shallow craters, and most were of recent origin, as indicated by their whiteness against the sandy-brown color of the painted surface of the TV camera housing. There was a definite concentration (10 to 100 times more) of these white craters on the side of the TV camera facing the LM, as compared with the other side of the camera. The number density of these craters peaked at a region approximately directly in line with the LM. In addition, protuberances on the camera (such as screwheads, support struts, etc.) left dark shadows on the camera paint, which again pointed toward the LM. After a detailed examination of the geometry involved and by taking into account the relative angles of the shadowing, the TV camera, the Surveyor spacecraft, and the LM position, it was readily shown that the LM was the most probable origin for these craters. It is, therefore, postulated that in the final moments of landing, the LM generated a dust shower that affected the Surveyor 3 spacecraft and sandblasted the camera surface that faced toward the LM. This is the most significant result obtained from the meteoroid-impact examination of the Surveyor 3 TV camera to date.

The preliminary examination of the polished tube at low magnifications revealed four craters that exhibited the characteristics of hypervelocity impacts. Subsequent detailed examination at higher optical magnifications has ruled out the possibility that one of these craters was caused by a hypervelocity impact. The examination revealed the crater to be a surface pit caused by polishing, and the other three craters have not yet been identified positively as impact pits. Further analysis of these craters is proceeding. The detailed scan has established that the surface of the tube is covered with polishing scratches and gouges, with the gouges being easily mistaken for lipped hypervelocity craters. There appears to be a marked concentration of surface effects on the same side of the tube as the deposit, and the significance of this correlation is currently being analyzed.

Both sections of the polished aluminum tube have been examined with a scanning electron microscope at up to 12 000-power magnification. At this power, it has been possible to observe the suspected micrometeoritic craters in sufficient detail and to examine the residue in some of them. Positive identification of the origin of the craters is still not possible at this time. Analysis of all the data obtained to date is in progress, and a comprehensive report will be made available at a later date.

Alpha-Particle Activity of the Spacecraft

Kraner, Schroeder, Davidson, and Carpenter (ref. 13–2) have suggested that diffusion of radon and thoron from the lunar soil and deposition of their radioactive daughter products on the lunar surface could result in the presence of significant quantities of these radioactive noble gases in the atmosphere of the Moon and could also result in the formation of a radioactive deposit on the surface of the Moon. By assuming a diffusion rate of 0.02 cm^2/sec and a lunar surface porosity of 0.25 and by using concentrations of radon and thoron measured in terrestrial surface layers, Kraner et al. calculated the equilibrium lunar surface activities to 4 disintegrations/cm^2/sec for radon and 1×10^{-2} disintegrations/cm^2/sec for thoron.

Yeh and Van Allen (ref. 13–3), by using alpha-particle measurements made by the Explorer 35 spacecraft, have shown that the alpha-particle emissivity of the Moon is not likely to exceed 0.1 of the value estimated by Kraner et al. This observed upper limit of alpha-particle emissivity of the Moon implies that the concentration of ^{238}U in lunar surface material is less than in average terrestrial crustal material but may be comparable to the uranium content of terrestrial basalt or chondritic meteorites.

Measurement of the surface alpha-particle activity of samples of the Surveyor 3 spacecraft provides further information on the radioactivity of the lunar surface layer. During the period that the Surveyor 3 spacecraft was on the Moon, a radioactive deposit of radon and thoron daughter products should have accumulated on the surfaces of the spacecraft by the mechanism described by Kraner et al. When samples removed from Surveyor 3 were returned to the LRL and released from quarantine, many of the radionuclides with short half lives would have decayed to leave the alpha-emitting nuclide ^{210}Po, with a half life of 138 days, supported by ^{210}Pb, with a half life of 22 yr. In the 30-month period that the Surveyor 3 spacecraft was on the lunar surface, ^{210}Po activity on the surface of the spacecraft would have reached 6 percent of the amount that would be in equilibrium with radon in the lunar atmosphere.

The alpha-particle activity of a section of the Surveyor 3 unpainted aluminum support tube was measured by using a 300-mm^2 gold-silicon surface-barrier detector, with the results recorded in a 256-channel spectrum. The sample was 1.2-cm-diameter tubing of aluminum alloy 2024, approximately 2.5 cm long. For background measurements, typical samples of the same aluminum alloy were used.

Preliminary results show no detectable alpha-particle activity above the background level. Considering the sensitivity of the method, this measurement indicates that the quantity of radon in the lunar atmosphere at the Surveyor 3 site is an order of magnitude less than the upper limit measured by Yeh and Van Allen.

References

13-1. Anon.: Surveyor III Mission Report. JPL Tech Rept. 32-1177, 1967. Part I — Mission Description and Performance. Part II — Scientific Results. (Also, Rennilson, J. J.: Surveyor III Mission Results. Part II — Scientific Results (Addendum). JPL Tech Rept. 32-1177, 1967.)

13-2. Kraner, H. W.; Schroeder, G. L.; Davidson, G.; and Carpenter, J. W.: Radioactivity of the Lunar Surface. Science, vol. 152, no. 3726, May 27, 1966, pp. 1235-1236.

13-3. Yeh, R. S.; and Van Allen, J. A.: Alpha-Particle Emissivity of the Moon: An Observed Upper Limit. Science, vol. 166, no. 3903, Oct. 17, 1969, pp. 370-372.

APPENDIX A
Glossary

achondrite — a stony meteorite devoid of rounded granules.

anhedral — having mineral grains lacking external crystals.

anorthosite — a granular, plutonic igneous rock composed almost exclusively of a soda-lime feldspar.

augite — one of a variety of pyroxene minerals, containing calcium, magnesium, and aluminum. Usually black or dark green in color.

bleb — small bit or particle of distinctive material.

breccia — a rock consisting of sharp fragments embedded in a fine-grained matrix.

bytownite-anorthite — calcium-rich varieties of plagioclase feldspar.

calcic — derived from or containing calcium.

chondrite — a meteoritic stone characterized by the presence of rounded granules.

chondrules — a rounded granule of cosmic origin.

clast — a discrete particle or fragment of rock or mineral, commonly included in a larger rock.

clinopyroxene — a mineral occurring in monoclinic, short, thick, prismatic crystals, varying in color from white to dark green or black (rarely blue).

cristobalite — an isometric variety of quartz that forms at high temperature (SiO_2).

dacite — an extrusive igneous rock composed of plagioclase and quartz, with other minerals.

detrital — pertaining to loose material that results directly from rock disintegration or abrasion.

diabase — an igneous rock of basaltic composition, but with slightly coarser texture.

diorite — a granular, crystalline igneous rock.

eucrite — a meteorite composed essentially of feldspar and augite.

euhedral — pertaining to minerals whose crystals have had no interference in growth.

exfoliation — the process of breaking loose thin concentric shells or flakes from a rock surface.

fayalite — an iron-rich variety of olivine (Fe_2SiO_4).

felsic — consisting of or chiefly consisting of feldspar or feldspar-type minerals, commonly with quartz. Also refers to light-colored rocks.

gabbro — a granular igneous rock of basaltic composition with a coarse-grained texture.

goniophotometric — pertaining to angle measurement by means of photography.

groundmass — the fine-grained or glassy mixture of a porphyritic rock in which the larger, distinct crystals are embedded.

holocrystalline — consisting wholly of crystals.

ilmenite — a mineral rich in titanium and iron; usually black with a submetallic luster.

lithic — of, relating to, or made of stone.

lithification — consolidation and hardening of fines into rock.

magma — molten rock material that is liquid or pasty.

magnetosphere — the region dominated by the magnetic forces of the Earth.

microlite — small lath-shaped minerals, commonly plagioclase feldspar, occurring as minute phenocrysts in basalt.

olivine — an igneous consisting of a silicate of magnesium and iron.

ophitic — a rock texture characterized by lath-shaped plagioclase crystals enclosed in augite.

peridotite — an igneous rock, composed largely of olivine and pyroxene with little or no plagioclase feldspar.

phenocryst — crystals larger than the crystalline matrix in which they occur.

picrite basalt — a basalt containing ferromagnesian minerals and a little feldspar.

pigeonite — a variety of pyroxene.

plagioclase — a feldspar mineral composed of varying amounts of sodium and calcium with aluminum silicate.

plutonic — pertaining to igneous rock that crystallizes at depth.

Poisson's ratio — ratio of elongation to diameter contraction.

porphyritic — a rock texture displaying mineral grains in a relatively fine-grained base.

porphyry — a rock with distinct mineral grains in a relatively fine-grained base.

pyroxene — a mineral occurring in short, thick, prismatic crystals or in square cross section; often laminated; and varying in color from white to dark green or black (rarely blue).

pyroxmangite — a pale red mineral, consisting essentially of manganese silicate, containing approximately 20 percent manganese.

regolith — the layer of fragmental debris that overlies consolidated bedrock.

sanidine — a feldspar mineral that is somewhat glassy (KAlSi$_3$O$_8$).

scoriaceous — having the characteristics of rough, vesicular, cindery, usually dark lava.

sodic — relating to or containing sodium.

spinel — a mineral that is noted for its great hardness MgAl$_2$O$_4$).

tholeiitic — pertaining to a composition of basalt having low olivine.

tridymite — a variety of quartz with minute, thin, tabular forms of crystallization (SiO$_2$).

troilite — a mineral that is native ferrous sulfide.

variolitic — pertaining to a fine-grained, basic rock, containing crystal forms made up of fibers of feldspar and augite in radial development.

vesicle — a small cavity in a mineral or rock, ordinarily produced by expansion of vapor in the molten mass.

vug — a small cavity in a rock.

APPENDIX B

Acronyms

A/D — analog to digital
ALSCC — Apollo lunar surface closeup camera
ALSEP — Apollo lunar surface experiments package
ALSRC — Apollo lunar sample return container
ASU — Arizona State University
ac — alternating current
BRN — Brown & Root-Northrop
CCIG — cold cathode ion gage
CM — command module
CSM — command and service module
DPS — descent propulsion system
dc — direct current
EOS — Electro-Optical Systems, Inc.
e.s.t. — eastern standard time
EVA — extravehicular activity
GASC — gas analysis sample container
g.e.t. — ground elapsed time
G.m.t. — Greenwich mean time
IR — infrared
JPL — Jet Propulsion Laboratory, California Institute of Technology
LESC — lunar environment sample container
LM — lunar module
LP — long period (seismometer)
LPD — landing point designator
LPX, LPY — long-period horizontal component (seismometers)

LPZ — long-period vertical component (seismometer)
LRL — Lunar Receiving Laboratory, NASA Manned Spacecraft Center
MIT — Massachusetts Institute of Technology
MSC — Manned Spacecraft Center
MSFN — Manned Space Flight Network
NASA — National Aeronautics and Space Administration
ORNL — Oak Ridge National Laboratory
PFK — perfluorokerosene
PRA — parabolic reflector array
PSE — passive seismic experiment
PTL — Photographic Technology Laboratory, NASA Manned Spacecraft Center
RCL — Radiation Counting Laboratory, NASA Manned Spacecraft Center
RCS — reaction control system
RTG — radioisotope thermoelectric generator
SIDE — suprathermal ion detector experiment
SIVB — Saturn IVB (rocket stage)
SP — short period (seismometer)
SPZ — short-period vertical component (seismometer)
SWC — solar-wind composition (experiment)
TV — television
USGS — U.S. Geological Survey
VFC — video film converter

www.ingramcontent.com/pod-product-compliance
Lightning Source LLC
Chambersburg PA
CBHW081722170526
45167CB00009B/3667